Remington Education

Pharmaceutics

Remington Education

The *Remington Education* series is a new series of indispensable guides created specifically for pharmacy students. Providing a simple and concise overview of the subject matter, the guides aim to complement major textbooks used.

The guides assist students with integrating the science of pharmacy into practice by providing a summary of key information in the relevant subject area along with cases and questions and answers for self assessment (as subject matter allows). Students will be given a practical way to check their knowledge and track the progress of their learning before, during, and after a course.

Key features of the texts include:

- Learning Objectives
- Tips and Tricks Boxes
- Key Points Boxes
- Illustrations
- Assessment Questions
- Further Reading

The *Remington Education* series is a must-have for all pharmacy students wanting to test their knowledge, gain a concise overview of key subject areas and prepare for examinations.

Covering all areas of the undergraduate pharmacy degree, the first titles in the series include:

- *Drug Information and Literature Evaluation*
- *Pharmaceutics*
- *Introduction to Pharmacotherapy*
- *Physical Pharmacy*
- *Law & Ethics in Pharmacy Practice*

Remington Education

Pharmaceutics

Shelley Chambers Fox

Clinical Associate Professor,
Department of Pharmaceutical Sciences,
Washington State University

University of the Sciences
Philadelphia
College of
Pharmacy

Pharmaceutical Press

Published by Pharmaceutical Press

1 Lambeth High Street, London SE1 7JN, UK

© Royal Pharmaceutical Society of Great Britain 2014

(**PP**) is a trade mark of Pharmaceutical Press

Pharmaceutical Press is the publishing division of the Royal Pharmaceutical Society

Typeset by OKS Group, Chennai, India
Printed in Great Britain by TJ International, Padstow, Cornwall

ISBN 978 0 85711 070 1

A catalogue record for this book is available from the British Library.

Library of Congress Cataloging-in-Publication Data has been requested

Contents

Preface

This text is written to provide students studying for a doctorate in pharmacy with an overview of pharmaceutics that connects this science to the practice of pharmacy. The presentation that follows is a result of the judicious consideration of the subject matter of pharmaceutics in light of what today's pharmacists do and what future pharmacists will do in their practices. These carefully selected, essential principles are presented in concise form accompanied by examples of their application.

The first six chapters present the fundamental principles relating to the behavior of solids, solutions and dispersed systems, solubility, stability, and the processes that move drugs from the dosage form to the portion of the body where the drug's receptor is located. The reader is introduced to the scientific basis of generic substitution in the chapter on bioavailability and bioequivalence. The remaining chapters present drug delivery organized by route of administration.

Each chapter reinforces the same basic principles of dosage form design: drug chemistry considerations, materials and methods to prepare a dosage form with the desired qualities and characteristics of the route in which these systems are administered to deliver a drug to its receptors.

I am grateful to my students whose intelligence and enthusiasm have shaped my teaching, and to my husband, Larry Fox, for his energetic support of this project.

About the author

Shelley Chambers Fox is a registered pharmacist with experience in pharmacy practice and academia. Dr Chambers received a bachelor of pharmacy and PhD from Washington State University where her graduate work was supported by an American Foundation for Pharmaceutical Education fellowship. She has taught pharmaceutics and extemporaneous compounding at Washington State University since 1991. Dr Chambers' scholarly interests include peptide chemistry and, more recently, quality assurance for compounded products and the development of methods to improve pharmacy education. In 2012 she was recognized with the Washington State University Distinguished Teaching Award. Dr Chambers lives on a small farm in eastern Washington State with her husband, a dairy scientist, and some of his dairy cow subjects.

Acknowledgments

Table 3.4 is reprinted from *Aulton's Pharmaceutics: The Design and Manufacture of Medicines*, Aulton, M. E., 'Disperse Systems', p. 71, Copyright (2007), with permission from Elsevier.

Figure 1.5 is reprinted from *Anesthesiology*, Vol. 21, Issue 1, Price, H., 'A Dynamic Concept of the Distribution of Thiopental in the Human Body', Copyright (1960), with permission from Wolters Kluwer Health.

Figure 2.3 is reprinted from *Advanced Drug Delivery Reviews*, Vol. 48, Issue 1, Vippagunta, S. R., Brittain, H. G. and Grant, D. J. W., 'Crystalline Solids', pp. 3–26, Copyright (2001), with permission from Elsevier.

Figure 10.3 is reprinted from *InnovAIT*, Vol. 5, Issue 3, Simpson, D., 'Glaucoma – In the GP's Blind Spot?', p. 10, Copyright (2012), with permission from SAGE.

Figure 15.4 is reprinted from *The Journal of Laryngology & Otology*, Vol. 114, Raghavan, U. and Logan, B. M., 'New Method for the Effective Instillation of Nasal Drops', pp. 456–9, Copyright JLO (1984) Ltd (2000).

Figure 15.6 is reprinted from *Otolaryngology – Head and Neck Surgery*, Vol. 130, Issue 1, Benninger, M. S., Hadley, J. A., Osguthorpe, J. D., Marple, B. F., Leopold, D. A., Derebery, J. M. and Hannley, M., 'Techniques of Intranasal Steroid Use', p. 20, Copyright (2004), with permission from SAGE.

Figure 17.1 is reprinted from *International Journal of Pharmaceutics*, Vol. 131, Issue 2, Moghimi, H. R., Williams, A. C. and Barry, B. W., 'A Lamellar Matrix Model for Stratum Corneum, Intercellular Lipids. II. Effect of Geometry of the Stratum Corneum on Permeation of Model Drugs 5-Fluorouracil and Oestradiol', pp. 117–129, Copyright (1996), with permission from Elsevier.

1

Introduction to dosage form design

Learning objectives

Upon completion of this chapter, you should be able to:

- Define dosage form, stability, excipient, pharmaceutical necessity, chemical stability, physical stability, microbiological stability, first pass effect, elimination and excretion.
- List the drug properties that must be considered in dosage form design.
- Describe the qualities that are built into an ideal dosage form/drug delivery system, and identify from a list of ingredients and an ingredient table, how the ingredient contributes to the manageable size of the dosage form, its palatability or comfort, its stability (chemical, microbial, or physical), the convenience of its use or the release of drug from the dosage form.
- List the properties of the routes of administration that must be considered in the design of dosage forms.
- Briefly outline the movement of an orally administered drug from its site of administration through its elimination from the body:
  - administration
  - liberation from the dosage form by (a) dissolution; (b) diffusion
  - absorption
  - distribution
  - metabolism
  - elimination.
- Distinguish between the following patterns of drug release from dosage forms:
  - immediate release
  - delayed release
  - sustained release
  - controlled release.
- Rank the different routes in terms of speed of onset of action.

Introduction

What is pharmaceutics and why do pharmacists study this subject? Historically pharmacists provided two services in the healthcare system: they extracted drugs from their natural sources and prepared or 'compounded' the drug into a convenient form for patient use, the dosage form.

Currently little dosage form compounding is required of most pharmacists, but the pharmacist in a compounding specialty practice has the need to design dosage forms that are both safe and effective. The pharmacist in charge of an admixture service must develop policies and procedures to ensure the products prepared remain sterile as well as physically and chemically stable. And our most pressing pharmaceutical task is to ensure that patients and caregivers are properly educated in the appropriate use of eye drops and inhalers, injections and transdermal patches such that the drugs contained in these dosage forms will provide the intended benefit. The products of biotechnology for rheumatoid arthritis and multiple sclerosis are injectable dosage forms used primarily by the patient in the home setting. New technologies to deliver drugs through the skin utilize handheld devices for iontophoresis and thermal ablation. Pharmacists serve as the dosage form experts both to other medical personnel and to our patients. The study of pharmaceutics provides the scientific foundation for the design and appropriate use of dosage forms and drug delivery systems.

Definitions

A *dosage form* is the form that we take our drug in – in other words, a tablet, a syrup, an ointment, an injection. The concept of a *drug delivery system* includes products that are designed to provide optimal control over the release of a drug to achieve enhanced safety or efficacy.

This text is about the physicochemical principles of dosage form and drug delivery system design, and the biological environment in which these systems are designed to perform to deliver a drug to its receptors.

Properties of the drug

Key Point

To design a dosage form for a drug, the following properties must be considered: water solubility; log *P*, a measure of lipophilicity; molecular weight; stability in solution; enzymatic degradation; and location of drug receptors.

Pharmaceutics is a science that applies both drug chemistry and drug biology to the problem of delivering drugs to their target tissues. During the preformulation process, key chemical and physical properties of the drug are studied in order to rationally design a delivery system and predict the fate of the drug *in vivo* after administration. Some aspects of drug chemistry that must be considered in the development of a dosage form include:

- *Water solubility:* The drug must be dissolved in aqueous body fluids before it can cross membranes.
- *Log P:* The drug must have sufficient lipophilicity to cross membranes.
- *Molecular weight:* Large molecular weight drugs have difficulty moving across membranes.
- *Stability in solution:* Drugs may hydrolyze, oxidize, photolyze or otherwise degrade in solution.
- *Enzymatic degradation:* Some drugs may be substantially degraded by enzymes before they reach their target.
- *Location of receptors:* Intracellular drug receptors or receptors in the brain or posterior eye are challenging to target.
- *Selectivity:* Some drugs cause serious toxicity as a result of poor selectivity for their target.

Anticancer drugs are notoriously nonselective in their cytotoxic effects; however, drug delivery systems may be designed to selectively target tumor cells with known biological features. Drug delivery systems can be developed to address the challenges that drugs present: poor solubility, large molecular weight, inaccessible targets and lack of selectivity. Appendix Table A.1 provides an overview of the preferred drug properties for different routes of administration.

Pharmaceutical example

If we consider the opiate pain reliever, oxycodone, its advantages and disadvantages as a drug for administration to the gastrointestinal tract are presented in Table 1.1.

Qualities of the ideal dosage form

Key Point

Pharmaceutical excipients and manufacturing processes are used to prepare dosage forms with the following qualities:

- one dose in a manageable size unit
- palatable or comfortable
- stable chemically, microbiologically and physically
- convenient/easy to use
- release the drug to the receptors in a timely fashion with minimal side effects and for the optimal duration.

Table 1.1 Analysis of oxycodone's physicochemical properties for use by the oral route

Design issue	Drug property	Analysis
Must be dissolved before drug can cross membranes	Water solubility of 100 mg/mL, maximum dose 80 mg	Drug is soluble in the contents of the gastrointestinal tract, even better if the patient takes 250 mL water with the tablet
Must have sufficient lipophilicity to cross membranes	Log P of 0.3 = log $\frac{[conc]_{oil}}{[conc]_{water}}$ Antilog(0.3) = 1.995	This drug will favor partitioning into lipid membranes over water by a 2:1 ratio
Large molecular weight drugs have difficulty moving across membranes	Formula weight of 315.37	Based on molecular weight, oxycodone should move easily through membranes
Drugs may hydrolyze, oxidize, photolyze, or otherwise degrade in solution	Stability, solid dosage form, stable to acid	Stability in the dosage form should not be an issue for this drug because it is in a solid dosage form. Stability in stomach acid is not an issue
Some drugs cannot be used by mouth because of substantial losses to enzyme degradation	Metabolized in the liver, 60–80% of the dose is delivered to the systemic circulation	Some drug is lost to enzyme deactivation but most of the drug makes it to the blood
Intracellular drug receptors or receptors in the brain or posterior eye are challenging to target	Receptors are in the brain	Oxycodone's lipophilicity is low for reaching the brain though still within the range. Its size (molecular weight) is within the optimal range
Some drugs cause serious toxicity as a result of poor selectivity of action	Not applicable	

Drugs and their delivery systems manufactured by the pharmaceutical industry or compounded by the pharmacist are prepared according to industry and compendial standards and have been carefully designed with each of the following qualities:

1. *One dose in a manageable size unit*
   Dosage forms should be formulated such that the dose is contained in a unit that can be counted or measured by the patient with reasonable accuracy. For example, vehicles such as syrups or water for injection can be used to prepare solution dosage forms in measureable units.
2. *Palatable or comfortable*
   Many drugs are bitter or salty, so we add sweeteners and flavors to mask their taste. Parenteral drugs (injections) and drugs for application to mucous membranes must be comfortable enough to prevent tissue damage

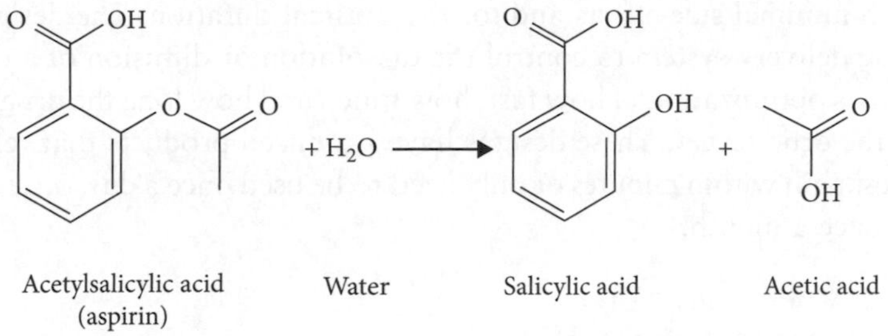

Figure 1.1   Hydrolysis of aspirin

or loss of drug from the site. We add tonicity agents and buffers to make them more comfortable and compatible with body tissues.

3. *Stable*
   Stability is defined as the extent to which a product retains the same properties and characteristics that it possessed at the time of its preparation or manufacture. There are three types of stability:
   - *Chemical stability:* The drug structure remains the same throughout the shelf life. Drugs in solution are particularly susceptible to chemical reactions that will change their structures (Figure 1.1).
   - *Physical stability:* The product retains the original physical characteristics of the dosage form, in particular the initial appearance and uniformity. For example, precipitation may occur in a solution dosage form or a suspended drug may settle and be difficult to redisperse.
   - *Microbiological stability:* Requirements will fall into one of two categories: (a) the dosage form must be sterile, or (b) the dosage form must resist microbial growth.

4. *Convenient/easy to use*
   Because most drug therapy takes place in the community, the dosage form should be portable and easily carried to work or play activities. The industry has developed a variety of new inhaler products that can be carried in the purse or pocket, as well as injection devices and portable infusion pumps for administration of injections in the community setting.
   Drugs that will be used for children or elderly patients should be available in liquid dosage forms that can be easily swallowed.
   Protein drugs cannot be formulated as oral dosage forms because they are quickly degraded in the gastrointestinal tract. These medications can be administered as injections but will require special training and patient motivation. Recent research has shown that protein drugs can be delivered via inhalation, and a number of peptide drugs are available as nasal aerosols for systemic absorption.

5. *Release of drug*
   To provide a therapeutic response, if the drug is not already in solution in the dosage form, it has to be released for absorption in a timely fashion

with minimal side effects and for the optimal duration. The design of a drug delivery system to control the dissolution or diffusion of a drug allows optimization of how fast, how much and how long the drug moves to the drug target. These designs have produced products that relieve chest pain within minutes or only need to be used once a day, once a week or once a month.

Pharmaceutical necessities

Pharmaceutical necessities are added to the drug and specific manufacturing processes are applied to build these properties into a delivery system.

> ### Definition
>
> *Pharmaceutical necessities* or *excipients* are non-drug ingredients added to the dosage form to build these properties into a delivery system (Appendix Table A.2).

Examples of pharmaceutical necessities include:

- preservatives
- buffers
- diluents
- solvents
- suspending agents.

In addition, specific manufacturing processes such as particle size reduction or sterilization are applied to provide a dosage form with these five qualities.

Pharmaceutical example

Let's consider these principles as they apply to the design of a tablet containing oxycodone and acetaminophen. This product is commercially available as a prompt release tablet that provides relatively rapid pain relief and must be taken every 4 to 6 hours. If we look at our table of excipients for oral solid dosage forms (Appendix Table A.3), we can identify the category of each of these ingredients in Endocet tablets. And if we think about each one we can see how the ingredient contributes to the qualities that we want to build into our dosage form (Table 1.2).

Table 1.2 Ingredients in immediate release oxycodone and their functions: Endocet (oxycodone and acetaminophen) tablets

Ingredient	Function	Quality
Croscarmellose sodium	Tablet disintegrant	Release: Promotes wicking of water into an oral solid so that it falls apart into particles. This causes the tablet to release drug faster because it exposes more surface to body fluids
Crospovidone	Tablet disintegrant	As above
Microcrystalline cellulose	Tablet binder	Ensures that the tablet remains intact after compression. Maintains one dose in a manageable size unit and physical stability of the tablet.
Povidone	Tablet binder	As above
Pregelatinized corn starch	Tablet binder	As above
Silicon dioxide (colloidal)	Glidant	Physical stability (uniformity of tablets). Improves flow properties and prevents caking of the powder mixture as it moves through tableting equipment so it can be accurately metered into the die and each tablet weighs the same amount and contains the same amount of drug
Stearic acid	Lubricant	Physical stability (uniformity of tablets). Prevents adhesion of the powder to tablet presses so powder can be accurately metered into the equipment and each tablet weighs the same amount and contains the same amount of drug
D&C Yellow No. 10	Dye	Makes the tablet distinctive so that pharmacists do not mistake it for a different product

Properties of the routes of administration

Key Point

Knowledge of each of the following route-related characteristics can be used to predict the rate and extent of drug travel to its target from a dosage form:

- accessibility, permeability and surface area of the absorbing membrane
- nature of the body fluids that bath the absorbing membrane
- retention of dosage form at the absorbing membrane long enough to release its drug
- location of enzymes or extremes of pH that can alter drug chemical structure before it is distributed to its target
- rate and extent of blood flow to the absorbing membrane and the distribution time from it.

To design a drug delivery system the pharmacist must apply an understanding of drug chemistry to the preparation of a product that will deliver the drug from the site of administration to the site of its activity. This requires an appreciation of drug travel in the body and the challenges and opportunities presented by the different routes of drug administration. To design dosage forms, recommend drug therapy or doses for a particular route of administration there are several key concepts to take away about each route:

- location and accessibility of the absorbing membrane – will the dosage form need to travel to reach it?
- permeability of the absorptive surface based on the number of layers and type of epithelial membrane
- the surface area of the absorbing membrane and the nature of the body fluids that bath it
- retention of the dosage form at the absorbing membrane long enough to release its drug
- enzymes or extremes of pH that can alter its chemical structure encountered by the drug before it is distributed to its target
- rate and extent of blood flow to the absorbing membrane and the distribution time from it.

Routes of administration

The routes by which we administer drugs can be divided into those that produce systemic drug effects and those that are site-specific (Tables 1.3 and 1.4).

Table 1.3 Systemic routes of administration

Route	Dosage forms – drug delivery systems	Administration	Absorbing membrane	Onset
Intravenous route	Aqueous solutions, emulsions	Places drug solution directly in a vein	Capillaries in target tissue	Within 60s
Intramuscular route	Aqueous solutions, oily solutions, suspensions	Places drug into one of the larger muscles: deltoid, vastus lateralis or gluteals	Capillaries in muscle and target tissue	15–30 minutes for aqueous solutions
Subcutaneous route	Aqueous solutions, suspensions, implants	Places drug under the layers of the skin	Capillaries in muscle and target tissue	15–60 minutes for aqueous solutions
Sublingual	Tablets, solutions, aerosols	Placed under the tongue	Sublingual mucosa	1–3 minutes
Buccal	Tablets, lozenges, chewing gum	Placed between the cheek and gum	Buccal mucosa	2–3 minutes
Oral	Tablets, solutions, suspensions	Swallowed	Mucosa of the small intestine	30–60 minutes for immediate release dosage forms
Rectal	Suppositories, ointments, creams, solutions, suspensions, foams	Placed in the rectal cavity	Rectal mucosa	30–60 minutes
Vaginal	Suppositories, creams, gels, tablets, rings	Placed in the vaginal cavity	Vaginal mucosa	Slow for systemic effects
Skin (trandermal)	Patches, ointments, gels	Applied to thinner skin surfaces (trunk, upper arm, abdomen)	Stratum corneum	1–4 hours
Nose	Aerosolized solutions or suspenions	Applied to the anterior turbinate area	Respiratory or olfactory mucosa	Within 10 minutes
Lung	Aerosolized solutions, suspensions or powders	Inhaled to bronchiolar and alveolar epithelia	Respiratory mucosa	Within 5 minutes

Table 1.4 Local routes of administration

Route	Dosage forms – drug delivery systems	Administration	Absorbing membrane	Onset
Buccal	Solutions, suspensions, lozenges	Applied to the mucosa of the oral cavity	Oral mucosa	Rapid for local effects
Rectal	Suppositories, ointments, creams, solutions, suspensions, foams	Placed in the rectal cavity	Rectal mucosa	Rapid for local effects
Vaginal	Suppositories, creams, gels, tablets, rings	Placed in the vaginal cavity	Vaginal mucosa	Rapid for local effects
Skin (local)	Ointments, creams, gels, lotions, aerosols, powders	Applied to the affected skin surface	Stratum corneum	Rapid for local effects
Eye (anterior segment)	Solutions, suspensions, ointments, gels, inserts	Applied to the conjunctival sac	Cornea, sclera	Within 10 minutes
Eye (posterior segment)	Solutions, suspensions, inserts	Injected into the vitreous humor	Retina	Varies by dosage form
Nose	Solutions, suspensions, aerosols	Applied to the anterior turbinate area	Respiratory mucosa	Rapid
Lung	Aerosolized solutions, suspensions or powders	Inhaled to bronchiolar and alveolar epithelia	Respiratory mucosa	Within 5 minutes

Definitions

- *Systemic drug administration* means that the drug is carried from the site of administration to a site of activity elsewhere and potentially exposes all body tissues to drug.
- *Site-specific drug administration* means that the drug is administered to the tissue in need of drug therapy (also known as local drug therapy) or is targeted to a particular site by specific interactions with that site, for example, by binding a cell-associated antigen or selective activation in the target cell.

Table 1.5 Onset and duration of anti-anginal effects of nitroglycerin from various dosage forms and routes

Route and dosage form	Onset	Duration
Sublingual tablet or spray	2 minutes	30 minutes
Buccal extended release tablet	2–3 minutes	3–5 hours
Oral extended release tablet	1 hour	Up to 12 hours
Intravenous solution (push)	1–2 minutes	3–5 minutes
Transdermal ointment	30 minutes	3 hours
Transdermal patch	60 minutes	12 hours

Table 1.5 illustrates how fast and how long a patient would experience the anti-anginal effects of nitroglycerin from various dosage forms and routes of administration.

Parenteral routes

Definition

- *Parenteral* means 'other than enteral (by mouth)' and is generally intended to mean the routes by which we inject drugs.

All drug products used by parenteral routes must be sterile and those products that will be used for multiple doses must also resist microbial growth.

The most commonly used parenteral routes are intravenous, intramuscular and subcutaneous. These routes can be used for drugs that would be destroyed by gastrointestinal acid and enzymes and to delay exposure to liver enzymes.

- The response to drug therapy by the intravenous route is generally rapid and predictable because there is no absorption step.
- Because subcutaneous injections are relatively easy to give, this route is usually chosen for drugs that require outpatient, parenteral therapy.

Enteral routes

The enteral routes include application to the oral cavity, oral dosage forms that are swallowed and administration to the rectum.

The preferred enteral route from the standpoint of ease of use is to swallow a dosage form for absorption into the blood supply of the small intestine. Although convenient and easy to explain to patients, this route provides slower and less predictable onset than the parenteral routes or administration of drug to the oral cavity.

The loss of active drug as it encounters the extremely low pH of the stomach, the array of enzymes in the small intestine and the liver is termed the *first pass effect*.

Rectal and vaginal routes

The rectal and vaginal routes are safe and technically simple but are not often preferred. They can be used for the administration of local or systemic medications.

Drugs applied to the skin

Drugs may be applied to the skin for local or systemic absorption.

- When intended for local effects, the response to an applied drug is usually rapid and confined to the area of application.
- Drugs applied transdermally are intended to penetrate all layers of the skin and circulate from the dermal vasculature in sufficient concentration to provide systemic effects.
- The transdermal route provides safe and convenient drug therapy of long duration.

Inhalation routes: nasal and lung

Drugs administered to the lungs must be provided in finely divided aerosol dosage form in order to travel through the branches and turns of the airways to reach the lower respiratory tract.

- The lungs provide a very large surface resulting in reproducible and rapid absorption.
- If the dose is sufficiently small, the effect of the drug may be confined to the lungs.
- Drug administered to the lungs can effectively provide systemic treatment if the dose is properly titrated.

Liberation, absorption, distribution, metabolism and elimination (LADME)

The application of a drug at the site of administration is only the start of its journey to its target tissues. The drug must be liberated from the dosage form, absorbed, and distributed to the site of action. At the same time that it is distributed to its receptor, the drug is delivered to the metabolic enzymes of the liver and filtered through the kidneys into the urine. The termination of

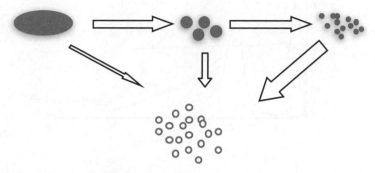

Figure 1.2 Drug dissolution from oral solid dosage form

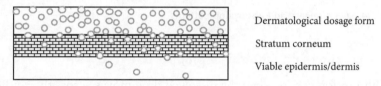

Dermatological dosage form

Stratum corneum

Viable epidermis/dermis

Figure 1.3 Drug diffusion from a topical dosage form

drug action through these latter two processes may be inconveniently rapid, requiring the patient to take a dose of drug frequently to sustain its effect. Advances in dosage form design based on knowledge of the biological fate of drugs have produced drug delivery systems optimized to the drug's therapeutic purpose and elimination rate.

Drug liberation

After administration, if the drug is not in an aqueous solution, it is released or liberated from the dosage form by either dissolution or diffusion or a combination of both processes.

- *Dissolution:* If the drug is administered in the form of an immediate release tablet, the tablet disintegrates into large granules in the fluids of the stomach, which in turn come apart into fine particles and finally result in the drug in solution (Figure 1.2).
- *Diffusion:* A drug administered as a trandermal patch must diffuse from an area of high concentration (the patch) to an area of low concentration (the skin) where it is dissolved in the lipid or water content of the skin (Figure 1.3).

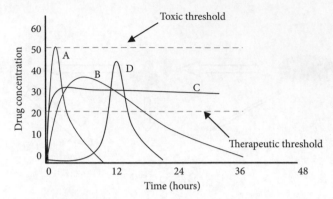

Key:
A Immediate release
B Sustained release
C Controlled release
D Delayed release

Figure 1.4 Drug release and absorption patterns

Drug absorption and patterns of drug release

Drug absorption occurs when the dissolved drug moves across the epithelial barriers it encounters and into the blood stream.

The step that is slowest (liberation or absorption) will determine the rate of drug appearance in the blood stream.

Broadly speaking we see four different patterns of drug absorption that are a result of the release design of the dosage form and the body's handling of the drug (Figure 1.4).

- *Immediate release* is characterized by a rapid rise in blood levels of the drug followed by a decline determined by the ability of the body to eliminate the drug.
- *Delayed release*, as the name implies, delays the appearance of the drug in the blood until the dosage form reaches a certain compartment of the gastrointestinal tract, where it is able to dissolve and release medication.
- *Sustained release* is characterized by a slow rise in drug concentrations in the blood, followed by a slow decline that provides therapeutic amounts long enough to reduce the frequency of administration compared with an immediate release dosage form.
- Controlled release is an absorption pattern with a slow rise in drug concentrations to a level that remains constant for a predictable interval.

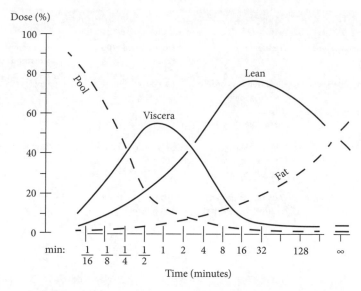

Figure 1.5 Blood flow primacy to the organs and tissues. (Reproduced with permission from selected reading 6)

Drug distribution

The drug is distributed primarily by transport with blood or lymph to the tissues, broadly speaking in order of their perfusion primacy (Figure 1.5, Table 1.6).

Some tissues are less accessible to drugs than others because of the reduced permeability of capillary endothelia in the retina, brain, spinal cord and enteric nervous system. The blood–brain barrier is one that is of great importance to drug development, in that many effects and side effects of drugs are mediated through receptors in the brain.

Drug elimination

The activity of a drug is terminated by its *elimination* either as intact drug via body wastes or by metabolic transformation to an inactive compound.

- *Excretion*, or the removal of intact drug, occurs through the kidneys into urine for water soluble drugs, the bile into feces for amphiphilic drugs and the lungs via expired air.
- Drug-metabolizing enzymes can be found in many tissues though the most significant populations are in liver, kidneys and intestine. These proteins catalyze reactions that transform fat soluble drugs into more water soluble forms or add large moieties such as amino acids to make an amphiphilic metabolite. These changes in drug structure facilitate the final elimination of the drug metabolite through urine or feces.

Table 1.6 Blood flow to body organs

Fluid compartment	Tissues	Flow rate per 100 g tissues	Time to complete drug distribution
Well perfused viscera	Brain, lungs, heart, liver, kidneys, adrenal, thyroid, gastrointestinal tract	50–400 mL/minute per 100 g tissue	<2 minutes
Intermediate perfusion: lean tissue	Muscle, skin, red marrow, non-fat subcutaneous tissue	5–12 mL/minute per 100 g tissue	60–90 minutes
Poorly perfused: fat	Adipose tissue, fatty marrow, bone cortex	1–3 mL/minute per 100 g tissue	300 minutes

Table 1.7 Characteristics of the oral (to small intestine) route

Questions	Answer	Challenge or opportunity?
What is the membrane that will absorb the drug from this route and how accessible is it?	Drug is absorbed through the small intestinal epithelium. The dosage form must be swallowed and be emptied from the stomach before absorption starts	Challenge
Is the absorbing membrane single or multiple layers and how permeable is it?	Single, excellent permeability	Opportunity
How large is its surface?	Large, 100 m^2	Opportunity
What is the nature of the body fluids that bathe the absorbing membrane?	Small intestinal fluids: pH 5.0–7.0, relatively large volume	Opportunity
Will the dosage form remain at the absorbing membrane long enough to release its drug?	Usually but not always. Some drugs are absorbed in a small window of the small intestine and are not completely released by the time they move	Neither
Will the drug encounter enzymes or extremes of pH that can alter its chemical structure before it is distributed to its target?	Yes, extremely low pH in the stomach, lots of enzymes: peptidases, esterases, cytochrome P450	Challenge
What is the blood flow primacy to the absorbing membrane and the distribution time from it?	Highly perfused <2 minutes	Opportunity

Pharmaceutical example

Let's consider these characteristics of the route of administration and travel to the drug receptor as they apply to the design of our tablet containing oxycodone and acetaminophen. Think about what you know about the anatomy and physiology of the gastrointestinal tract as it applies to each of the route-related characteristics for dosage form design and whether the characteristic represents a challenge or an opportunity for drug delivery (Table 1.7).

Questions and cases

1. A pharmaceutical necessity is:

   (a) The active ingredient

   (b) An ingredient required by the Food and Drug Administration

   (c) An ingredient that enhances the size, stability, palatability or release of a drug

   (d) An ingredient that prevents drug elimination

2. Syrup dosage forms must be:

   (a) Sterile

   (b) Resistant to microbial growth

   (c) Sterile and resistant to microbial growth

   (d) Prepared as powders for reconstitution just prior to use

3. Liberation from a dosage form refers to:

   (a) Movement of the drug across the epithelial barriers it encounters and into the blood stream

   (b) The application of a dosage form to an absorptive site

   (c) Movement of the drug from interaction with elements of the dosage form to interaction with molecules of biological fluid

   (d) Termination of drug activity

4. Which of the following drugs is likely to move through membranes most rapidly?

   (a) Levodopa, log P −1.8, molecular weight (MW) 197

   (b) Amphotericin, log P 0.8, MW 924

   (c) Citalopram, log P 3.5, MW 324

   (d) Amikacin, log P −7.4, MW 585

5. A patient has a prescription for ondansetron orally disintegrating tablets. The reason that the drug may be prepared in this dosage form includes:

   (a) Need for rapid absorption

   (b) Need for infrequent administration

   (c) Need to avoid first pass metabolism

   (d) Need to avoid renal excretion

   (e) Need to administer a small dose

6. Oxytocin is a peptide drug containing nine amino acids. Because oxytocin increases trust and bonding behaviors and reduces fear, there is interest in using the drug to treat children with autism. Which of the following formulation approaches would be useful to prevent oxytocin's destruction by gastrointestinal peptidases?

(a) Preparation as an oral suspension

(b) Preparation as an extended release tablet

(c) Preparation as a buffered solution

(d) Preparation as a buccal tablet

7. The lactam ring in penicillin G is hydrolyzed in solution. This is an example of:

(a) Poor water solubility

(b) Formation of a precipitate

(c) Physical instability

(d) Chemical instability

(e) Microbial instability

8. After application to the affected area of skin, fluticasone moves out of the cream dosage form and into the skin. The mechanism of release of fluticasone from the cream can be described as:

(a) Distribution

(b) Dissolution

(c) Diffusion

(d) Disintegration

9. The pattern of drug absorption characterized by a slow rise in drug levels in the blood followed by a slow decline is termed:

(a) Immediate release

(b) Delayed release

(c) Sustained release

(d) Controlled release

10. The correct ranking of routes of administration in terms of fastest speed of onset to slowest speed of onset of action is:

(a) Sublingual, oral, intravenous, transdermal

(b) Transdermal oral, sublingual, intravenous

(c) Intravenous, sublingual, oral, transdermal

(d) Intravenous, oral, sublingual, transdermal

11. Research the chemical and biological properties of the following drugs using Drugbank (http://www.drugbank.ca/) or another drug database: ibuprofen, amoxicillin, and lidocaine. Chemical stability information is available in the American Hospital Formulary Service Drug Information. What challenges do these drugs present to the formulator in terms of dissolution in body fluids, movement through membranes, stability in dosage forms, loss of drug due to metabolism, access to receptors, or selectivity of the target?

Ibuprofen

Design issue	Drug property	Analysis
Drug must be dissolved before it can cross membranes		
Drug must have sufficient lipophilicity to cross membranes		

Design issue	Drug property	Analysis
Large molecular weight drugs have difficulty moving across membranes		
Drugs may hydrolyze, oxidize, photolyze or otherwise degrade in solution		
Some drugs may be substantially degraded by enzymes before they reach their target		
Intracellular drug receptors or receptors in the brain or posterior eye are challenging to target		
Some drugs cause serious toxicity as a result of poor selectivity of drug action		

Amoxicillin

Design issue	Drug property	Analysis
Drug must be dissolved before it can cross membranes		
Drug must have sufficient lipophilicity to cross membranes		
Large molecular weight drugs have difficulty moving across membranes		
Drugs may hydrolyze, oxidize, photolyze or otherwise degrade in solution		
Some drugs cannot be used by mouth because of substantial losses to enzyme degradation		
Intracellular drug receptors or receptors in the brain or posterior eye are challenging to target		
Some drugs cause serious toxicity as a result of poor selectivity of drug action		

12. Research the inactive ingredients in the following dosage forms. Determine the function of the excipients using Appendix Table A.2. Categorize them as contributing to the manageable size of the dosage form, its palatability or comfort, its stability (chemical, microbial, or physical), and the convenience of its use or the release of drug from the dosage form. Note that ingredients may be included in many dosage forms for more than one purpose.

Adrenaline chloride injection multidose vial (King, other manufacturers)

Ingredient	Category	Function
Sodium bisulfate		
Chlorobutanol		
Water for injection		

Lumigan (bimatoprost) ophthalmic solution

Ingredient	Category	Purpose
Benzalkonium chloride		
Sodium chloride		
Dibasic sodium phosphate		
Citric acid		
Purified water		

Donepezil orally disintegrating tablet (Zydus)

Ingredient	Category	Purpose
Crospovidone		
Magnasweet		
Magnesium stearate		
Mannitol		
Silicon dioxide (colloidal)		
Sucralose		

Depakene (valproic acid) oral solution

Ingredient	Category	Purpose
Cherry flavor		
FD&C Red No. 40		
Glycerin		
Methyl and propyl parabens		

Ingredient	Category	Purpose
Sorbitol		
Purified water		
Sucrose		

13. Discuss the following route of administration with a group of classmates. From what you know about anatomy and physiology, determine the answer to the questions below about the route. Do you think the characteristic represents a challenge or an opportunity for drug delivery? (You may conclude that it is neither or equivocal.) Appendix Table A.4 has some data that may be helpful.

Skin (local treatment) route

Questions	Answer	Challenge or opportunity?
What is the membrane that will absorb the drug from this route and how accessible is it?		
Is the absorbing membrane single or multiple layers and how permeable is it?		
How large is its surface?		
What is the nature of the body fluids that bath the absorbing membrane?		
Will the dosage form remain at the absorbing membrane long enough to release its drug?		
Will the drug encounter enzymes or extremes of pH that can alter its chemical structure before it is distributed to its target?		
What is the blood flow primacy to the absorbing membrane and the distribution time from it?		

Selected reading

1. USP Convention. Pharmaceutical compounding – nonsterile preparations, Chapter <795>. United States Pharmacopeia, 34th edn, and National Formulary, 29th edn. Baltimore, MD: United Book Press, 2011.

2. VF Patel, F Liu, MB Brown. Advances in oral transmucosal drug delivery, J Contr Rel 2011; 153: 106–116.
3. NR Mathias, MA Hussain. Non-invasive systemic drug delivery: developability considerations for alternate routes of administration, J Pharm Sci 2010; 99: 1–20.
4. X Liu, B Testa, A Fahr. Lipophilicity and its relationship with passive drug permeation, Pharm Res 2011; 28: 962–977.
5. HY Rojanasakul, LY Wang, M Bhat, et al. The transport barrier of epithelia: a comparative study on membrane permeability and charge selectivity in the rabbit, Pharm Res 1992; 9: 1029–1034.
6. HL Price. A dynamic concept of the distribution of thiopental in the human body, Anesthesiology, 1960; 21: 40–45.
7. F Broccatelli, G Cruciani, LZ Benet, TI Oprea. BDDCS Class prediction for new molecular entities. Mol Pharmaceut 2012; 9: 570–580.

2

Intermolecular forces and the physical and pharmaceutical properties of drugs

Are you ready for this unit?

- Review relationships between
  - amounts: grams, moles, milligrams, millimoles
  - concentrations: molarity, molality, percent, weight/weight, weight/volume, volume/volume
  - density, specific gravity
- Review for each of the organic functional groups:
  - bond polarity
  - acidic, basic or neutral
- Given a drug structure, determine whether it is an acid or a base.

Learning objectives

Upon completion of this chapter, you should be able to:

- Define crystal, crystalline habit, polymorph, hydrate, efflorescence, cocrystals, amorphous solid, drug salt, drug derivative, trituration, micronized, granulation, adsorption, hygroscopic, deliquescent, surface energy, molecular solution, solubility, log partition coefficient, dissolution and intrinsic dissolution rate.
- Explain the effect of solid form on melting point, solubility and dissolution rate.
- List the consequences of surface interactions on the pharmaceutical properties of drugs.
- Explain the effect of particle size on dissolution rate, surface energy, wettability and powder flow.
- Describe how powder flow can be improved.
- Given a drug's melting point, predict whether its solubility will be affected by attractive forces within the crystal.
- Given a drug structure, predict whether the drug's functional groups are likely to confer water solubility or octanol solubility (membrane permeability).
- Given a drug's log partition coefficient, predict the partitioning preferences of a drug.
- Explain the factors that affect dissolution rate of solids in liquids.

Introduction

As we use solid drugs and pharmaceutical necessities to make dosage forms, we need to be able to mix them to uniformity, push them through tableting machines or capsule fillers and ensure that they release drugs for absorption through dissolution. Powder mixing and flow and the dissolution of solids all depend on the characteristics of the solids used. In this chapter we want to consider the characteristics of solids that affect the preparation and performance of solid dosage forms. These are:

- melting point
- arrangement of the molecules of drug in the solid
- particle size, shape and surface energy
- solubility
- dissolution rate.

Solid state, forces between ions and molecules, and physical properties of drugs

Key Point

> The physical properties of a drug solid may be modified by preparation of the drug as a salt, a derivative, a polymorph, a hydrate, or a cocrystal.

There are several attractive forces between molecules or ions that can exceed the repulsive forces and result in the solid state of matter. Some of the stronger forces that exist between ions and molecules are presented in Table 2.1.

Types of attractive forces

Electrostatic forces are those between oppositely charged ions. These forces tend to be quite strong compared with the other attractive forces. Ions may also interact with oppositely charged dipoles.

A *dipole* is an unequal sharing of electrons between atoms in a covalent bond resulting in one atom that is relatively negative and one that is relatively positive (Figure 2.1).

Van der Waals forces are attractive forces that involve permanent or induced dipoles; these forces are also referred to as dipole–dipole, hydrogen bonds or induced dipole interactions.

Hydrophobic interactions are created in polymers such as proteins by the preference for hydrophobic areas of the molecule to be sequestered away from polar water molecules.

Table 2.1 Attractive forces between molecules		
Attractive force	**Description**	**Example**
Ion–ion	Interaction between anion and cation	Na^+ and Cl^-
Ion–dipole	Interaction between an area of electron density and a cation or an area of electron deficiency and an anion	Na^+ and the oxygen of water
Dipole–dipole	Interaction between an area of electron density created by unequal sharing of electrons in a covalent bond and an area of electron deficiency in adjacent molecule	$C^{\delta+}$ of a carbonyl and $O^{\delta-}$ of the carbonyl on an adjacent molecule
Hydrogen bond (a type of dipole–dipole attraction)	Interaction between a molecule with a hydrogen atom and a strongly electronegative atom (O, N, F)	$H^{\delta+}$ and $O^{\delta-}$ of water
Hydrophobic Interactions	Interaction between two areas of hydrocarbon in a molecule	Folding of proteins to place amino acids with hydrophobic side chains on the inside of the polymer

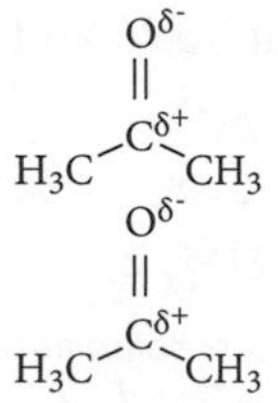

Figure 2.1 Dipoles in organic compounds

Repulsive forces

There are also repulsive forces between ions and molecules encountered when the electron clouds of two atoms become close enough to enter the same space and repel each other. At a distance of 0.3–0.4 nm the attractive forces will equal the repulsive forces and the volume occupied by the two atoms is most stable.

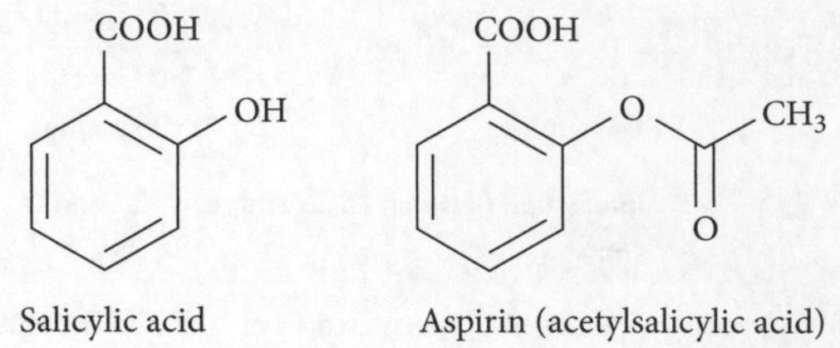

Salicylic acid   Aspirin (acetylsalicylic acid)

Figure 2.2 Melting point and solubility

Melting point

Key Point

- Melting point is a measure of of the strength of interactions between molecules in a solid and of drug–drug attractive forces.
- Crystalline forms have higher melting points than amorphous forms.

A *crystal* is an orderly arrangement of molecules that permits optimal attractive interactions between adjacent molecules within the solid. This creates the diffraction pattern we observe, as X-rays are reflected from the regular array of atoms.

- The lower melting point a solid has, the less energy is required to disrupt interactions within the solid to become a liquid. The lower attractive forces between molecules in a solid with a lower melting point will also confer greater solubility in any solvent.
- The higher melting point a solid has, the more energy is required to disrupt interactions within the solid and the poorer the solubility.

Pharmaceutical example

The effect of attractive forces on the solubility of drug molecules is most easily observed in structurally similar drugs (Figure 2.2 and Table 2.2).

Polymorphs

Definition

Polymorphism is the phenomenon whereby molecules arrange themselves in more than one pattern within a crystal.

Table 2.2 Melting points and water solubilities of structurally similar drugs

Compound	Melting point (°C)	Water solubility
Acetylsalicylic acid (aspirin)	135	1 g/300 mL
Salicylic acid	157–159	1 g/460 mL

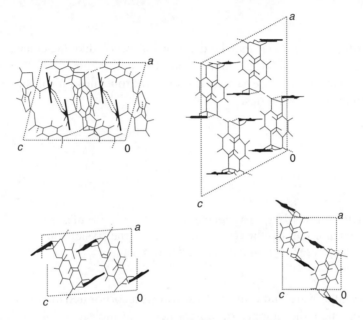

Figure 2.3 Four polymorphs of sulfathiazole. (Reproduced with permission from selected reading 1)

Polymorphism can influence the melting point, solubility, dissolution rate and bioavailability of a drug, and processing properties such as powder flow and compressibility. About 36% of drugs exhibit polymorphism.

- There will be one polymorph in which the intermolecular attractive forces are best aligned and it will be the most thermodynamically stable. This form of the drug is the highest melting and the least soluble in any solvent (Figure 2.3).
- Less stable polymorphs can spontaneously convert to the more stable form over time.
- Because polymorphs have different physical properties they offer the opportunity to select a lower melting point and therefore a more soluble form of a drug for use in a solid dosage form.
- Polymorphism can complicate the preparation of suspensions because as the drug moves in and out of solution it will ultimately re-form in the most thermodynamically stable and least soluble form.

Crystalline habit

> *Definition*
>
> *Crystalline habit* is a term used to describe the shape of the drug crystal that can be appreciated by visual inspection.

- A drug's crystal habit may be described as needle-like (acicular), columnar, pyramidal, prismatic, plate-like and tabular.
- The shape of the drug particle is important to its flow, dispersibility, and aerodynamic properties.

Solvates

> *Definitions*
>
> *Solvates* are orderly arrangements of drug molecules that include solvent molecules in the crystalline lattice. If the crystallization solvent is water, these forms are called *hydrates*.

The solvent or water molecule in the solvate is incorporated in an exact molar ratio such that the number of waters per drug molecule will be included into the name of the solid, for example, magnesium sulfate heptahydrate or amoxicillin trihydrate.

- It has been estimated that about 30% of drugs form hydrates.
- Hydrates may lose their waters of crystallization and become sticky, a process called *efflorescence*.

Cocrystals

> *Definition*
>
> *Cocrystals* are crystals composed of an active drug species and another organic molecule.

Another approach to the modification of the physical properties of a solid drug is the preparation of cocrystals.

- The other molecule in the crystal, referred to as a coformer, forms hydrogen bonds with the drug molecule stoichiometrically providing a lower melting and more soluble solid than a pure crystalline form of the active drug. The coformer must be generally recognized as safe (GRAS) for human consumption; compounds serving as coformers include drugs such as caffeine, vitamins such as nicotinamide or endogenous molecules such as adenine or cytosine.
- Cocrystals may be more stable than polymorphs of the drug thus reducing concerns about polymorph instability in dosage forms.
- A number of cocrystals of poorly soluble drugs – carbamazepine, theophylline, furosemide and itraconazole – have been developed but none are commercially available.

Amorphous solids

Definition

Amorphous solids have no orderly arrangement of molecules within the solid; they are simply a pile of molecules.

- The minimal interactions between molecules in an amorphous solid are easily disrupted by solvent. Thus the amorphous form of a drug will always have a lower melting point and a better solubility than the crystalline form(s).
- Amorphous solids may be formed when a drug solution is cooled so rapidly that solute molecules lose mobility before assuming their lattice positions as occurs in freeze drying procedures.
- Amorphous solids are generally porous and low density, rendering them susceptible to sorption of vapor from the atmosphere and subsequent chemical or microbial instability.

Pharmaceutical example

Orally disintegrating tablets, a solid dosage form made to dissolve in the small volume of saliva in the mouth, are often made with the amorphous forms of the drug and excipient solids. Each dosage form must be packaged individually to prevent adsorption of water from the atmosphere.

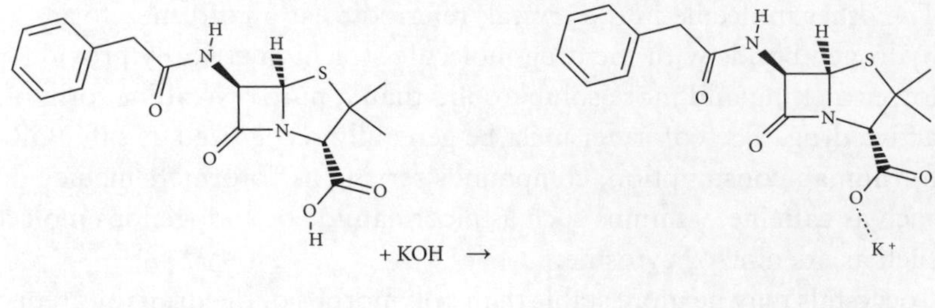

+ KOH →

Penicillin free acid → Potassium penicillin
FW 334.39; water solubility 210 mg/1000 mL → FW 372.48; water solubility 1 g/0.5 mL

Figure 2.4 Drug salts – penicillin

Drug salts

Approximately 75% of drugs are weak bases and about 20% are weak acids. By virtue of their ability to accept or donate protons, these drugs are capable of existing in both a water soluble (charged) and a lipid soluble (uncharged) form. This property has importance implications for their formulation and biological fate, and the process of drug absorption illustrates this well. To be absorbed (to move across biological membranes towards drug receptors), a drug must be in solution in the aqueous body fluids.

> *Definition*
>
> Most drug dosage forms, whether solid or liquid, are prepared with the charged, water soluble form of the drug referred to as a *drug salt* (Figure 2.4).

- Drug salts are prepared from the reaction of the weak base drug with strong acid, or the weak acid drug with strong base.
- A drug may exist as different drug salts, morphine hydrochloride and morphine sulfate, which will have different physical properties such as solubility and melting point (Figure 2.5).
- These physical properties can affect drug absorption and, as a result, different salt forms of the same drug are not regarded by the Food and Drug Administration (FDA) as 'pharmaceutically equivalent'.

Pharmaceutical example

While both forms of metoprolol can treat high blood pressure equally well, metoprolol tartrate tablets should not be dispensed in lieu of metoprolol succinate tablets because of differences in the biopharmaceutical behavior of the two forms.

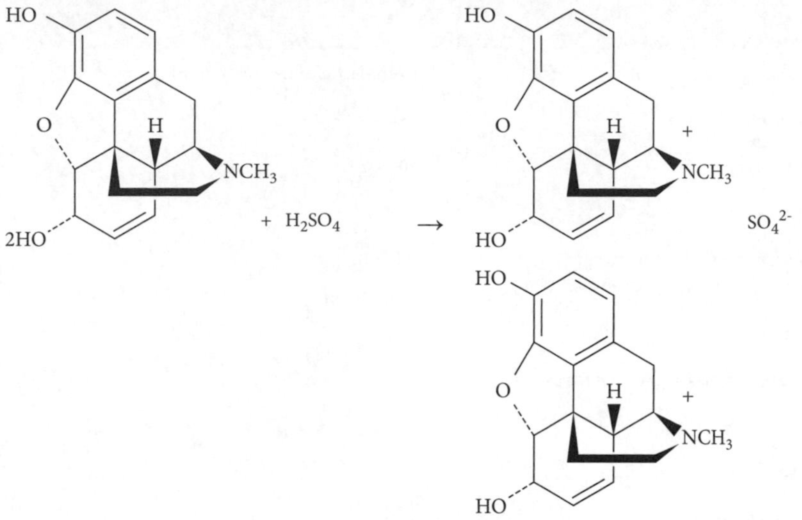

2 Morphine base → (Morphine)$_2$ sulfate

FW 285.34; water solubility 1 g/5000 mL → FW 668.83; water solubility 1 g/15.5 mL

Figure 2.5 Drug salts – morphine

Drug derivatives

> ### Definition
>
> *Drug derivatization* involves the covalent attachment of a moiety that can be cleaved *in vivo* to produce the active form of the drug. Generally this means the formation of an ester or amide bond between the drug molecule and the derivatizing group that can be removed by esterases or peptidases.

As with drug salts, drug derivatives will have different physical properties and significantly larger formula weights than the original drug and are not 'pharmaceutically equivalent' to dosage forms containing underivatized forms of the same drug.

Surface interactions and the pharmaceutical properties of drugs

Particle size

The particle size of a drug solid will have significant effects on the drug's pharmaceutical properties such as dissolution rate, stability and powder flow. Particle sizes that are encountered in dosage forms can range from 0.1 μm (dry powder inhalers) to 1000 μm in diameter (Table 2.3).

Table 2.3 Particle sizes in commercially manufactured dosage forms

Dosage form	Drug particle size range (µm)
Ophthalmic suspensions	≤10
Oral suspensions	10–50
Parenteral suspensions	0.5–25
Parenteral dispersions for intravenous use (emulsions, liposomes, nanoparticles) USP	0.1–0.5
Aerosols for the lung	1–5
Tablets and capsules, immediate release	≤50
Topical aerosols	50–100
Topical emulsions	≤50
Topical suspensions	10–50

A drug powder may be expected to have a range of particle sizes which can be described by a frequency distribution plot. A pharmaceutical powder can then be characterized by its mean, d_{ave} or median, d_{med} particle diameter. Particle size of powders can be measured with light and electron microscopy, sieving, sedimentation and centrifugation, Coulter counter, quasi-elastic light scatter and column chromatography. In the compounding pharmacy, sieving is traditionally used because of its simplicity. Table 2.4 provides the classification of powders by sieve number (openings per linear inch) and width of mesh opening.

Reducing the particle size of the active ingredient:

- increases surface area, which increases dissolution rate and bioavailability
- improves mixing or blending of several solid ingredients if they are reduced to approximately the same size
- minimizes segregation
- contributes to greater dose uniformity.

The increased surface area created by milling:

- increases susceptibility to degradation due to air, moisture or light
- promotes air adsorption and inhibits wettability
- increases attractive forces between particles that cause particles to aggregate and flow poorly.

In a compounding pharmacy milling is accomplished on a small scale using a mortar and pestle. Many of the powders used in pharmacies are prepared in fine grind by the manufacturer and the pharmacist's task is to ensure that all particles in a solid mixture are uniform. The process of grinding a powder in a

Table 2.4 Classification of particle size

Descriptive category	Mesh (openings/inch)	Width of mesh opening (μm)	Applications
	2	9520	Granules
Very coarse	8	2380	Granules
	10	2000	Granules
Coarse	20	850	Granules
	30	600	Granules
Moderately coarse	40	425	Compounded divided
	50	300	Powders and suspensions
Fine	60	250	Compounded divided
Very fine	80	180	Powders and suspensions
	120	125	Compounded divided
	200	75	Powders and suspensions
Smallest sieve	325	45	
Micronized	NA	1–5	Commercially available micronized powders

Adapted from reference 6.

mortar is called *trituration*. Limited research in the area of extemporaneous preparation of powders indicates that conventional trituration in a mortar can produce particles with a median diameter (d_{med}) of 53.52 (±0.62) micrometers.

Adsorption

Definition

Adsorption describes a tendency for materials (molecules, ions, particles) to locate at a surface in a concentration different than that found in the surrounding medium. The phenomenon of adsorption occurs as a result of unequal forces at surfaces or interfaces.

Adsorption is more commonly observed in liquid dispersions (molecular, colloidal and coarse).

Solid particles may demonstrate a preference for surfaces of plastics or glass. This behavior can be observed when mixing powders extemporaneously in sealed plastic bags as may be done to prepare antineoplastic capsules or simply as a method to produce diffusive mixing.

The pharmacist should be aware that excessive loss of one powder over another can change the concentration of the dilution and the resulting dosage form.

Gases including water vapor will adsorb to drug and excipient solid particle surfaces creating concerns for the stability of hydrolyzable or oxidizable drugs, or interfering with proper wetting of solids.

Hygroscopicity and deliquescence

Pharmacists must note whether drug powders under their care are hygroscopic or deliquescent.

Key Points

- *Hygroscopic* powders take up water from the atmosphere.
- Typical relative humidity for drugs and associated excipients stored without special precautions is between 25% and 75% at 25°C.
- Drugs or excipients with polar surface groups (good water solubility) will adsorb a few molecular layers of water on their surfaces, particularly if their surface area is large (particle size is small).
- Metal salts (sodium, potassium, lithium, calcium and magnesium) are prone to hygroscopicity, some forming both stochiometric and non-stochiometric hydrates.
- *Deliquescent* powders will sorb water vapor from the environment and gradually form a solution that impacts the chemical stability of the drug or excipient.

Table 2.5 lists important drugs and excipients with hygroscopic and deliquescent tendencies. The pharmacist in a compounding pharmacy may find taking the weight of a hygroscopic solid each time it is opened useful to monitoring its water sorption status.

Manufacturers of drug and excipient powders will provide information on the label or in the catalog about the hygroscopicity of their compounds. Once opened, the container of a hygroscopic powder should be stored in a tightly closed, glass container with a bed of desiccant. If the solid is refrigerated, it should be allowed to come to room temperature before the container is opened. Solid dosage forms may also have some tendency to sorb water and are packed in glass or in polypropylene containers with desiccants. The pharmacist should instruct the technical staff to return the desiccant to the original container of dosage forms for which they are provided.

Table 2.5 Hygroscopic and deliquescent drugs			
Drugs		**Excipients**	
Albuterol sulfate	Hygroscopic	Acetic acid, glacial	Hygroscopic
Ammonium chloride	Hygroscopic	Aspartame	Hygroscopic
Bacitracin	Hygroscopic	Bentonite	Hygroscopic
Bismuth subnitrate	Hygroscopic	Boric acid	Hygroscopic
Calcium chloride (anhydrous or dihydrate)	Hygroscopic	Carbomer	Hygroscopic
Calcium pantothenate	Hygroscopic	Carboxymethylcellulose, sodium	Hygroscopic
Diphenhydramine HCl	Deliquescent	Histamine dihydrochloride	Deliquescent
Hydroxyzine diHCl	Deliquescent	Neostigmine bromide	Deliquescent
Triflupromazine HCl	Deliquescent	Valproate sodium	Deliquescent

Particle interactions and surface energy

The ability to separate particles in a dosage form or to ensure that they flow through equipment that measures a single dose is essential to the uniformity of dispersed and solid dosage forms. Molecules on the surface of a particle have a net inward force exerted on them from the molecules in the interior of the particle. If we attached a string at the surface of the particle, the force per unit length we would have to exert to counter the inward pull is called the surface energy when the solid is immersed in air or interfacial tension (millinewtons/metre or dyne/centimetre) when the solid is immersed in a liquid. In fact this is very difficult to measure because a solid particle will have many faces (surfaces) each of which can have slightly different surface tension.

Key Points

- *Micronized* particles or those with diameters in micrometers will have high surface energy relative to larger particles of the same drug by virtue of their high surface area.
- Particles with high surface energy will reduce their surface area through aggregation.

Powder flow and its improvement

The flow of powders is extremely important to the production of solid dosage forms with a uniform dose in each tablet or capsule. Particles must pack uniformly into the die (cavity) of the tablet machine or the body of a capsule to produce a constant volume to mass ratio. Ease of flow is also essential for the proper function of the dry powder inhalers, which use the patient's air stream, rather than propellant to draw the drug and powder carrier out of the inhaler and into the patient's respiratory tract.

For drug particles to penetrate to the lower respiratory tract, they must be 5 μm in diameter or smaller. Their large surface and small size generally gives them poor flow properties.

Powder flow can be improved by creating uniform particles with a spherical shape, granulation, mixing with excipients with good flow properties, and use of flow enhancers (lubricants, glidants and antiadherents) in the commercial production of tablets.

Definition

Granulation is a process used in the manufacture of tablets and capsules in which drug and other excipient powders are prepared in larger, spherical granules to improve their flow and the metering of the solids into a tablet die. The optimal size of granules for this purpose is between 500 and 800 μm.

Solubility and dissolution

Key Points

- Water solubility of a drug is dependent on the intermolecular attractive interactions between drug–drug, water–water and drug–water.
- Water has strong attractive forces for other water molecules.
- Drug molecules that have a formal electric charge or polarized covalent bonds can form hydrogen bonds with water.
- Dissolution rate is increased by higher surface area exposed to the solvent and greater solubility of the drug.

Definitions

- A *molecular solution* is a homogeneous mixture of the molecules of one substance with another. This implies that the molecules of the solute (dissolved substance) are interacting with the molecules of the solvent and not with one another as might occur in crystals.
- *Dissolution* is the process of solute molecules mixing with solvent molecules and can be studied mathematically as a rate.
- *Solubility* is the extent to which a drug dissolves in a given solvent at a given temperature. Close attention to the factors that influence drug solubility allows the pharmacist to prepare and maintain drug solutions required for patient care.

Melting point, solubility and dissolution rate

- What is it that determines that one drug will form a molecular solution and another a coarse dispersion?
- Thermodynamics tell us that there are two major driving forces for any reaction:
  - disorder (entropy) S
  - enthalpy (heat energy) H
- Drug solutions are created by a combination of disorder and attractive forces that occur when a solute mixes with a solvent.

When a drug moves from orderly interactions of molecules in a crystalline solid to less orderly interactions with liquid solvent, the entropy term favors formation of a solution. However, it is the enthalpy term, the favorable energetics that some drugs derive from interacting with water molecules, that makes them water soluble while others have no attraction for water molecules and remain in the solid state.

$$DD + WW \rightarrow 2\,DW$$

- A solution begins with a drug solid, DD, in which the molecules of drug are interacting with other drug molecules in the solid, and a solvent, in this case water, in which molecules of water are interacting with other molecules of water, WW.
- To form a solution from these two separate entities, all molecules of drug will need to move out of the solid and interact with molecules of solvent, DW.
- For a drug to be water soluble, it must replace the attractive interactions with other drug molecules in its solid form with attractive interactions with water molecules.
- If DD and WW attractive forces predominate, the drug will form a coarse dispersion.
- Desirable DW interactions occur when water molecules can replace the hydrogen bonding with other water molecules with hydrogen bonding with polar groups on the drug molecule.
- To provide the reader with perspective on the impact of attractive forces on solubility of dissimilar drugs, Yalkowsky estimates that each 100°C increase in melting point above 25°C (room temp) corresponds to at least a 10-fold decrease in solubility (Table 2.6).

Table 2.6 Effect of melting point on drug solubility	
Melting point	**Impact on drug solubility**
>300°C	Major factor
>200°C	Probable factor
100–200°C	Equivocal
<100°C	Not a factor

Information from SH Yalkowsky (ed), Techniques of Solubilization of Drugs, New York: Marcel Dekker, 1981.

Predicting water solubility from desirable drug–water interactions

Drug solutes that are able to form favorable interactions with water molecules include polar solutes such as:

- Nonelectrolytes with bonds in which the electrons are not equally shared (OH, NH, SH)
- Strong electrolytes (inorganic and drug salts) that carry a formal charge:
  - The sodium and potassium salts of inorganic anions are all quite water soluble as are the chloride and acetate salts of inorganic cations.
  - Not all inorganic salts are water soluble. Magnesium, calcium and aluminum carbonates, phosphates and hydroxides form suspensions in water.
  - The pharmacist must be alert to the fact that a combination of two water soluble salts, calcium chloride and sodium carbonate, may precipitate insoluble calcium carbonate.
  - Preparation of a salt form of a drug (a strong electrolyte) creates a species that dissociates readily from its counter ion to form attractive interactions with water.
- Weak electrolytes that ionize (become charged) incompletely:
  - Drugs that are weak organic acids and bases donate or accept protons incompletely in water.
  - Ionization produces a formal charge on the molecule.
  - Ionization is a reversible process that, if the pH or counter ion concentration of the solution is altered, can result in formation of the uncharged form of the drug and precipitation from solution.
  - If we increase the extent of ionization using strong acid or base, the water solubility of the drug molecule is enhanced considerably. (See strong electrolytes above.)
  - Not all salts of drug acids and bases are water soluble. Drug salts formed by the reaction of a drug with another weak acid or base do not dissociate appreciably and will have lower water solubility than drugs salts formed with strong acid or base.

Common ion effect on solubility

Hydrochloride and sodium salts of weak drug bases and acids may be considered preferable to other ions because they are physiologically abundant and would not be expected to cause adverse effects. However, the abundance of these ions in the blood or gastrointestinal tract may suppress the solubility of the drug they are intended to enhance through the common ion effect. The common ion effect is the displacement of an ionic equilibrium by the addition of more of one of the ions involved. Dissolution of an ionic compound can be depicted as follows:

$$CA_{(s)} \rightleftharpoons C^+_{(aq)} + A^-_{(aq)}$$

where C is the cation and A is the anion.

And the equilibrium expression for the dissolution would be:

$$K_{eq} = \frac{[C^+_{(aq)}][A^-_{(aq)}]}{CA_{(s)}}$$

If we were to add more of the anion, A^-, the law of mass action tells us that the equilibrium will be driven to the left or back towards the solid form. For example, addition of chloride ion to an aqueous solution of terfenadine hydrochloride will suppress its solubility by driving the solubility equilibrium back towards terfenadine hydrochloride solid.

$$\overset{\leftarrow\ Cl^-_{(aq)}}{R_3NHCl_{(s)} \rightleftharpoons R_3NH^+_{(aq)} + Cl^-_{(aq)}}$$

Partitioning behavior of drug molecules: Log P

Definition

Log partition coefficient or *log P* is a parameter that estimates whether a drug will prefer to interact with water or prefer to escape water and interact with an immiscible organic solvent such as octanol. It is the log of the drug concentration in the solvent divided by the drug concentration in water.

Partition coefficient can be measured experimentally by dissolving a known amount of drug in water, mixing the solution well with octanol and allowing enough time for the mixture to reach equilibrium in partitioning between the two layers. The drug concentration in the aqueous layer is quantified and used to calculate the concentration in the octanol phase. The two concentrations are expressed as a ratio:

$$P = \frac{[\text{Drug}]_o}{[\text{Drug}]_w} \quad \text{and} \quad \log P = \log \frac{[\text{Drug}]_o}{[\text{Drug}]_w}$$

Pharmaceutical example

A log P of 2 indicates that there are 100 molecules of drug in octanol for every 1 molecule of drug in the water phase.

Key Points

- Log partition coefficient is the log of the drug concentration in octanol divided by the drug concentration in water.
- Log P provides the pharmacist an appreciation of the drug's partitioning preference.
  - A positive log P tells us that more drug has moved into octanol
  - A negative log P tells us that the drug has partitioned more into water.
- The log P of an ionizable drug is understood to be for the unionized form.
- Log D is the partition coefficient for the combination of ionized and unionized forms of a drug at a specified pH.
- Using the *United States Pharmacopoeia* definition of water solubility (greater than 3.3% in water), compounds with a log P less than 0.5 are considered water soluble.
- Compounds with a log P of 0.5 or greater are considered water insoluble. Although log P is used in some instances to predict water solubility, its significance lies in its ability to predict partitioning into biological membranes.

Measurement of solubility

The accurate prediction of water solubility from drug structure is the pharmaceutical 'holy grail'. While the practicing pharmacist has access to water solubility data, it is at times unsatisfactory, using descriptive terms rather than precise values and is not specific about the form of the drug to which the solubility applies. However, equations that estimate water solubility from a molecule's structure are insufficiently accurate to replace measurement techniques. Solubility is measured in the lab using excess solid drug and a known volume of solvent at a specified temperature. The solvent is allowed to interact with the drug solid until the number of molecules going into solution is equal to the number returning to the solid state.

Sources of information on drug solubility are:

- *The Merck Index*
- *The United States Pharmacopoeia*

Table 2.7 Solubility definitions	
Descriptive terms	**Parts of solvent for 1 part of solute**
Very soluble	Less than 1
Freely soluble	From 1 to 10
Soluble	From 10 to 30
Sparingly soluble	From 30 to 100
Slightly soluble	From 100 to 1000
Very slightly soluble	From 1000 to 10 000
Practically insoluble, or insoluble	More than 10 000

From L Felton (ed.) Remington: Essentials of Pharmaceutics. Pharmaceutical Press, London, 2012, reproduced with permission.

- *Martindale's Extra Pharmacopoeia*
- *AHFS Drug Information*
- *DrugBank* (http://www.drugbank.ca/), a database provided by David Wishart at the University of Alberta
- *ChemIDplus* (http://chem.sis.nlm.nih.gov/chemidplus/) from the United States National Library of Medicine
- *PubChem Compound* (http://pubchem.ncbi.nlm.nih.gov/) from the United States National Library of Medicine.

Methods of expressing solubility

When searching for solubility information on older drug compounds, the pharmacist will encounter descriptive terms. The meaning of the descriptive terms used for the expression of solubility is presented in Table 2.7.

Key Point

- Although the concentrations of most drugs are expressed using weight in specified milliliters of *solution*, the solubility of most drugs is expressed using weight in specified milliliters of *solvent*.

Dissolution of solids

Dosage forms that present the drug in solid form such as tablets and capsules, or incompletely dissolved form such as suspensions must dissolve to release drug for absorption. An important parameter for understanding how quickly a drug will be available for absorption is the *intrinsic dissolution rate*, the amount of a drug that dissolves per unit time and unit surface area.

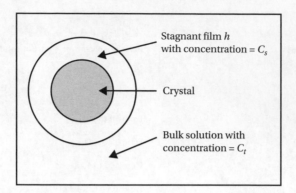

Figure 2.6   Dissolving particle

Measurement of intrinsic dissolution rate requires the use of a nondisintegrating disc that exposes a known surface of drug to the dissolution medium. Drugs with water solubility less than 0.01 mg/mL typically will demonstrate dissolution rate limited absorption. Rapid dissolution for a dosage form is defined as ≥85% of the amount of drug in the dosage form dissolved within 30 minutes in a volume of ≤900 mL dissolution medium at 37°C.

To understand the relationship of dissolution and drug solubility, consider the following mathematical description of dissolution rate. If we have a spherical particle dissolving in a bulk medium, the rate of transfer of mass from the particle to the dissolution medium can be described by the Noyes Whitney equation (Figure 2.6):

$$\text{Dissolution rate} = \frac{dm}{dt} = \frac{DA(C_s - C_t)}{h}$$

where D is the diffusion coefficient of drug in dissolution medium; A is the area exposed to solvent; $C_s - C_t$ is the saturation solubility of the drug minus the concentration in the bulk medium – in other words, the concentration gradient; and h is the the width of the boundary layer around the dissolving particle.

In the body, the drug is rapidly carried away from the dissolving particle, making the concentration of drug in the bulk body fluids close to zero. The Noyes Whitney equation is often simplified to:

$$\frac{dm}{dt} = \frac{DAC_s}{h}$$

If these different variables affect how fast a drug gets into the blood stream (dissolution rate limited absorption) then we are interested as pharmacists in how we might affect these parameters.

Questions and cases

1. Ionic compounds would be expected to have:
    (a) High melting points due to strong attractive forces between ions
    (b) Low melting points due to strong attractive forces between ions
    (c) High melting points due to weak attractive forces between ions
    (d) Low melting points due to weak attractive forces between ions

2. A polymorph describes:
    (a) A solid with solvent molecules incorporated in its crystalline structure
    (b) A solid that takes up water from the environment
    (c) A solid that forms distinct orderly arrangements between molecules
    (d) A solid that sorbs water vapor from the environment and gradually forms a solution

3. A hygroscopic solid refers to:
    (a) A solid with solvent molecules incorporated in its crystalline structure
    (b) A solid that takes up water from the environment
    (c) A solid that forms distinct orderly arrangements between molecules
    (d) A solid that sorbs water vapor from the environment and gradually forms a solution

4. The melting points (mp) of three beta agonist homologues are given below:

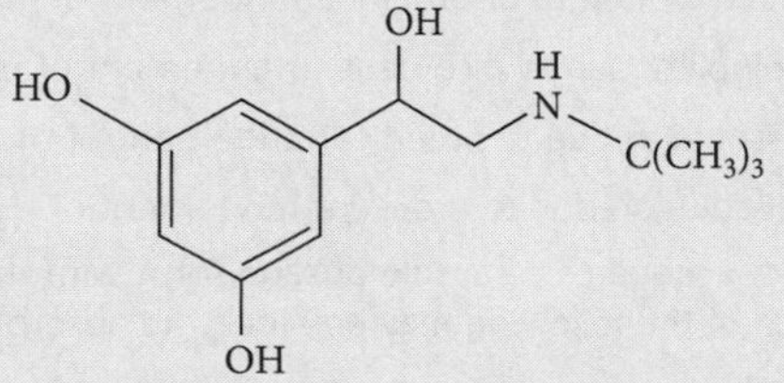

Terbutaline mp 119-122°C

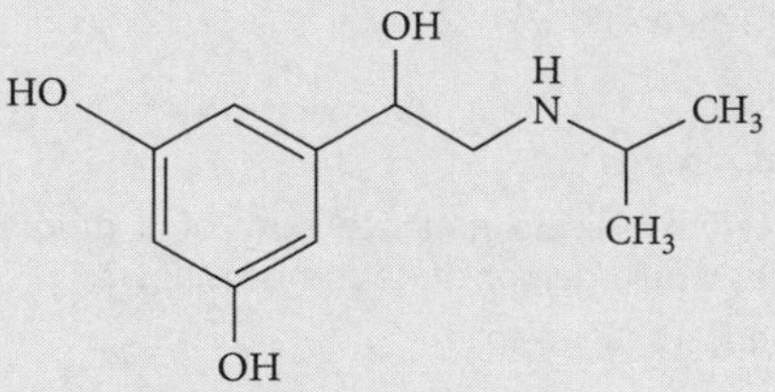

Metaproterenol mp 100°C

Isoproterenol mp 155.5°C

What is their expected order of solubility in water, from *highest* to *lowest*?

(a) terbutaline > metaproterenol > isoproterenol

(b) metaproterenol > terbutaline > isoproterenol

(c) isoproterenol > metaproterenol > terbutaline

(d) isoproterenol> terbutaline > metaproterenol

5. Reduction of the particle size of a drug enhances:

(a) Drug stability

(b) Drug dissolution rate

(c) Powder flow

(d) Particle segregation

6. A pharmacist purchased a bottle of procaine hydrochloride crystal. Which of the following describe this solid?

(a) Poor solubility due to disorderly arrangement of molecules within the solid

(b) Good solubility due to disorderly arrangement of molecules within the solid

(c) Slow dissolution due to orderly arrangement of molecules within the solid

(d) Rapid dissolution due to orderly arrangement of molecules within the solid

7. The pharmacist prepares a capsule dosage form with the procaine hydrochloride crystal. Which of the following may enhance the absorption rate from this dosage form?

(a) Milling

(b) Use of diffusive mixing

(c) Granulation

(d) Use of shear mixing

8. The pharmacist would like to enhance flow of a drug powder. Which of the following particles would demonstrate the best flow?

(a) A large spherical shaped solid

(b) A small rod-shaped solid

(c) A large needle shaped solid

(d) A small cylindrical shaped solid

9. A hydrophobic drug with some attraction for alcohol molecules would demonstrate which of the following?

(a) A contact angle greater than 90°

(b) Complete wetting

(c) A contact angle of 0°

(d) Partial spreading on the solid surface

10. Drug derivatives are:

(a) Different arrangements of drug molecules within a crystal

(b) A drug molecule that has been chemically altered to change its physical properties

(c) A drug molecule that interacts with water molecules in its crystal

(d) A drug molecule that hydrogen bonds with a conformer within a crystal

11. You are looking in a catalog to buy sodium phosphate monobasic for compounding. It is available as a powder and a granular form. Which of the following characteristics describes the granular form?

(a) Greater surface area, poor wettability, better flow

(b) Greater surface area, smaller surface energy, poorer flow

(c) Greater surface area, faster dissolution, tendency to aggregate

(d) Smaller surface area, faster dissolution, better flow

(e) Smaller surface area, slower dissolution, better flow

12. You have a patient who has a painful, sprained ankle. Which of the following advice will hasten the dissolution of the ibuprofen in the tablet he has purchased?

Ibuprofen

(a) Take the tablet without water to increase its concentration at the gastrointestinal membrane surface

(b) Take the tablet with food so it remains in the acidic environment of the stomach longer

(c) Crush the tablet first and take with lots of water

(d) Take the tablet with acidic fluids so it is ionized as it dissolves

13. Warfarin is an anticoagulant drug used to treat deep venous thrombosis and other cardiovascular thrombotic disorders. It has a narrow therapeutic index and must be carefully dosed and monitored to prevent bleeding. The first available warfarin product was Coumadin, the crystalline sodium salt of warfarin. In 1980, a generic version of the drug was introduced that was amorphous warfarin sodium. How might these two forms of warfarin be different? Would you observe a difference in the extent of warfarin solubility, the rate of warfarin dissolution or both? What effect might these differences have on patients who were switched from Coumadin to generic warfarin?

14. Search through catalogs and the web to find compounds that can serve as desiccants for your hygroscopic or deliquescent drug powders. List the names of these compounds. What do they have in common? Make a plan for how you will know that it is time to replace your desiccant in your desiccant containers used to protect drug powders.

Selected reading

1. SR Vippagunta, HG Brittain, DJW Grant (2001). Crystalline solids. Adv Drug Del Rev 48: 3–26.
2. NR Goud, S Gangavaram, K Suresh, et al. Novel furosemide cocrystals and selection of high solubility drug forms. J Pharm Sci 2012; 101: 664–680.
3. H Nakamura, Y Yanagihara, H Sekiguchi, et al. Effect of mixing method on the mixing degree during the preparation of triturations. J Pharm Soc Japan. 2004; 124: 127–134.
4. H Nakamura, Y Yanagihara, H Sekiguchi, et al. Effect of particle size on mixing degree in dispensation. J Pharm Soc Japan 2004; 124: 135–139.
5. AW Newman, SM Reutzel-Edens, G Zografi. Characterization of the "hygroscopic" properties of active pharmaceutical ingredients, J Pharm Sci 2007, 97: 1047–1059.
6. Chapter <786> Particle size distribution estimation by analytical sieving, United States Pharmacopeia, 34. USP Convention, Rockville, MD, 2011.
7. RD Odegard, Minimizing the probability of out-of-specification preparations: results that make you say ... hmmm! Int J Pharm Compounding 2008; 12: 130–135.

3

Dispersed systems

Learning objectives

Upon completion of this chapter, you should be able to:

- Differentiate between molecular solutions, colloidal dispersions and coarse dispersions.
- Define colloid, surfactant, micelle, surface or interfacial energy, emulsions, suspension, foam, flocculated and deflocculated.
- Describe the behavior of dispersed particles (or droplets) in terms of their surface or interfacial energy.
- Explain the following approaches to increasing the physical stability (uniformity) of dispersed systems in terms of interfacial behavior:
  - emulsification with surfactants and polymers
  - wetting of hydrophobic powders
  - flocculation with polymers, electrolytes and surfactants.
- Describe the difference between an oil-in-water and a water-in-oil emulsion, and explain the use of the hydrophilic lipophilic balance system to select an emulsifier.
- Describe the difference between flocculated and deflocculated particles in terms of their sedimentation and redispersion behavior.
- Explain how the density, viscosity, and flow properties of vehicles affect the physical stability (uniformity) of dispersed systems.
- Define viscosity, plastic, pseudoplastic and dilatant.
- Distinguish between Newtonian and non-Newtonian fluids.
- Given the behavior of a specific liquid, determine whether it is Newtonian, plastic or pseudoplastic.

Introduction

If we take a solid drug and mix it with water and it 'disappears' into the solvent, we say that it has formed a solid in liquid solution. At a molecular level individual molecules of this drug have ceased to interact with one another as they do in a crystal and are surrounded entirely by individual solvent molecules. If we take another solid drug and mix it with water and over time we can still see solid material in the solvent, we have a coarse dispersion that in pharmacy is referred to as a suspension. These hydrophobic drug molecules prefer interacting with each other to interacting with molecules of solvent and this behavior has significant impact on how we formulate and handle suspensions. There is a third set of interactions that can be observed in a mixture between a solid drug or pharmaceutical necessity and a solvent that may be considered intermediate in nature between a molecular solution and a suspension. In these colloidal dispersions we find a network of overlapping solute molecules interacting with molecules of solvent. The behavior of colloidal dispersions can be explained by these simultaneous interactions between colloid and colloid, and colloid and solvent. This chapter focuses on the behavior of colloids and coarse dispersions. Chapter 4 considers molecular solutions in detail.

Behavior of molecular solutions, colloidal dispersions, and coarse dispersions

Table 3.1 will assist you in differentiating between molecular solutions, colloidal dispersions, and coarse dispersions.

Colloidal dispersions

- Mixtures of colloids are considered dispersed systems, although in some cases, their lack of homogeneity may not be initially appreciated. Colloidal dispersions include solutions of hydrophilic polymers, suspended particles of colloidal dimensions and micellar solutions.
- The distinguishing feature of colloidal dispersions is the size of their dispersed phase and the resulting large surface that they present to the continuous phase.

> ### Definition
>
> A *colloid* is a subvisible particle or macromolecule between 1 and 500 nm (0.5 μm) in diameter.

- Hydrophilic polymers form colloidal solutions with water, which appear to be molecular solutions until their light scattering and non-Newtonian flow properties are examined.

Pharmaceutical example

Colloidal solutions of hydrophilic polymers are used as gel bases for drugs to be applied to the skin and as vehicles for drug solutions or suspensions that require viscosity (Table 3.2).

Table 3.1 Molecular solutions, colloidal solutions and coarse dispersions

	Molecular solutions	Colloidal solutions	Coarse dispersions
Example solute	Glucose	Solute: methylcellulose	Water insoluble drugs
Molecular weight	180 Da	180 000–300 000 Da	Not applicable
Diameter	<1 nm	1–500 nm; solutes have a large surface to volume ratio	>5000 nm
Molecular weight	Solute is small, <1000 Da	Solutes large, >1000 Da	Have a dispersed phase and a continuous phase
Mobility	Solutes exhibit Brownian motion	Exhibit Brownian motion	Too large to exhibit Brownian motion
Diffusion	Diffuse easily from an area of high concentration to an area of low concentration	Diffuse from an area of high concentration to an area of low concentration	Unable to diffuse
Effect on light	Do not scatter light	Scatter light	Subject to the forces of gravity and will noticeably settle over time
Flow	Demonstrate Newtonian flow	Exhibit non-Newtonian flow	Exhibit non-Newtonian flow
Behavior of solutes	Solute separated from other solute molecules by solvent molecules	Solutes tend to associate in solution	Particles tend to aggregate
Examples	Syrups, elixirs, solutions for injection, nasal, otic and ophthalmic use	Colloidal solutions include gels and vehicles for suspensions	Coarse dispersions include suspensions, emulsions, foams and aerosols

Information from selected reading 2.

- Particles of colloidal dimensions may be wettable with solvent but not dissolve. They form suspensions in which the particle size is small enough (colloidal) to prevent sedimentation.

Pharmaceutical example

Aluminum hydroxide/magnesium hydroxide antacid gels are examples of these colloidal suspensions.

- *Surfactants* are amphiphilic molecules that have a hydrophobic segment and a hydrophilic segment (Table 3.3). At concentrations above their critical micellar concentration, surfactant molecules assemble into spherical associations called *micelles*. Micelles are large enough to exhibit colloidal behavior.

Table 3.2 Properties of hydrophilic colloidal viscosity enhancing agents

Viscosity agent	Concentration	Viscosity (mPa · s)	Flow properties
Acacia	30% w/v	100	<25% Newtonian >25% non-Newtonian; viscosity reduced by salts and increasing temperature
Alginic acid	2% w/w	2000	Gels in the presence of physiological concentration of polyvalent cations
Carbomer 934	0.5% w/v neutralized	30 500–39 400	Plastic flow; solution is liquid at low pH but gels at pH >7.0
Carboxymethyl-cellulose sodium	1% w/v	10–12 000	Pseudoplastic; gels in the presence of physiological concentrations of polyvalent cations
Gelatin	6.67% w/v	2.7–4.8 (60°C)	Pseudoplastic; thixotropic, gels below 40°C
Hydroxy ethylcellulose	2% w/v	2–20 000	Pseudoplastic; viscosity is concentration and formula weight dependent
Hydroxy propylcellulose	2% w/v	150–6500	Pseudoplastic; viscosity is concentration and formula weight dependent
Hypromellose	2% w/v	3–100 000	Pseudoplastic; viscosity is concentration and formula weight dependent. Gels at 50–90°C
Methylcellulose	2% w/v	2–75 000	Pseudoplastic; viscosity is concentration and formula weight dependent; gels at 50–60°C
Polaxamer 188 (Pluronic)	0.2–5%	Not applicable	Pseudoplastic, thixotropic; liquid at room temperature but gels at body temperature
Polyvinyl alcohol	4% w/v	4.0–65.0	Viscosity is concentration dependent; gels in the presence of borax
Povidone	10% w/v	1.3–700	Pseudoplastic; viscosity is concentration and formula weight dependent
Sodium alginate	1% w/v	20–400	Gels in the presence of physiological concentration of polyvalent cations
Tragacanth	1% w/v	100–4000	Pseudoplastic
Xanthan gum	1% w/v	1200–1600	Pseudoplastic; mucoadhesive

Information from selected reading 3.

Pharmaceutical example

Colloidal parenteral delivery systems use polymers or lipids to prepare drug carriers of colloidal dimensions, generally between 10 and 100 nm in diameter, which protect proteins and nucleic acid based macromolecules from enzymatic degradation and glomerular filtration.

Table 3.3 Functions of surfactants

Function	HLB no.
Antifoaming	1–3
Emulsifiers (water-in-oil)	3–6
Wetting agents	7–9
Emulsifiers (oil-in-water)	8–18
Solubilizers	15–20
Detergents	13–16

HLB = hydrophilic lipophilic balance.
Adapted from selected reading 2.

Coarse dispersions

Coarse dispersions include those multiphase mixtures in which lack of uniformity can be appreciated by visual inspection (Table 3.4). They are thermodynamically unstable and will reduce the surface of dispersed phase exposed to continuous phase by either coalescence or sedimentation.

Their formulation, preparation, and use require attention to their inherent physical instability through reduction of surface energy, manipulation of viscosity and application of shear stress (shaking) to disperse or redisperse.

- Suspensions – mixtures of solid in liquid – may be used orally, topically, subcutaneously or intramuscularly.
- Emulsions – mixtures of liquid dispersed in liquid – are used topically, parenterally, and occasionally orally.
- Aerosols – mixtures of liquids or solids in gases – are used topically and applied to the nose or the lungs.
- Coarse dispersions tend to separate into their two phases to minimize the interfacial (surface) energy of the two phase system unless the design of the dosage form includes a mechanism to reduce the energy at the surface

Table 3.4 Types of coarse dispersions

Dispersed phase	Continuous phase	Examples
Liquid	Gas	Liquid aerosols
Solid	Gas	Powder aerosols
Gas	Liquid	Foams
Liquid	Liquid	Creams (emulsions)
Solid	Liquid	Suspensions

Adapted from selected reading 4.

or to prevent the close approach of the dispersed phase droplets or particles.

Interfacial (surface) behavior of the dispersed phase

The tendency of coarse dispersions to separate into two phases is an attempt to minimize surface or interfacial energy by minimizing the surface area of each phase exposed to the other. The coarse dispersions are systems in which the forces of attraction between molecules of the same phase (cohesional forces) exceed the forces of attraction between the molecules of the different phases (adhesional forces). This imbalance results in a boundary between the two phases that the system will seek to minimize. Pharmaceutical scientists use the term *surface* to describe a boundary between a liquid or solid and a gas and the term *interface* to describe a boundary between two liquids or a solid and a liquid.

Liquid–liquid interfacial behavior

If we consider first the boundary between two immiscible liquids, oil and water, the layering behavior we observe at the interface is a result of the negligible attractive forces between molecules of oil and molecules of water compared with the attraction between two water molecules or two oil molecules (Figure 3.1).

The net effect is that the molecules at the interface experience an inward force or pull and the interface or surface is contracted (minimized). We call this force per unit length of boundary the *interfacial tension*, which takes the units mN/m. As a result, work in the form of shaking must be applied to the two liquid phases to create the increase in surface (ΔArea) necessary for a 'uniform' dispersed system. Even so the droplets of dispersed liquid will assume a spherical shape since this shape has the smallest surface area per unit volume. If we allow the dispersed liquids to rest, they will eventually minimize their exposed interface by forming two separate layers thereby minimizing the energy of the system.

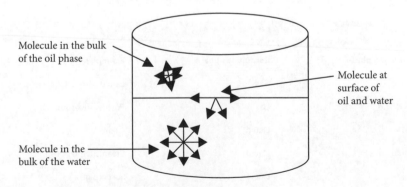

Figure 3.1 Interfacial tension of two immiscible liquids

> *Definition*
>
> *An emulsion* is a mixture of two immiscible liquids (usually oil and water), which to remain uniformly dispersed, requires formulation approaches that will reduce interfacial tension rather than interfacial (surface) area.

- There are two types of emulsions:
  - oil-in-water emulsions in which the oil is dispersed in a water continuous phase (o/w)
  - water-in-oil emulsions in which water droplets are dispersed in an oil continuous phase (w/o).
- The interfacial energy of two immiscible liquids can be reduced by the use of a class of pharmaceutical excipients called emulsifying agents. These ingredients position themselves at the interface between the two liquids and lower interfacial tension either through chemical interaction or physical barrier properties.

Emulsifying agents

Many emulsions will remain homogeneously dispersed without shaking if an appropriate emulsifier is included in the formulation. Emulsifying agents include surfactants, film forming polymers, and solid particulate film formers.

- Surfactants or surface active agents are a class of emulsifying agents that have a polar section that orients itself into the water phase of an emulsion and a hydrophobic section that orients itself into the oil phase of the emulsion (Figure 3.2).
- Film forming polymers (Figure 3.2) are only weakly surface active but create a multilayered, physical barrier of polymer molecules at the interface between dispersed droplets and the continuous phase and prevent coalescence into separate phases.
- The film forming polymers are generally significantly less toxic than surfactants.
- Solid particles of colloidal dimensions (Figure 3.2) find their way to the interface between the two phases and create a physical barrier to the coalescence of two immiscible liquids.
- The particulate film formers do not dissolve but they must be wetted by the continuous phase.

Pharmaceutical example

The main application for solid particulate film formers are the sun screens.

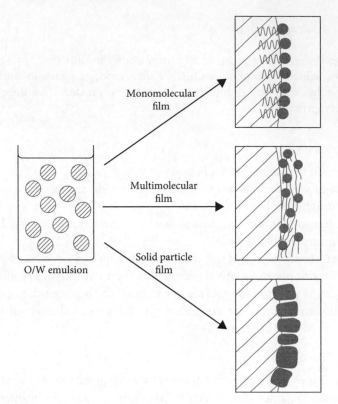

Figure 3.2 Types of films formed by emulsifying agents at the oil–water interface. Orientation shown for o/w emulsions.

The hydrophilic lipophilic balance numbering system

The preference of surfactants for the water or the oil phase has been quantified using the hydrophilic lipophilic balance (HLB) numbering system. The numbering system can be used to identify surfactants useful for several applications (Table 3.5).

HLB numbering system and emulsifier selection:

- Surfactants that orient more of the molecule into a water continuous phase have HLB numbers between 8 and 18 and will make oil-in-water (o/w) emulsions.
- Surfactants that orient more of the molecule into the oil continuous phase have HLB numbers between 3 and 6 and will make water-in-oil (w/o) emulsions.
- Surfactants can be blended to produce an optimal HLB.
- Surfactants have the potential to cause some irritation to mucous membranes and the concentration of surfactant used by the oral or

Table 3.5 Emulsifying agents

Example	Routes	HLB no.
Nonionic surfactants		
Glyceryl monooleate	TO	4.1
Sorbitan trioleate (Span 85)	TO	1.8
Sorbitan monooleate (Span 80)	TO	4.3
Sorbitan monolaurate (Span 20)	TO	8.6
Polyoxyethylene lauryl ether (Brij 30)	T	9.5
Polyoxyethylene sorbitan monooleate (Tween 80)	TOP	15.0
Polyoxyethylene sorbitan monolaurate (Tween 20)	TOP	16.7
Polyoxyl 8 stearate	T	11.1
Polyoxyl 40 stearate	TOP	16.9
Polyoxyl 50 stearate	T	17.9
Polyoxyl 35 castor oil (Cremophor EL)	TOP	12–14
Polyoxyl 40 Hydrogenated castor oil (Cremophor RH 40)	TOP	14–16
Polaxamer 181 (Pluronic)	TOP	1–7
Polaxamer 124 (Pluronic)	TOP	12–18
Polaxamer 407 (Pluronic)	TOP	18–23
Anionic surfactants		
Sodium lauryl sulfate	T	40
Sodium oleate	TO	18
Cationic surfactants		
Cetyltrimethylammonium bromide	T	40
Zwitterionic surfactants		
Triethanolamine oleate	T	12
Lecithin	TOP	7–10
Hydrophillic colloids (film forming polymers)		
Methylcellulose	TO	10.5
Acacia	TO	8.0
Gelatin	TO	9.8
Tragacanth	TO	13.2

(continues)

Table 3.5 *(continued)*		
Example	**Routes**	**HLB no.**
Colloidal solids (solid particulate film formers)		
Bentonite (hydrated aluminum silicate)	T	o/w
Veegum (Mg-Al silicate)	T	o/w
Magnesium oxide	T	w/o
Magnesium trisilicate	T	w/o
Titanium dioxide	T	o/w, w/o
Talc	T	

HLB = hydrophilic lipophilic balance; T = topical; O = oral; P = parenteral; o/w = oil-in-water; w/o = water-in-oil.
Information from selected reading 3.

ophthalmic routes must be relatively low compared with what is used topically.

- Some surfactants can only be used topically.

Gas–liquid dispersions

Definition

A *foam* is a coarse dispersion of a gas in a liquid.

Before it emerges from the container, the foam consists of a liquid propellant uniformly mixed with another liquid by surfactant emulsifiers. The propellant is an ingredient that is a gas at room temperature and atmospheric pressure but is easily liquefied by lowering the temperature or increasing the pressure. When placed in a sealed container, the propellant establishes an equilibrium between its vapor and liquid phases that creates pressure inside the container. When the mixture is released from the container, the propellant becomes a gas creating a foam from the surfactant liquid mixture. Pure liquids are unable to create foams and must be blended with surfactants. Foam products are prepared at pressures between 240 and 275 kPa (35–40 psi) at 20°C and generally contain 6–10% propellant. The most commonly used propellant in pharmaceutical foams is a blend of propane and isobutane.

Solid–liquid interfaces

Solid particles dispersed in a liquid medium with no significant attraction to the molecules of liquid develop interfacial tension (energy) as a result of the net inward pull on the molecules at the surface of the particle. The surface or

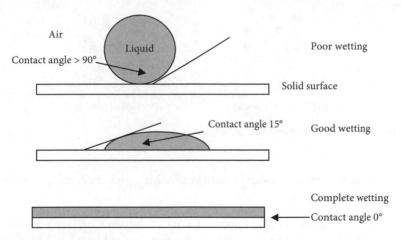

Figure 3.3 Wetting and contact angle

interfacial energy of a solid is measured by placing a drop of liquid on a drug particle and measuring the contact angle (Figure 3.3).

- If the surface energy between a hydrophobic solid and a hydrophilic solvent is high, the solid will wet poorly. The solvent will remain as a spheroidal droplet with a contact angle greater than 90°.
- The surface energy between a solid and liquid with adequate attraction will demonstrate good wetting. The solvent will spread partially over the solid forming a 'contact angle' of 90° or less.
- The surface energy between a solid and a liquid with strong attraction will be low, resulting in complete wetting. The solvent displaces adsorbed gases and forms a monolayer of liquid on the solid with a contact angle of 0°.
- If the liquid is water, the solid can be considered hydrophilic.

This imbalance results in two major behaviors that complicate the preparation of a uniform suspension: problems with wettability and caking.

Wetting of hydrophobic powders

> *Definition*
>
> *The ability of a liquid to displace air and spread on a solid surface is called wetting* and can be quantified in terms of the angle made by the liquid with the solid surface at the point of contact.

- A hydrophobic solid will trap air on to its surface rather than allowing the hydrophilic vehicle to spread on the solid surface.

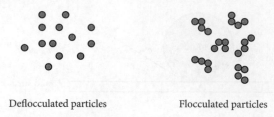

Deflocculated particles   Flocculated particles

Figure 3.4 Deflocculated and flocculated particles

Problems with wettability can be managed with a pharmaceutical ingredient called a wetting agent.

- Some wetting agents are simply solvents with lower surface tension that are still miscible with water such as glycerin, propylene glycol or alcohol.
- The other class of ingredients that can be used as wetting agents are the surfactants. These molecules reduce the interfacial energy between the hydrophobic solid and hydrophilic solvent by providing a molecular section with which each phase can associate.

Use of flocculating agents to manage caking in suspensions

The tendency of a suspension to cake may not be apparent initially. If we were to constantly shake or stir a suspension, the system would maintain the large surface area of dispersed particles exposed to the liquid phase. If we allow the suspension to rest, over time hydrophobic particles that have been dispersed in a hydrophilic vehicle will respond to gravity by dropping to the bottom of the container. The collection of particles at the bottom of a suspension may aggregate to reduce the surface exposed to the dispersion medium and can become sufficiently dense (caked) that the particles cannot be easily redispersed.

- One solution to caking is to allow the particles to associate in loose aggregations called flocs (Figure 3.4). A suspension with particles in these large, loose associations is termed *flocculated*.
- Individual particles that have a strong attraction for one another and form compact masses at the bottom of the container are *deflocculated*.
- Addition of flocculating agents prevents the close approach of individual particles, and creates loose collections of particles that are easily broken up by shaking but sediment rapidly.
- Flocculating agents include polymers, electrolytes and surfactants (Table 3.6).
  - Polymers such as the hydrophilic colloids used as film forming emulsifiers create bridges between the particles. The bridges are broken by shaking and reformed at rest (Figure 3.5).

Table 3.6 Flocculating agents

Flocculating agent	Routes
Film forming polymers	
Methylcellulose	TO
Acacia	TO
Gelatin	TO
Tragacanth	TO
Aliginate	TO
Electrolytes	
Sodium acetate	TOP
Mono-, di-, trisodium phosphate	TP
Mono-, di-, trisodium citrate	TOP
Surfactants	
Lecithin	TOP

T = topical; O = oral; P = parenteral.

— Many particles dispersed in water will have a net charge on their surface to provide more favorable interactions with water molecules. Particles with a surface charge will attract ions of the opposite charge. The addition of a low concentration of an ionic compound creates loose networks of particles and ions that can be easily redispersed by shaking (Figure 3.6).

— Surface active agents will distribute themselves at the surface of particles and reduce the surface energy and tendency to aggregate. In addition either because they are ionic or because they are polymeric, they will hold particles in flocs and prevent cake formation (Figure 3.7).

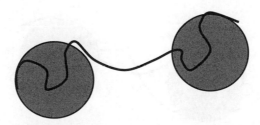

Figure 3.5 Polymeric flocculating agents

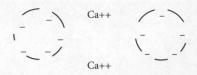

Ca++

Ca++

Figure 3.6 Ionic flocculating agents

Use of density and viscosity to slow sedimentation in coarse dispersions

Stokes' law provides us with the variables that influence the sedimentation of spherical particles in a dilute dispersion:

$$v = \frac{gd^2(p_1 - p_2)}{18\eta}$$

where v is the velocity of sedimentation, g is the acceleration due to gravity, d is the diameter of the particle, p_1 is the density of the particle, p_2 is the density of the dispersion medium, and η is the viscosity of the dispersion medium.

- Larger particles sediment faster and small, individual particles will sediment more slowly.
- Particle or droplet sizes encountered in pharmaceutical disperse dosage forms:
  - coarse suspension prepared in the pharmaceutical manufacturing sector: between 1 and 50 μm.
  - extemporaneous suspensions given the particles sizes that can be achieved in the mortar: between 50 and 400 μm.
  - droplet size of parenteral emulsions: between 0.2 and 0.5 μm.
  - droplet size of topical emulsions: from 0.5 to 500 μm in diameter.

Strategies to slow sedimentation rate

- Reducing the particle size of the dispersed solid which as noted above can result in caking.
- Use of denser and more viscous dispersion media. The density and viscosity of various vehicles for the preparation of oral, topical and parenteral products is shown in Table 3.7.

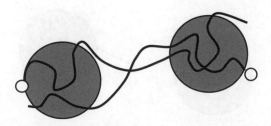

Figure 3.7 Surfactant flocculating agents

Table 3.7 Properties of liquid vehicles

Vehicle	Density (g/mL)	Viscosity RT (cp or mPa · s)	Surface tension, γLV (mN/m or dynes/cm)
Alcohol USP	0.812–0.816	1.22 (100%)	22.75 (100%)
Canola oil	0.913–0.917	77.3–78.3	
Corn oil	0.915–0.918	37–39	
Cottonseed oil	0.916	70.4	35.4
Corn syrup		1380.6	
Glycerin	1.2636	1500	63.4
Isopropyl alcohol	0.783–0.787	2.43	
Mineral oil, heavy	0.845–0.905	110–230	35
Olive oil	0.910–0.915	100	35.8
Peanut oil	0.915	35.2	37.5
Polyethylene glycol 400	1.120	105–130	44
Propylene glycol	1.035–1.037	58.1	40.1
Sorbitol 70%	1.293	110.0	
Sucrose 85% w/v	1.31	124.7	
Water	0.9997 (1)	0.89	71.97

Information from selected reading 3.

- — Vehicles that are either pure liquids or molecular solutions will exhibit Newtonian flow (Figure 3.8).
- — The combination of molecular solutions (syrups) with colloidal solutions of hydrophilic polymers can provide a suspension vehicle with suitable density, viscosity and flow properties.
- If the dispersed phase of an emulsion is less dense than the continuous phase, sedimentation will not occur, however, viscosity enhancing agents can prevent the coalescence of these dispersions.

Flow and viscosity

A distinguishing characteristic of the disperse systems, both colloidal and coarse, is that the dispersed molecules or particles have a large surface to volume ratio. To reduce their interfacial area (and therefore interfacial energy) they will tend to associate resulting in flow properties that differ from pure liquids or molecular solutions. To understand how dispersed systems flow, let's look first at how pure liquids flow.

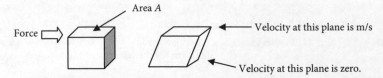

Figure 3.8 Newtonian flow. If we apply a force at the top of a cube of liquid, it will flow at a velocity that is related to the applied force and the area of the fluid layer. Note also that since the force is applied at the top of the liquid, there is a velocity gradient that decreases to zero at the bottom plane of the liquid

Newtonian flow

Newton's law states that the force it takes to move area A is equal to the resistance, η, times the change in velocity over the distance, d:

$$\frac{\text{Force}}{\text{Area}} = \frac{\eta \Delta \text{ velocity}}{\Delta \text{ distance}}$$

Another term for $\dfrac{\text{Force}}{\text{Area}}$ is sheer stress

$$\frac{\Delta \text{ velocity}}{\Delta \text{ distance}} = \text{flow rate} = \text{shear rate}$$

If we were to study a pure liquid like glycerin, vary the force that we apply to a known surface, A, we would find that the resulting rate of flow would be linearly related to the force/unit area as shown in Figure 3.9.

The line depicting Newtonian flow reveals that as force is increased, flow rate increases proportionately and the slope, which is the resistance to flow or *viscosity* of the liquid, remains constant.

Key Points

- Liquids that demonstrate Newtonian flow include pure liquids such as glycerin or water or alcohol, and molecular solutions, such as syrup or 0.9% sodium chloride or 5% dextrose.
- The rate of flow of a Newtonian liquid is proportionate to the force applied. The viscosity of a Newtonian liquid remains constant.

Non-Newtonian flow

The dispersed pharmaceutical systems do not flow proportionately to the applied force. The association of particles or molecules in a dispersion produces the following different flow patterns.

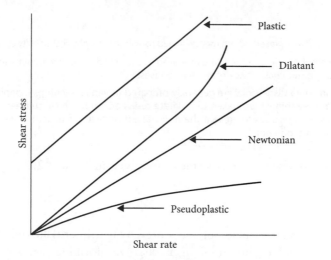

Figure 3.9 Correlation of rate of Newtonian flow with force per unit area. Note that the slope (viscosity) of the pseudoplastic liquid decreases as force is increased

- *Plastic flow* is demonstrated by those liquid mixtures that do not flow at low values of force. Then when the force reaches the yield value the liquid will flow proportionate to the applied force.
  - A plastic liquid is one that demonstrates high viscosity until the force reaches the yield value, at which point the viscosity becomes constant. In other words, plastic liquids do not flow until they reach the yield value.
  - Plastic flow is exhibited by concentrated, flocculated suspensions, and by semisolid emulsions (creams).
- *Pseudoplastic flow* is demonstrated by 'shear-thinning' liquids that begin to flow as soon as force is applied but as greater force is applied, the viscosity actually decreases.
  - Fluids that exhibit pseudoplastic flow include liquid emulsions and aqueous dispersions (colloidal solutions) of hydrophilic polymers such as tragacanth, methylcellulose, povidone and sodium alginate.
  - The long, high molecular weight molecules in solution become entangled as they assemble to reduce their surface and the entanglement increases the viscosity of the liquid. When force (shaking) is applied, the molecules disentangle and the viscosity of the liquid is reduced.
- *Dilatant flow* is demonstrated by a liquid that increases its viscosity as the force applied increases.

Key Points

- The flow of a dispersed system may be varied to enhance its physical stability or uniformity.
- Many emulsion dosage forms are prepared as semisolids that flow minimally and thus do not permit coalescence of the two phases.
- Solid in liquid dispersions are generally prepared in vehicles with pseudoplastic flow so that they thin when shaken to facilitate redispersion but have sufficient viscosity when shaking ceases to prevent the redispersed particles from settling before the dose can be poured.

Ingredients for viscosity enhancement are given in Table 3.2.

Gels

When linear, water soluble polymers are covalently crosslinked they swell extensively in water but no longer dissolve, forming semisolids called hydrogels. These crosslinked compounds include polycarbophil, crospovidone, croscarrmellose and carbomer. Carbomer is a crosslinked acrylic acid polymer that produces highly viscous gels when the pH is increased to between 6 and 11. It is used for oral and topical formulations.

Key Points

- The physical stability of a coarse dispersion can be improved by careful attention to the characteristics of the dispersion medium as well as the dispersed phase.
- The high interfacial tension of the dispersed phase can be reduced with surfactants, and physical barriers to coalescence formed using polymers or film forming particles.
- The dispersion medium should be prepared with suitable viscosity and density to prevent or reduce sedimentation or coalescence of the dispersed particles or droplets.
- In suspensions, the sedimentation rate may be reduced by reducing the particle size of the dispersed solid, followed by treatment of the particles with surface active agents to reduce high surface tension and the resulting tendency to cake.
- Alternatively the particles may be flocculated in a controlled manner with electrolytes, hydrophilic colloids or surfactants to allow loose associations that are easy to redisperse.
- Solid in liquid dispersions can be prepared in vehicles with pseudoplastic flow so that they thin when shaken to facilitate redispersion but have sufficient viscosity to prevent the redispersed particles from settling before the dose can be poured.

Questions and cases

1. Colloidal solutions scatter light because:

(a) They are larger than molecular solutes

(b) They exhibit Brownian motion

(c) They associate in solution

(d) (a) and (b)

(e) (a), (b) and (c)

2. High interfacial tension at a solid–liquid interface would cause:

   (a) The solid particles to repel each other

   (b) The solid particles to pull together

   (c) The solid particles to move randomly

   (d) The solid particles to interact with solvent molecules

3. A pharmacist has two emulsifying agents, one with an HLB of 5 and the other with an HLB of 18. To make a water in oil emulsion, the pharmacist should:

   (a) Use the emulsifying agent with the HLB of 5 because it will orient mostly in the oil phase

   (b) Use the emulsifying agent with the HLB of 18 because it will orient mostly in the oil phase

   (c) Use the emulsifying agent with the HLB of 5 because it will orient mostly in the water phase

   (d) Use the emulsifying agent with the HLB of 18 because it will orient mostly in the water phase

4. To create a foam when the formulation is expelled from the container:

   (a) The pressure on the surfactant ingredient must be decreased

   (b) The pressure on the surfactant ingredient must be increased

   (c) The pressure on the propellant ingredient must be decreased

   (d) The pressure on the propellant ingredient must be increased

5. You are preparing a ketoprofen suspension. The drug has a log P of 3.2 and a water solubility of 51 mg/L. You add a suspending vehicle to the powder and when you attempt to stir the mixture, the powder bounces off your stirring rod attached to several air bubbles. This observation indicates that:

   (a) The interfacial (surface) tension between water and ketoprofen is high

   (b) The interfacial (surface) tension between water and ketoprofen is low

   (c) The surface tension between air and ketoprofen is high

   (d) The surface tension between air and water is low

6. A pharmacist has prepared a flocculated suspension. This liquid will most likely demonstrate:

   (a) Slow rate of sedimentation

   (b) Good uniformity

   (c) Loosely packed sediment

   (d) Particles exist as separate entities

7. A dispersion of mineral oil in water:

   (a) Will require viscosity agents to prevent sedimentation of the oil

   (b) Will require an increase in the density of the continuous phase to prevent sedimentation

   (c) Will require reduction of droplet size to prevent sedimentation

   (d) Will not sediment but will coalesce

8. What is the slope of the line defined by Newton's Law?

   (a) Sheer stress

   (b) Sheer rate

   (c)   Velocity

   (d)   Viscosity

9. Hydrophilic colloidal solutions thin as the force applied to them increases because:

   (a)   Hydrophilic colloids are rigid until shaken

   (b)   Hydrophilic colloids are highly associated until shaken

   (c)   Hydrophilic colloids are highly flexible until shaken

   (d)   Hydrophilic colloids are highly dissociated until shaken

10. A pharmacist needs a vehicle that will thin as it is forced through a needle. This vehicle would be described as:

   (a)   Newtonian

   (b)   Plastic

   (c)   Pseudoplastic

   (d)   Dilatant

11. Viscosity enhancing agents are used to slow the loss of drug solutions and suspensions from the eye. If we assume that blinking is a form of shear stress, what kind of vehicle would be best for ophthalmic solutions and why?

12. Below are the formulations of the sugar-free suspension structured vehicle, NF, Soft hand cream, and Carbomer Gel. Determine the function of the excipients using Appendix Table A.2 and other resources. Categorize them as contributing to the manageable size of the dosage form, its palatability or comfort, its stability (chemical, microbial, or physical), the convenience of its use or the release of drug from the dosage form. Note that ingredients may be included in many dosage forms for more than one purpose.

Sugar-free suspension structured vehicle, NF

Ingredient	Category	Quality
Xanthan gum		
Saccharin sodium		
Potassium sorbate		
Citric acid		
Sorbitol		
Mannitol		
Glycerin		
Purified water		

Soft hand cream

Ingredient	Category	Quality
Methylparaben		
Propyl paraben		
Stearic acid		

Ingredient	Category	Quality
Triethanolamine		
Glycerin		
Mineral oil		
Purified water		

Carbomer aqueous jelly

Ingredient	Category	Quality
Carbomer 934		
Triethanolamine		
Methylparaben		
Propyl paraben		
Purified water		

13. The flow diagram in Figure 3.9 presents shear rate (flow rate) on the x axis and shear stress (force) on the y axis. Given Newton's law this means that the slope of the line formed reflects the viscosity of the liquid under study. Some authors will present flow diagrams with shear rate (flow rate) on the y axis and shear stress (force) on the x axis. Why is the latter presentation more scientifically 'proper'? What advantage is there in presenting flow diagrams with shear rate on the x axis and shear stress on the y axis?

Selected reading

1. W Im-Emsap, J Siepmann, O Paeratakul. Disperse Systems, in Modern Pharmaceutics, 4th edn. GS Banker and CT Rhodes (eds), Marcel Dekker, New York, 2002.
2. Remington: The Science and Practice of Pharmacy, 20th edn, AR Gennaro (ed.). Lippincott Williams & Wilkins, Baltimore, 2000.
3. Handbook of Pharmaceutical Excipients, 6th edn, RC Rowe, PJ Sheskey, ME Quinn (eds). Pharmaceutical Press, London 2009.
4. Pharmaceutics: The Design and Manufacture of Medicines, 3rd edn, ME Aulton (ed.), Churchill Livingston, London, 2007.
5. USP Convention. United States Pharmacopeia, 34th edn, and National Formulary, 29th edn. Baltimore, MD: United Book Press, 2011.

4

Properties of solutions and manipulation of solubility

Are you ready for this unit?

- Review moles, chemical equilibria, ionization, acid, base, pH and pK_a.
- Given a drug structure determine whether it is an acid or a base.
- Review molecular solutions, solubility, polymorphs, solvates and amorphous solids.

Learning objectives

- Define electrolyte, colligative property, isotonic, hypertonic, hypotonic, van't Hoff factor (*i*), cosolvent, dielectric constant, lyophilization and critical micellar concentration.
- Explain the difference between osmolality and osmolarity and calculate the osmolarity of a drug solution using the USP equation.
- Given a solute structure and relevant tables, determine the number of moles of particles (*i* or van't Hoff factor) that will result when a mole of the solute is dissolved in water.
- Calculate the amount of sodium chloride required to make a drug solution isotonic given the sodium chloride equivalent, *E* value.
- Predict whether a drug solution will be acidic or basic given the drug's structure and labeled contents.
- Given a drug's structure and pK_a, predict whether a drug will be water soluble or membrane soluble at a given pH.
- Calculate the pH of precipitation for a drug given concentration, formula weight, pK_a and relevant formulas.
- Explain how a buffer pair is chosen to resist pH changes.
- List the cosolvents that are used in pharmaceutical dosage forms and calculate the percent cosolvent required to dissolve a known drug concentration in a water/cosolvent blend.
- Describe how amorphous solids, drug derivatization, surfactants and cyclodextrins affect solubility.

Introduction

Why solubility is important

Perhaps the most obvious reason that solubility is important in pharmacy is that many dosage forms are solutions: syrups, elixirs, parenteral solutions or solutions for the eye, ear, nose, respiratory tract or skin. Most of these dosage forms are solutions of drug in water. In the case of parenteral intravenous solutions it is essential to patient safety that the drug remains in solution, meaning that the pharmacist must have a good understanding of the factors that will affect drug solubility.

The second reason that solubility is important in the practice of pharmacy is that those dosage forms that are not solutions must dissolve in aqueous body fluids before they can move across membrane barriers in the body. Formation of a solution, then, is the initial step in delivering many drugs to the site of action, affecting both the rate and extent of drug absorption.

While our initial interest as pharmacists may be in the water solubility of a drug, we will learn that our drug will also need adequate lipid solubility in order to diffuse across the lipid bilayers of the body's membranes. Thus we find that drug molecules are a collection of compounds whose affinity for water and for lipid fall within a reasonably well-defined range. A highly water soluble drug, that is, a *hydrophilic* compound, must also have enough lipid solubility to move from the site of administration to the receptors and out of the body through eliminating organs. A lipid soluble drug, that is, a *lipophilic* compound, must have enough water solubility to dissolve in body fluids at the site of administration and be carried with body fluids to the site of action. It is essential then, to understand what properties of drug molecules contribute to their water and lipid solubilities.

Colligative properties of molecular solutions

In a molecular solution the molecules of solute interact with the molecules of solvent. This interaction permits the solute to influence the properties of the solvent. The degree of influence of a solute on its solvent depends on the type of solute (Table 4.1):

- Nonelectrolyte solutes do not ionize in water.
- *Electrolytes* are compounds that dissociate into ions in solution:
  - Strong electrolytes are those compounds that dissociate completely in solution such as mineral acids, strong bases and most salts, including drug salts.

Table 4.1 Characteristics of solutes in aqueous solutions

Characteristics	Nonelectrolytes	Strong electrolytes	Weak electrolytes
Ionizability	No ions	Completely ionized	Weakly ionized; ionization pH dependent
Conductivity	Do not conduct current	Conduct strong current	Conduct weak current
Colligative properties	Depend on number of molecules in solution	Depend on number of ions in solution	Depend on the degree of ionization
Pharmaceutical excipient examples	Dextrose, mannitol, sorbitol, glycerin	HCl, NaOH, NaCl, KCl	Acetic acid, citric acid, benzoic acid
Drug examples	Carbamazepine, prednisone, lovastatin	Drug salts: morphine sulfate, penicillin potassium, propranolol hydrochloride	Morphine, penicillin, propranolol

Pharmaceutical example

$NaCl \rightarrow Na^+ + Cl^-$

- Weak electrolytes ionize to a very small extent in water such that the ratio of ionized form to unionized form is a very small number.

Pharmaceutical example

For acetic acid with a pK_a of 4.74:

$CH_3COOH + H_2O \rightleftarrows CH_3COO^- + H_3O^+$

free acid   ions

$$\text{Equilibrium constant} = \frac{[\text{products}]}{[\text{reactants}]} = \text{extent of ionization}$$

$$K_a = \frac{[CH_3COO^-][H_3O^+]}{[CH_3COOH]} = \text{antilog}[-pK_a] = \text{antilog}[-4.74] = 0.0000182$$

The alteration of the properties of pure water and other solvents by solutes depends on the concentration of dissolved particles (species) rather than the molar concentration. Referred to as *colligative properties*, these alterations include:

- the lowering of vapor pressure
- elevation of boiling point
- depression of freezing point
- production of osmotic pressure.

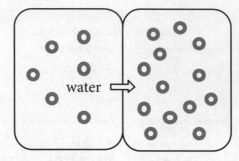

Figure 4.1 Osmotic pressure – water will move across a semipermeable membrane to equalize the concentrations of dissolved particles on either side

The *colligative properties* are so named because they are 'tied together' by the common underlying relationship to the concentration of dissolved particles. The effect of a solute on one colligative property can be used to predict the effect on other colligative properties. For example, the effect of solute particle concentration on freezing point depression can be used to predict the effect on osmotic pressure.

- Osmosis is a phenomenon created by particles in solution which are prevented from diffusing throughout a solution by a semipermeable membrane.
- Water moves to the side containing the greater concentration of *dissolved particles* until the concentrations are equal (Figure 4.1).
- A change in the concentration of dissolved particles will change each of the colligative properties in a predictable way. For example as the concentration of dissolved particles increases, the freezing point will be more depressed and the osmotic pressure of the solution will be increased.
  - One mole of nonelectrolyte in water produces one mole of dissolved particles.
  - The effect of electrolytes on the colligative properties of solutions is related to the number of ions rather than the number of moles dissolved in the solution.
- The contributions of different solutes to altering the property are additive. That is to say, the addition of dextrose to a drug solution in water will increase the osmotic pressure of the drug solution because the dissolved particles of both solutes contribute to the osmotic pressure.
  - *Any* dissolved solute will contribute to the osmotic pressure: sodium, chloride, proteins, glucose.

The osmotic pressure of drug solutions

The relationship between freezing point depression and the osmotic pressure of aqueous solutions has been useful to pharmaceutical scientists to develop a system for adjusting the osmotic pressure of drug solutions that are injected or applied to mucous membranes. The van't Hoff equation is the basis for these calculations:

$$\Delta T_f = iK_f m$$

where ΔT_f is the depression of the freezing point of water caused by the solution, i is the deviation from ideal behavior created by solute interactions (the van't Hoff factor), K_f is 1.86, and m is the concentration of solute in moles per kg solvent (molality). A nonelectrolyte such as dextrose has an i factor of 1 in dilute solutions. The freezing point of blood has been determined to be –0.52°C, so for the nonelectrolyte, dextrose monohydrate (MW 198), the concentration of a solution that will be iso-osmotic with blood is:

$$\Delta T_f = iK_f m$$

Rearranging: $m = \dfrac{\Delta T_f}{iK_f}$

$$\frac{\text{moles dextrose}}{\text{kg}} = \frac{0.52°\text{C}}{(1)(1.86)} = 0.279 \text{ moles/kg}$$

$$\frac{(0.279 \text{ moles})}{(\text{kg water})} \frac{(198 \text{ g})}{(\text{moles dextrose})} = 55 \text{ g dextrose/kg water} \sim 5\% \text{ (w/v)}$$

Solute structure and number of dissolved particles

Van't Hoff recognized that the effect of *electrolytes* on the colligative properties of solutions was related to the number of ions rather than the number of moles dissolved in the solution. Thus a 1 mol/kg solution of sodium chloride has about twice the freezing point lowering capability of a 1 mol/kg solution of dextrose. However, the van't Hoff factor for sodium chloride is not 2, but rather 1.823 at isotonic concentrations. The osmotic activity of the two ions is dampened by interactions with other ions in solution:

Theoretically:

$$NaCl \rightarrow Na^+ + Cl^- \text{ two particles}$$

1 mole = 2 moles particles

Actually:

$$NaCl \rightarrow 82.3\% \, Na^+ + Cl^- + 17.7\% \, NaCl$$

1 mole = 1.646 moles + 0.177 moles = 1.823.

The i factor corrects for the deviation from ideal behavior that is observed in solutions of pharmaceutical interest.

For divalent–divalent ions that only weakly dissociate, such as zinc sulfate, the i factor is reduced significantly as the concentration of zinc sulfate

Table 4.2 van't Hoff factors for different ionic classes

Class	Examples	$L_{iso} = K_f i$	i factor
Nonelectrolyte	Dextrose	1.86	1.0
Weak electrolyte	Weak acids or bases, boric acid, citric acid, acetic acid	2.0	1.053
Divalent–divalent salts, 2 ions	$ZnSO_4$, $MgSO_4$	2.0	1.053
Univalent–univalent salts, 2 ions	NaCl, dipivefrin hydrochloride, cephalothin sodium	3.4	1.79
Univalent–divalent salts, 3 ions, anion polyvalent	Atropine sulfate, ticarcillin disodium, dibasic sodium phosphate, morphine sulfate	4.3	2.26
Divalent–univalent salt, 3 ions, anion univalent	$CaCl_2$, magnesium gluconate, hydroxyzine dihydrochloride	4.8	2.53
Univalent–trivalent salts, 4 ions, anion polyvalent	Sodium citrate, sodium phosphate	5.2	2.74
Trivalent–univalent salts, 4 ions, anion univalent	Aluminum chloride, ferric chloride	6.0	3.16
Tetraborates	Sodium borate	7.6	4.0

Adapted from FM Goyan et al. J Am Pharm Assoc, Pract Ed. 5: 99: 1944.

increases (theoretical $i=2$, measured$=1.24$) If the i factor for the electrolyte of interest has not been measured, the estimates in Table 4.2 may be used.

Pharmaceutical example

For the univalent–univalent electrolyte sodium chloride (FW 58.5) the concentration of a solution that will be iso-osmotic with blood is:

$$m = \frac{\Delta T_f}{iK_f}$$

$$\frac{\text{moles NaCl}}{\text{kg}} = \frac{0.52°C}{(1.823)(1.86)} = 0.153 \text{ moles/kg}$$

$$\frac{(0.153 \text{ moles})}{(\text{kg water})}\frac{(58.5 \text{ g})}{(\text{moles NaCl})} = 8.95 \text{ g NaCl/kg water} \sim 0.9\% \text{ (w/v)}$$

Molality (moles per kilogram solvent, g MW (molecular weight)/kg) is useful in the measurement of a solution which will be frozen and therefore will change its volume, however, the preferred concentration convention for room temperature solutions is molarity (moles/liter solution) or g MW/L solution. While molality can be converted to molarity using the density of the solution, in practice the difference is ignored because, especially in solutions with concentrations less than 0.1 mol/L, the difference is small.

Osmolality and osmolarity

- The USP designates the units for expressing the osmotic pressure of a solution across a semipermeable membrane as osmoles per kg or mosmoles per kg of water (*osmolality*).
- One osmole equals 1 mole of dissolved particles, meaning for a nonelectrolyte, 1 mole equals 1 osmole and by definition would lower the freezing point of water by 1.86°C.
- The osmolality of blood and other body fluids is between 285 and 310 mosmol/kg.
- Drug solutions typically have osmotic concentrations in the milliosmolal range.
- *Osmolarity*, the number of osmoles or milliosmoles per liter of solution is calculated and cannot be directly measured.

The USP provides the following equation for calculation of the theoretical osmolar concentration of multicomponent drug solutions:

$$\text{Osmolarity} = \sum i_i C_i$$

where i_i is the van't Hoff factor of each component and C_i is the molar concentration.

Pharmaceutical example

We can calculate the osmolarity of a 0.9% sodium chloride solution using this equation:

$$\frac{(9\text{ g})}{(L)}\frac{(1\text{ mole})}{(58.5\text{ g})} = 0.1538\text{ mol/L NaCl}$$

Osmolarity = (1.823)(0.1538 mol/L) = 0.28038 osmol/L or 280 mosmol/L

Pharmaceutical example

To illustrate the calculation for a multicomponent solution, the osmolarity of 1 g of ampicillin sodium (Figure 4.2, FW 371.4, van't Hoff factor for a univalent–univalent salt from Table 4.2 is 1.79) in 10 mL of normal saline would be:
Molarity of ampicillin 1 g/10 mL:

$$\frac{(1\text{ g})}{(10\text{ mL})}\frac{(1000\text{ mL})}{(L)}\frac{(1\text{ mole})}{(371.4\text{ g})} = 0.269\text{ mol/L ampicillin}$$

Calculating its osmolarity:

Osmolarity = (1.79)(0.269 mol/L) = 0.482 osmol/L or 482 mosmol/L

The osmolarities are additive:

280 mosmol/L + 482 mosmol/L = 762 mosmol/L.

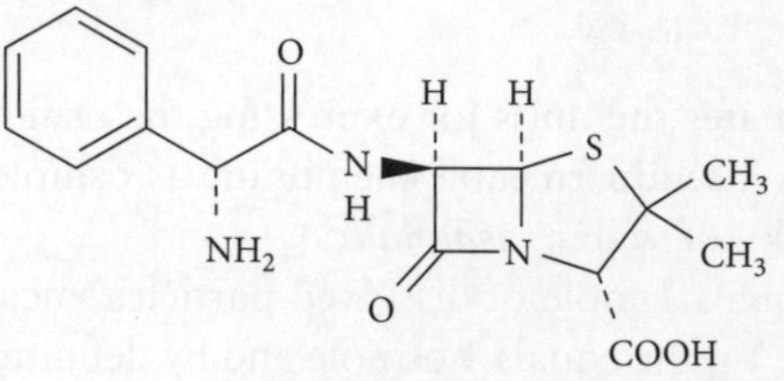

Figure 4.2 Ampicillin. (From Sweetman S. Martindale: The Complete Drug Reference, 27th edn (online), published by Pharmaceutical Press, 2013.)

Tonicity and isotonicity

- The term 'tonicity' is used to describe the osmotic pressure exerted across a membrane in the body.
- A solution that exerts the same pressure as body fluids is termed *isotonic*, and theoretically would have the same number of dissolved particles as body fluids.
- *Hypertonic* solutions have more dissolved particles than body fluids.
- *Hypotonic* solutions have fewer dissolved particles than body fluids.
- Drug solutions that are hyper- or hypotonic tend to cause irritation and pain when injected or applied to mucous membranes as a result of the osmotic pressure gradient between the solution and tissue fluids.
- Clinicians usually strive to keep the osmolarity of drug solutions administered through peripheral veins at or below 600 mosmol/L.
- Drugs with higher osmolarities must be administered via central veins where the blood flow is high enough to dilute them quickly.
- Hypertonic or hypotonic ophthalmic or nasal preparations cause tearing or nasal discharge, which flushes away applied drug and reduces its medicinal action.
  - We usually strive to make solutions for mucous membranes equivalent to 0.9% sodium chloride, which is considered isotonic.
  - The USP states that the eye can tolerate osmotic concentrations equivalent to 0.6–2% sodium chloride without significant discomfort.

Calculating tonicity of drug solutions using E value

- The preparation of isotonic solutions requires the use of units that can be conveniently measured in the pharmacy.
  - The *E* value system was developed to facilitate the addition of tonicity agent (an ingredient added to increase the osmotic pressure of a solution) measured by weight as a solid.
  - The USP volume method was developed by White and Vincent to facilitate the addition of a tonicity agent as an isotonic solution.

E value

Definition

The *E* value is defined as the weight of sodium chloride equivalent to 1 g of drug or other solute.

Related term: $\dfrac{\text{grams of NaCl}}{1 \text{ g of drug}}$

- Use when you will add the tonicity agent (an ingredient added to increase the osmotic pressure of a solution) as a solid measured by weight.
- The *E* values of many drugs have been calculated based on measured van't Hoff factors and tabulated for convenient use (Table 4.3).
- If the pharmacist needs to compound an isotonic product for a drug not previously tabulated, an *E* value can be calculated after consulting the appropriate reference books from the following formula:

$$E = \frac{17 \, K_f i}{\text{FW}}$$

where FW is the formula weight of the substances in grams, *i* the van't Hoff factor and $K_f = 1.86$. If the solute is a drug salt or a hydrate, the formula weight must include the weight of the counter ion or waters of crystallization. The van't Hoff factor may be estimated from Table 4.2 based on the drug's structure.

Pharmaceutical example

Rx   cefazolin sodium 5 mg/mL
Disp 10 mL

Sig : gtts i ou q h × 6 hours then q 4 hours × 7 days.

Consulting a reference such as *DrugBank* or *Merck Index*, we find that the formula weight of cefazolin sodium is 476.5.

What is *i*, the van't Hoff factor for cefazolin sodium?

If we consult the structure of cefazolin (Figure 4.3) we see that it has one carboxylic acid group that forms a univalent anion paired with the univalent cation, sodium. From Table 4.2, *i* = 1.79.

$$E = \frac{17(1.86)(1.79)}{476.5} = 0.11878 \sim 0.12$$

To use an *E* value, whether calculated or taken from a table, the pharmacist can use the definition, the weight of sodium chloride equivalent to 1 g of drug or other solute, to write the related term

Table 4.3 Isotonic values

Drug (1.0 g)	van't Hoff factor, i*	E value	USP volume (mL)
Atropine sulfate	2.79	0.13	14.3
Boric acid	0.96	0.50	55.7
Cocaine hydrochloride	1.68	0.16	17.7
Cromolyn sodium	2.2	0.14	15.6
Dextrose H_2O 12.3	1.0	0.16	17.7
Ephedrine sulfate	2.99	0.23	25.7
Epinephrine bitartrate	1.84	0.18	20.2
Mannitol	0.88	0.17	18.9
Morphine sulfate	2.86	0.14	15.6
Neomycin sulfate 20	2.30	0.12	12.3
Oxymetazoline hydrochloride	1.91	0.22	24.4
Phenylephrine hydrochloride	1.84	0.32	32.3
Pilocarpine nitrate	1.95	0.23	25.7
Procaine hydrochloride	1.76	0.21	23.3
Scopolamine hydrobromide	1.67	0.12	13.3
Sodium bisulfite	1.99	0.61	67.7
Sodium chloride	1.823	1	111.0
Sodium metabisulfite	3.88	0.67	74.4
Sodium phosphate (dibasic, 7 H_2O)	2.47	0.29	32.3
Sodium phosphate (monobasic, H_2O)	1.72	0.43	44.3
Sulfacetamide sodium	1.91	0.23	25.7
Tetracaine hydrochloride	1.68	0.18	20.0
Tobramycin sulfate	5.8	0.13	14.3
Vancomycin hydrochloride	1.60	0.05	5.6
Zinc sulfate (7 H_2O)	1.23	0.15	16.7

Adapted from selected reading 2 and the United States Pharmacopeia, XXI edn, page 1339. Rockville, MD: USP Convention, 1985.

*Calculated from $L_{iso}=1.86\ i$ or $\Delta T_f = iK_f m$ as tabulated in selected reading 2.

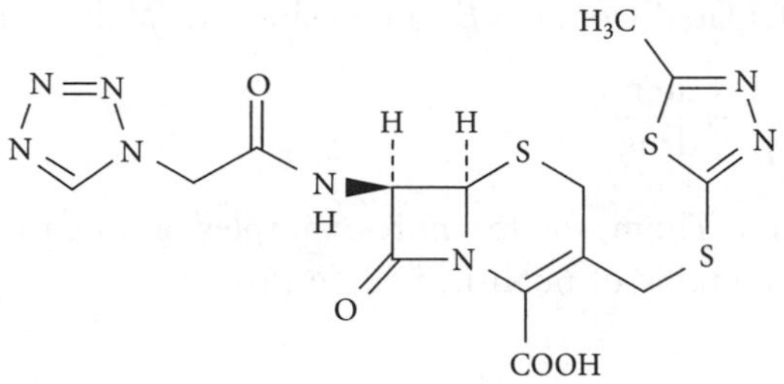

Figure 4.3 Cefazolin. (From Sweetman S. Martindale: The Complete Drug Reference, 27th edn (online), published by Pharmaceutical Press, 2013.)

$$\frac{\text{grams of NaCl}}{1 \text{ g of drug}}$$

This can be used to convert grams of drug to grams of sodium chloride. How much sodium chloride should be added to make the cefazolin solution isotonic?

1. Amount of sodium chloride equivalent to drug:
For cefazolin:
5 mg/mL $\times$ 10 mL = 50 mg = 0.05 g cefazolin E value = 0.12

$$\frac{(0.05 \text{ g cef}) (0.12 \text{ g NaCl})}{1 \text{ g cef}} = 0.006 \text{ g NaCl}$$

2. How much sodium chloride to make the volume of solution isotonic?

$$\frac{0.9 \text{ g NaCl}}{100 \text{ mL}} = \frac{y \text{ g NaCl}}{10 \text{ mL}} \qquad y = 0.09 \text{ g NaCl}$$

3. How much sodium chloride to add?

0.09 g NaCl

−0.006 g contributed by the cefazolin

0.084 g NaCl to add

USP volume

Definition

The USP volume is the volume in milliliters of isotonic solution that can be prepared by adding water to 1 g of drug.

Related term: $\dfrac{\text{mL water}}{1 \text{ g of drug}}$

- Use when you add tonicity agent as an isotonic solution.

- It can be calculated from the E value using the White-Vincent equation:

Related term: $\dfrac{\text{mL water}}{1 \text{ g of drug}}$

You are referred to Thompson for more examples of how to use E value and USP volume and a table of published values.

Manipulation of solubility

Drug solubility may be enhanced or decreased to achieve the design features necessary in an optimized dosage form. If the pharmacist finds that a drug's water solubility is inadequate to prepare a solution dosage form, there are several approaches that can be used to improve the drug's solubility. Of these approaches both pH and drug derivatization can be used to either increase or decrease the water solubility of a drug depending on the formulation's needs.

Approaches to the manipulation of drug solubility:

- pH: use of strong acid or base or buffers to prepare the ionized form of an acid or base
- use of cosolvents
- solid state manipulation
- solubilizing agents: surfactants, cyclodextrins
- drug derivatization

Solubility and pH

About 75% of drugs are weak organic bases and about 20% are weak acids. When acidic or basic drugs are in solution, either in dosage forms or in the body, they dissociate into ions. The extent to which they ionize is determined by the pH of the dosage form or the body fluid. The ionized form is significantly more water soluble while the unionized form is able to move through membranes. We can manipulate pH in dosage forms to increase the degree of ionization thus increasing the apparent solubility of the drug. If we know the pH of the body fluids we can predict the amount of drug that is available to cross membranes.

Drugs that are weak acids and bases ionize in solution when they are dissolved in water by donating or accepting protons and, in the absence of strong acid or base or buffer, the weak acid or base will determine the pH of the solution.

Drugs that are weak bases, B:

Drugs that are bases include amines.

$$R - NH_2 \quad R - NH - R \quad R - N - R$$
$$R$$

When the unionized form is dissolved in water:

$$B: + H_2O \rightleftharpoons B:H^+ + {}^-OH$$

When the ionized form (a salt or conjugate acid form) is dissolved in water:

$$B:H^+ + H_2O \rightleftharpoons B: + H_3O^+$$

K_b tells us the extent of the ionization or dissociation of the base form: i.e

$$K_b = \frac{\text{Products}}{\text{Reactants}} = \frac{[B:H^+][OH^-]}{[B:]}$$

K_a tells us the extent of the ionization or dissociation of the acid form: i.e

$$K_a = \frac{\text{Products}}{\text{Reactants}} = \frac{[B:][H_3O^+]}{[B:H^+]}$$

Drugs that are weak acids, HA:

Drugs that are acids include carboxylic acids, imides, sulfonamides, phenols, thiols and enols (β-dicarbonyls).

When the unionized form is dissolved in water:

$$HA + H_2O \rightleftharpoons H_3O^+ + {}^-A$$

When the ionized form (a salt or conjugate base form) is dissolved in water:

$$A^- + H_2O \rightleftharpoons HA + {}^-OH$$

K_a tells us the extent of the ionization or dissociation of the acid form: i.e

$$K_a = \frac{\text{Products}}{\text{Reactants}} = \frac{[{}^-A][H_3O^+]}{[HA]}$$

K_b tells us the extent of the ionization or dissociation of the base form: i.e

$$K_b = \frac{[{}^-OH][HA]}{[A^-]}$$

For weak organic acids and bases like drugs, both K_a and K_b are always very small numbers.

pK_a

- We usually use the equilibrium constant, K_a in a slightly different form.

$pK_a = -\log K_a$

While pK_a makes the numbers easier to work with, it also insulates us from the magnitude of the changes. For the conjugate acid, ammonium chloride:

$$K_a = 5.75 \times 10^{-10}$$

$$pK_a = -\log K_a = 9.24$$

Predicting whether a drug solution will be acidic or basic

The extent of ionization of weak acids and weak bases is pH dependent. This means that we can push this reaction towards the more soluble, ionic form by changing the pH with strong acid or base:

For a drug that is a weak base, B:, the unionized form in water forms hydroxyl ions and makes the solution basic:

$$B: + H_2O \rightarrow B:H^+OH^-$$

In the reaction above water acts as a weak acid. We can drive the ionization reaction to completion with *strong* acid:

$$B: + HCl \rightarrow B:H^+Cl^-$$

When the ionized form (the salt or conjugate acid form) is dissolved in water, it donates protons and makes the solution acidic:

$$B:H^+ + H_2O \rightleftarrows B: + H_3O^+$$

And for a drug that is a weak acid, HA, the unionized form in water donates protons and makes the solution acidic:

$$HA + H_2O \rightarrow A^- + H_3O^+$$

We can drive the ionization reaction to completion with strong base:

$$HA + NaOH \rightarrow Na^+A^- + H_2O$$

When the ionized form (the salt or conjugate base form) is dissolved in water, it forms hydroxyl ion and makes the solution basic:

$$A^- + H_2O \rightleftarrows HA + {}^-OH$$

Key points

- When we put a weak electrolyte into water, its ionization is what determines the pH of the solution.
- When we put weak electrolytes into a vehicle with a pH determined by a buffer, or a strong acid or base, the pH of the vehicle determines the extent of ionization.

Penicillin G

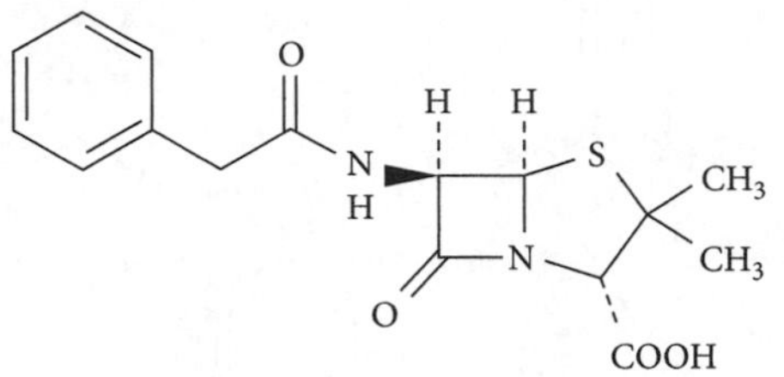

Penicillin G water solubility 210 mg/1000 mL

Figure 4.4 Penicillin G. (From Sweetman S. Martindale: The Complete Drug Reference, 27th edn (online), published by Pharmaceutical Press, 2013.)

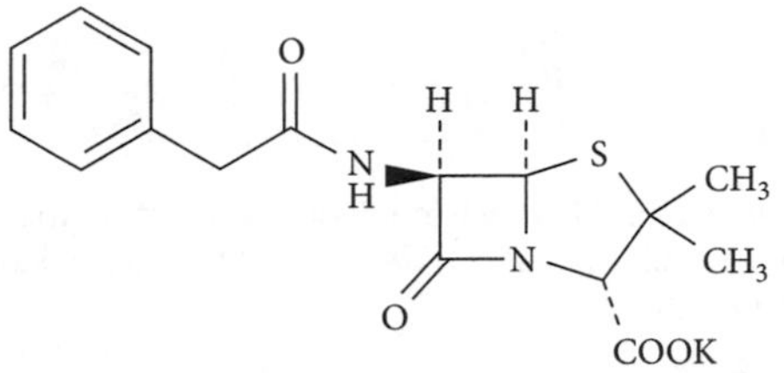

Conjugate base solubility 1 g/0.5 mL

Figure 4.5 Conjugate base of penicillin G. (From Sweetman S. Martindale: The Complete Drug Reference, 27th edn (online), published by Pharmaceutical Press, 2013.)

Pharmaceutical example

Let's consider three examples of drug conjugate bases and acids, and look at how strong acids or bases determine the extent of their ionization.

Consider the free acid form of penicillin G (Figure 4.4).

The conjugate base of penicillin G is shown in Figure 4.5.

The pK_a is 2.76. The K_a is $-\log K_a = 1.74 \times 10^{-3}$

The K_a is equilibrium constant of the reaction in which the proton on pencillin G's carboxylic acid is lost to water. Let HA represent penicillin G and A^- represent its conjugate base:

$$HA + H_2O \rightleftarrows A^- + H_3O^+$$

If K_a is 1.74×10^{-3} we can tell there is more HA in the solution than A^- when penicillin G is dissolved in water from the equation for equilibrium constant:

$$K_a = \frac{[A^-][H_3O^+]}{[HA]} = \text{extent of ionization}$$

Because the K_a is a fraction, there are more reactants, in other words, there is more HA in the solution. If we wanted the ionization of the acid to go to completion, we can use strong base such as sodium or potassium hydroxide:

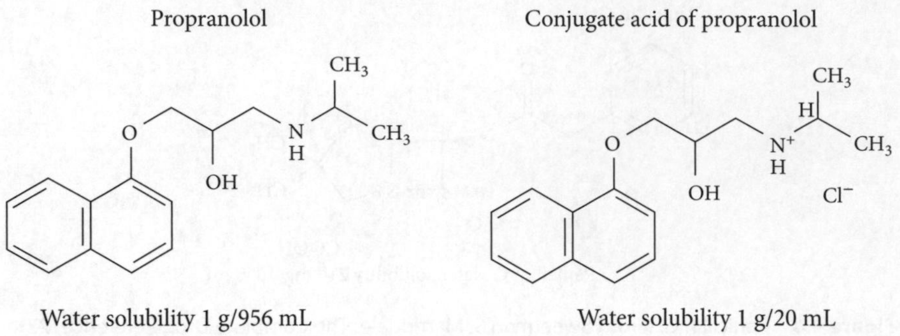

Propranolol

Conjugate acid of propranolol

Water solubility 1 g/956 mL

Water solubility 1 g/20 mL

Figure 4.6 Propranolol and its conjugate acid. (From Sweetman S. Martindale: The Complete Drug Reference, 27th edn (online), published by Pharmaceutical Press, 2013.)

$HA + KOH \Rightarrow K^+A^- + H_2O$

If you dissolve *potassium* penicillin G in water would you expect the pH to be acidic or basic? If we write the equilibrium reaction of potassium penicillin ionizing in water we can make a reasoned prediction.

$K^+A^- + H_2O \rightleftarrows HA + K^+OH^-$

The reaction of the conjugate base with water generates hydroxyl ion which makes the solution basic. The pH of reconstituted K penicillin G injection is about 8, slightly basic.

Pharmaceutical example

The pK_a of propranolol (Figure 4.6) is 9.45. The K_a is $-\log K_a = 3.55 \times 10^{-10}$ If B: represents propranolol and B:H$^+$ represents its conjugate acid, which of the reactions below does this equilibrium constant describe?

$B: + H_2O \rightleftarrows B:H^+ + {}^-OH$ (A)

$B:H^+ + H_2O \rightleftarrows B: + H_3O^+$ (B)

Because it is a K_a, the constant describes reaction B, the loss of a proton from the conjugate acid form of propranolol. If we want the ionization of the base (the addition of a proton) to go to completion, we would add strong acid, such as HCl to drive the reaction to completion:
The pH of propranolol HCl injection is 3.1, quite acidic.

$B: + HCl \Rightarrow B:H^+Cl^-$

If the drug manufacturer makes the intravenous injection of propranolol using propranolol HCl, is the injection likely to be acidic or basic? Again if we write the equilibrium reaction of propranolol HCl ionizing in water we can make a reasoned prediction.

$B:H^+Cl^- + H_2O \rightleftarrows B: + H_3O^+Cl^-$

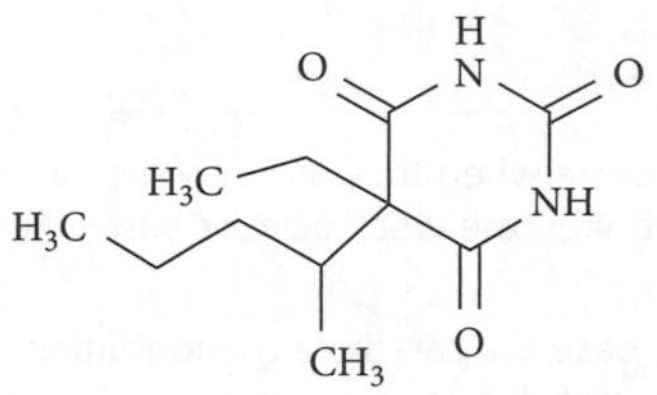

Figure 4.7 Pentobarbital (Nembutal). (From Sweetman S. Martindale: The Complete Drug Reference, 27th edn (online), published by Pharmaceutical Press, 2013.)

An imide   + H⁺   + H⁺

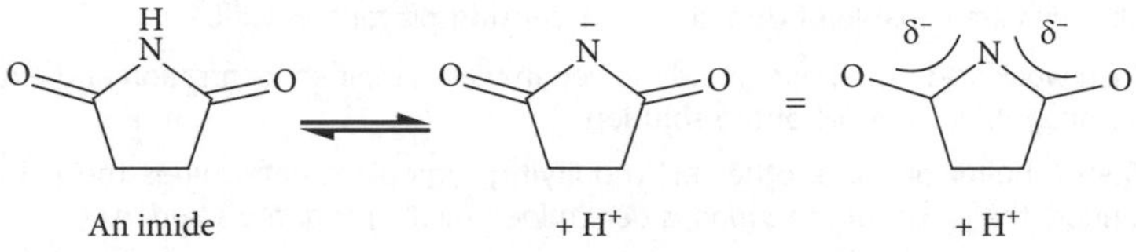

Figure 4.8 Formation of enol form of an imide. (From Sweetman S. Martindale: The Complete Drug Reference, 27th edn (online), published by Pharmaceutical Press, 2013.)

Pharmaceutical example

Pentobarbital is an example of a molecule that many students find difficult to identify as an acid or a base. We learned that carboxylic acids give protons to water with relative ease and that the nonbonding electrons of amines pick up protons from water quite readily. At first glance we might say that this molecule is neutral, that it is multiple amides in a six-membered ring (Figure 4.7).

However, these nitrogens flanked by carbonyls form the imide functional group which is found in a number of important drug molecules. The imide and β-dicarbonyl compounds rearrange to form an enol that is more readily identifiable as a proton donor and therefore an acid (Figure 4.8).

The free acid form of pentobarbital is very slightly soluble in water (0.1 to 1 mg/mL). If a drug manufacturer wanted to make an injectable solution of this drug with a concentration of 50 mg/mL, the equilibrium may be driven towards the charged form with sodium hydroxide. The sodium salt of pentobarbital is very soluble in water (>1000 mg/mL). Note, however, that the pK_a of pentobarbital is 8.0. It is a much weaker acid than penicillin G. This means to produce enough of the ionized form of this drug to dissolve 50 mg in 1 mL of water, the pH would have to be very high. In fact the pH of pentobarbital sodium injection is 9.5.

Pharmaceutical example

If you mix propranolol HCl injection and pentobarbital sodium injection, what will happen to the solubility of each of these drugs? Again if we write the reaction equation we can make a reasoned prediction:

$B{:}H^+Cl^- + Na^+A^- \rightleftarrows B{:} + HA + NaCl$

Water soluble forms $\rightleftarrows$ less water soluble forms
These two drugs will reduce the water solubility of one another and in fact are not compatible in the same syringe.

Key Points

- Weak acids and bases ionize when they are dissolved in water and, in the absence of strong acid or base or buffer, the weak acid or base will determine the pH of the solution.

- If we use strong acid or base, we can drive the ionization reaction to completion thus preparing water soluble, salt forms.

- Salt forms are significantly more soluble than the uncharged (unionized) form of the drug.

- The salt forms are also referred to as conjugate acids or conjugate bases.

- Salt forms are capable of donating or accepting protons as well.

- When weak acids and bases are dissolved in water, their concentration and extent of *ionization* drives the pH of the solution.

- When a buffer or some other pH modifying ingredient determines the pH of the solution, the *extent of ionization is determined by the pH of the solution*.

How are pK_a and pH useful to us?

- Predict the relative amounts of ionized or unionized drug at a given pH.
- Help us in choosing vehicles for drugs that will enhance their solubility since the ionized form will be more water soluble (see Table 4.4).
- Calculate the pH of drug solutions.
- Predict the advisability of mixing drug solutions with acidic and basic pHs.
- Predict whether drugs will accumulate in body compartments that have different pHs than serum (urine, breast milk). The unionized form moves through lipid membranes; the ionized form remains in compartmental fluids.

Table 4.4 Sweetened, flavored vehicles for oral liquids and their pH

Vehicle	pH
Acacia syrup	5.0
Aromatic eriodictyon syrup	8.0
Cherry syrup	3.5
Glycyrrhiza syrup	6.5
Orange syrup	2.5
Ora-Sweet	4.2
Ora-Sweet sugar free	4.2
Raspberry syrup	3.0
Simple syrup	7.0
Wild cherry syrup	4.5

Predicting whether a drug will be water soluble or membrane soluble at a given pH

In this last unit on pH concepts we will see that

The Henderson Hasselbach equation allows us to calculate the degree of ionization of a weak acid or base at a given pH:

$$pH = pK_a + \log\frac{[\text{base}]}{[\text{acid}]}$$

When we are concerned with the relative amounts of ionized and unionized forms of a drug, it can be helpful to write the Henderson Hasselbach equation more specifically:

$$pH = pK_a + \log\frac{[B:]}{[B:H^+]} \quad pH = pK_a + \log\frac{[A^-]}{[HA]}$$

If we rearrange the Henderson Hasselbach equation, we can calculate the relative amounts of acid and base and observe several important trends (Table 4.5).

- At pHs below the pK_a, the acid form predominates. This should make sense because in acidic solutions there are a lot of protons available to be picked up. They will find any base and drive the equilibrium towards the acid form.
  - For a drug that is an acid (HA), HA predominates. HA moves readily through membranes but is less soluble in water.

Table 4.5 Relative amounts of base and acid forms at pH units above and below the pK_a

pH – pK_a	log [base]* [acid]	[base] ** [acid]	% B: or A⁻	% HA or B:H⁺
−4	−4	1/10 000	0.01	99.99
−3	−3	1/1000	0.1	99.90
−2	−2	1/100	0.99	99.01
−1	−1	1/10	9.09	90.91
0	0	1	50	50
1	1	10/1	90.91	9.09
2	2	100/1	99.01	0.99
3	3	1000/1	99.90	0.10
4	4	10 000/1	99.99	0.01

*More specifically $\log\frac{[B:]}{[B:H^+]}$ or $\log\frac{[A^-]}{[HA]}$.

**More specifically [B:] or [A⁻].

- For a drug that is a base (B:), $B:H^+$ predominates. $B:H^+$ does not move through lipid membranes easily but has better solubility in water.
- At pH = pK_a, the amounts of acid and base forms are equal:

$[HA] = [A^-]$

$[B:] = [B:H^+]$.

- This interesting fact is extremely important to understanding how buffers work and we will consider it in more detail later.
- Equal amounts of weak acids and their conjugate bases produce buffer solutions that resist changes in pH.
- Buffers can be used to maintain drug solutions at the pH of maximum solubility or stability.
- Weak bases and their conjugate acids are used extensively as buffers in research applications but less often for pharmaceutical purposes.
- At pH > pK_a the basic form predominates. This again makes sense because in basic solutions there are few protons and many hydroxyl ions. This drives the equilibrium towards the loss of protons and the conjugate base species.
  - For a drug that is an acid (HA), A^- will predominate. A^- does not move through lipid membranes easily but has better solubility in water.
  - For a drug that is a base (B:), B: will predominate. B: moves readily through membranes but is less soluble in water.

The Henderson Hasselbach equation can also help us determine the relative amounts of charged and uncharged drug at different pHs encountered in the body.

Pharmaceutical example

Consider, for example, a nursing mother would like to take pseudoephedrine as a decongestant. She wants to know if the pseudoephedrine could affect the nursing baby. The pH of plasma is 7.4 and the pH of breast milk is 7.0. Pseudoephedrine is a weak base with a pK_a of 10.3. What is the ratio of B: to $B:H^+$ at pH 7.4 and pH 7.0?

$$pH = pK_a + \frac{\log[B:]}{[B:H^+]}$$

$$7.4 = 10.3 + \frac{\log[B:]}{[B:H^+]} = 7.4 - 10.3 = -2.9 = \frac{\log[B:]}{[B:H^+]} \frac{[B:]}{[B:H^+]} = \frac{1}{794}$$

$$7 = 10.3 + \frac{\log[B:]}{[B:H^+]} = 7 - 10.3 = -3.3 = \frac{\log[B:]}{[B:H^+]} \frac{[B:]}{[B:H^+]} = \frac{1}{1995}$$

These calculations tell us that in the breast milk there is more of the charged form of pseudoephedrine, $B:H^+$ than in plasma. Because this is the form of pseudoephedrine that cannot cross membranes, the drug is effectively trapped in the milk in a higher concentration than in plasma. This means that the baby could receive pseudoephedrine via the breast milk and the pharmacist should advise that the patient use nasal spray decongestant instead (Figures 4.9 and 4.10).

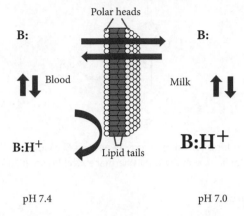

Figure 4.9 Effect of pH on concentration of basic drugs in different body fluids

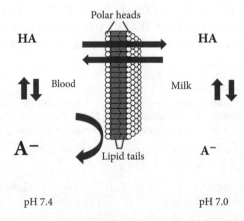

Figure 4.10 Effect of pH on concentration of acidic drugs in different body fluids

Calculating the pH of precipitation

We can calculate the pH of precipitation using a variation of the Henderson Hasselbach equation:

For a weak acid, HA:

$$pH_P = pK_a + \frac{\log\ [S_{A^-} - S_{HA}]}{[S_{HA}]}$$

where pH_P is the pH of precipitation, S_{A^-} is the solubility of the conjugate base and S_{HA} is the solubility of the free acid. If the conjugate base is not dissolved in the solution of interest at its solubility, S_{A^-} can be replaced by the actual or desired concentration of the drug in the solution.

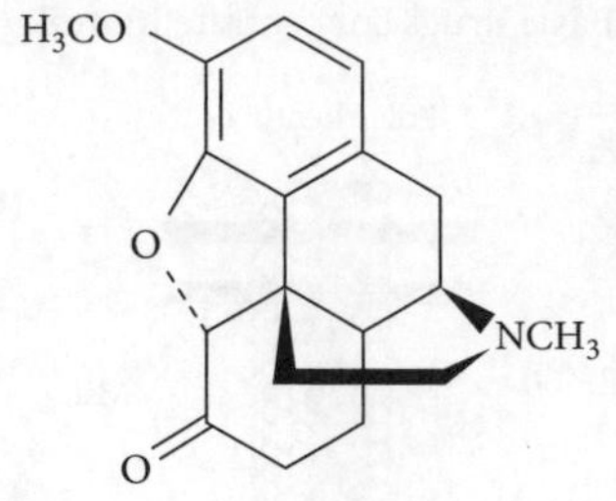

pKa 8.3

Figure 4.11 Hydrocodone (Vicodin). (From Sweetman S. Martindale: The Complete Drug Reference, 27th edn (online), published by Pharmaceutical Press, 2013.)

For a weak base, B:

$$pH_P = pK_a + \log\frac{[S_{B:}]}{[S_{BH^+} - S_{B:}]}$$

where pH_P is the pH of precipitation, $S_{B:}$ is the solubility of the free base, and S_{BH^+} is the solubility of the conjugate acid. If the conjugate acid is not dissolved in the solution of interest at its solubility, S_{BH^+} can be replaced by the actual or desired concentration of the drug in the solution.

Pharmaceutical example

You need to make hydrocodone injection using the hydrochloride salt (FW 335.8) at a concentration of 10 mg/mL. Hydrocodone (FW 299.4) (Figure 4.11) has a solubility of 6.87 mg/mL. Will you be able to prepare the injection at physiological pH (7.4)?

$$\frac{\text{Moles B:}}{L} = \frac{(6870\text{ mg})}{(L)}\frac{(1\text{ g})}{(1000\text{ mg})}\frac{(1\text{ mole})}{(299.4\text{ g})} = \frac{(0.0229\text{ mol/L B:})}{L}$$

$$\frac{\text{Moles B:H}^+}{L} = \frac{(10\text{ mg})}{(\text{mL})}\frac{(1\text{ g})}{(1000\text{ mg})}\frac{(1000\text{ mL})}{(1\text{ L})}\frac{(1\text{ mole})}{(335.8\text{ g})} = \frac{(0.0298\text{ mol/L B:H}^+)}{L}$$

$$pH_P = \frac{8.3 + \log[0.0229]}{[0.0298 - 0.0229]} = 8.3 + 0.52 = 8.8$$

You should be able to prepare the injection at pH 7.4 since it will not precipitate until pH 8.8.

Many drug molecules have more than one acidic or basic functional group; acids capable of donating two or more protons are called polyprotic acids and bases capable of accepting more than one proton are called polyprotic bases. See selected reading 7 for a presentation of similar calculations with these molecules.

Buffers

The other important use of the Henderson Hasselbach equation is the preparation of buffers. We use buffers in dosage forms to maximize drug stability or solubility, improve patient comfort or improve absorption.

Buffers are solutions of weak acids and their conjugate bases that resist changes in pH. Weak bases and their conjugate acids may also be used but they are less common. Buffers resist changes in pH at pHs where there are substantial amounts of both the acid form and the conjugate base forms. In general adequate amounts of both forms can be found within 1 pH unit of the acid's pK_a. Consider acetic acid (pK_a 4.74):

At pH 3.7, there are about 10R–COOH for every 1R–COO–Na$^+$

$$10R–COOH + 1R–COO–Na^+ + 1NaOH \rightarrow 9R–COOH$$

$$+ 2R–COO–Na^+ + H_2O$$

At pH 4.7, there are about 10R–COOH for every 10R–COO–Na$^+$

$$10R–COOH + 10R–COO–Na^+ + 1NaOH \rightarrow 9R–COOH$$

$$+ 11R–COO–Na^+ + H_2O$$

At pH 5.7, there are about 1R–COOH for every 10R–COO–Na$^+$

$$1R–COOH + 10R–COO–Na^+ + 1NaOH \rightarrow 11R–COO–Na^+ + H_2O$$

In the range of +1 to −1 pH unit of the pK_a, there are enough of both acid and base forms to consume any H$^+$ or OH$^-$ ions added to the solution and prevent any large changes in pH. Appendix Table A.6 presents buffer pairs commonly used in drug dosage forms. A 1:1 ratio of the acid and base will produce a buffer solution in which pH = pK_a.

Buffer selection

The buffer's pH is determined by the pK_a and the ratio of weak acid and conjugate base, but the buffer's capacity will depend on the concentration of the buffer components. The buffer concentration should be sufficient to maintain the pH in the dosage form without affecting the pH of the body when the dosage form is administered.

- Pharmaceutical buffers are usually used in concentrations between 0.05 and 0.5 molar.
- The pharmacist must consider the route of administration in the choice of buffer ingredients. Boric acid buffers are toxic parenterally; can only be used in the eye. Buffers made from amines have a disagreeable odor and cannot be used in oral products.
- The buffer pair selected should have a pK_a approximately equal (±1 pH unit of the pK_a) to the pH required for the dosage form.
- The Henderson Hasselbach equation may be used to calculate the ratio of conjugate base to weak acid required to produce the desired pH.
- A number of buffer formulas have been developed that may be used if the compounding pharmacist needs a buffer with a specific pH or one

Table 4.6 Polarity of various solvents		
Solvent	Dielectric constant (ε)	Log P
Water	78.5	−4.00
Glycerin	42.5	−1.76
Methanol	32.5	−0.77
Propylene glycol	32.1	−0.92
Ethanol	25	−0.32
Isopropanol	18.3	0.3
Polyethylene glycol 400	12.4	−0.88
Chloroform	4.8	2.3
Diethyl ether	4.34	0.9
Olive oil	3.1	
Mineral oil	0	>6

that is isotonic. Please see selected reading 2 for formulas and more details on buffer selection.

Manipulation of solubility with cosolvents

Key Points

- A *cosolvent* is any solvent that is miscible with water.
- Cosolvents are generally mixed with water to produce a blend that is less polar than water and that is capable of dissolving a hydrophobic drug at the required concentration.
- Occasionally a cosolvent or cosolvent blend is used without water to produce a solution dosage form that is miscible with body fluids.
- Blends with water are preferred because undiluted cosolvent is generally not palatable when taken by mouth, not comfortable when injected or dropped in the eye and too viscous to easily move through a needle.
- Cosolvents are also used in injectable drug solutions where they may precipitate when diluted with aqueous injectables and can precipitate in body fluids on injection.
- The polarity of solvents can be expressed using the *dielectric constant*, a measure of the ability of a solvent to separate two oppositely charged ions (Table 4.6).
  - The higher the dielectric, the more polar the solvent.
- The relationship between log solubility and fraction cosolvent is approximately linear and can be used to estimate the amount of cosolvent required to dissolve a drug at a specified concentration using the equation:

$$\log S_{mixed} = \log S_{water} + \sigma f c$$

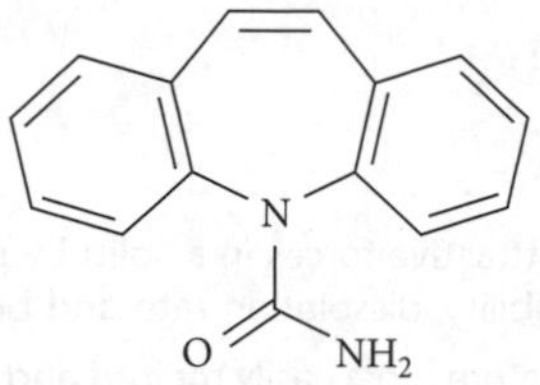

Figure 4.12 Carbamazepine FW 236. (From Sweetman S. Martindale: The Complete Drug Reference, 27th edn (online), published by Pharmaceutical Press, 2013.)

Table 4.7 Carbamazepine solubility parameters

Parameter	g/mL	Log Y	Fraction cosolvent (x)
Water solubility	0.0000177	−4.75	0
Alcohol solubility	0.0435	−1.36	1
Desired concentration	0.02	−1.7	?

Pharmaceutical example

Consider the following drug (Figure 4.12):
(sol 0.0177 mg/mL water, log P 2.7)
If we want to make a solution that is 100 mg/5 mL carbamazepine, we can use the equation on page 92 to determine what fraction of the vehicle should be alcohol and what fraction water. We need to look up the following data about carbamazepine and convert to logs (Table 4.7):
To calculate slope, we use our two data points:

$$\text{Slope} = \frac{\Delta y}{\Delta x} = \frac{y_1 - y_2}{x_1 - x_2} = \frac{-1.36 - 4.75}{1 - 0} = 3.39$$

Our equation is

$$-1.7 = -4.75 + (3.39)fc$$

Adding 4.75 from each side:

$$4.75 - 1.7 = -4.75 + 4.75 + (3.39)fc$$

$$3.05 = (3.39)\,fc$$

Dividing both sides by 3.39:

$$\frac{3.05}{3.39} = \frac{(3.39)fc}{(3.39)} = 0.899$$

The fraction of cosolvent required to dissolve 0.1 g carbamazepine in 5 mL solution is 0.899 or nearly 90%. This is a high concentration for an oral solution given that cosolvents are less palatable than aqueous solutions and must be sweetened with artificial sweeteners such as saccharin. In the case of alcohol, this is the same as 180 proof! In some cases the palatability of cosolvent solubilized drug solutions may be improved by blending alcohol with glycerin or propylene glycol. Because this particular product is intended for children with seizure disorders, we should think about another approach to formulating a liquid dosage form.

Solid state manipulation

Key Points

- Reduction of the strong attractive forces in a solid by preparation of a lower melting polymorph improves solubility, dissolution rate and bioavailability.
- The least orderly and therefore most easily melted and dissolved form of a drug is the amorphous solid.
- The methods used to prepare amorphous solids do not allow the molecules of drug enough time to form orderly arrays. These include rapid precipitation from solvent, rapid cooling of a melt of the drug, spray drying and freeze drying.

To spray dry solids, a suspension of drug is sprayed into a stream of hot air so that each droplet dries into an individual solid particle. Particles are very uniform, porous and generally amorphous. Freeze drying or *lyophilization* begins with an aqueous solution of the drug which is rapidly frozen and then exposed to a powerful vacuum that draws off the water in vapor form. In addition to being amorphous, a freeze dried solid will occupy the same volume as the original solution, is light and porous and has a large surface area. Drug products presented as powders for reconstitution are prepared using lyophilization.

Solubilizing agents

Key Points

Surfactants

- Surfactants are molecules with hydrophilic and hydrophobic functional groups arranged in clusters, i.e. hydrophilic functional group at one end of the molecule and hydrophobic functional group at the other end (Figure 4.13).
- Surfactants align themselves at surfaces between air and water or interfaces between immiscible liquids such as oil and water, in order to remove the hydrophobic end of the molecule from water.
- When added to water in sufficient concentration, the *critical micellar concentration* (CMC), surfactants will form spherical aggregates called micelles (Figure 4.14).
- Use of a surfactant at or above its CMC increases the apparent solubility of hydrophobic drugs by allowing them to remove themselves from the water into the micelle.
- Since solubilization depends on the presence of micelles, it does not occur below the CMC and drugs will be precipitated by dilution below the CMC.
- Surfactants that increase apparent drug–water solubility are not generally used in oral product formulations because the amount of surfactant needed may cause side effects.

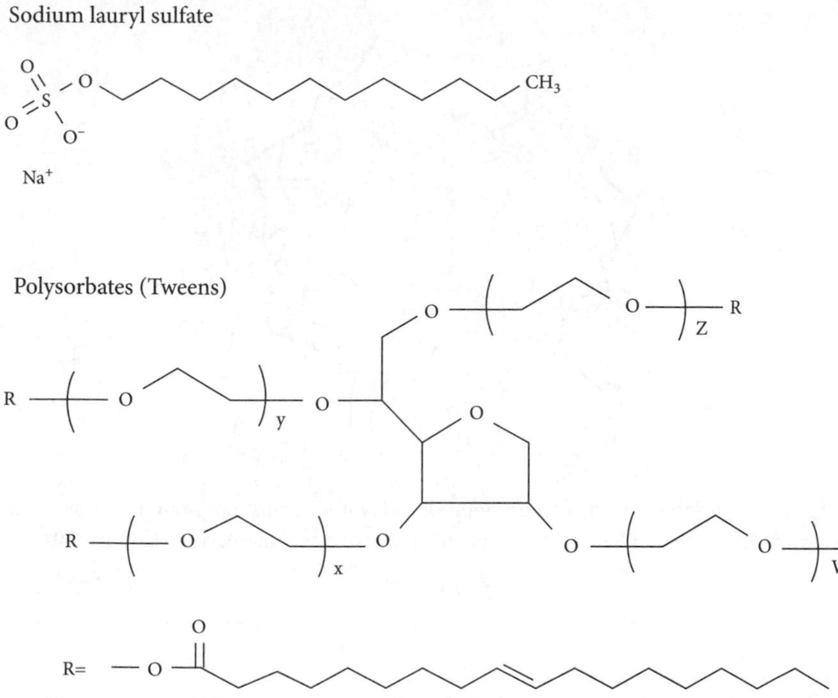

Sodium lauryl sulfate

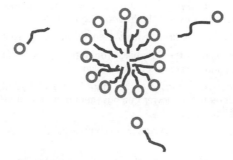

Polysorbates (Tweens)

Figure 4.13 Anionic and nonionic surfactants. (From Sweetman S. Martindale: The Complete Drug Reference, 27th edn (online), published by Pharmaceutical Press, 2013.)

Figure 4.14 Micelle and surfactant monomers

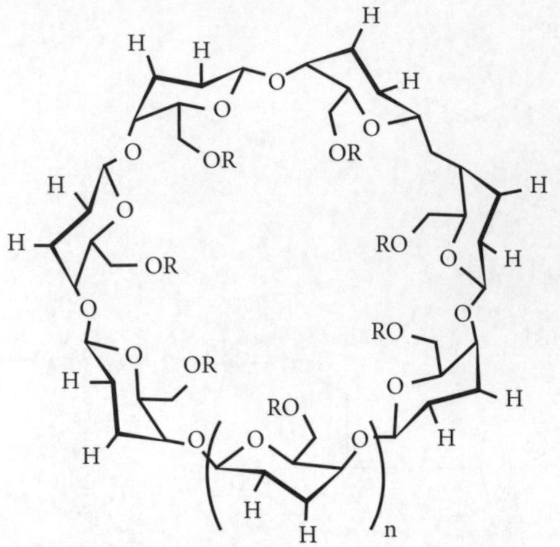

Figure 4.15 Cyclodextrin structure. (Reproduced with permission from L Felton (ed.), Remington: Essentials of Pharmaceutics. p. 691; published by Pharmaceutical Press, 2012.)

Key Points

Cyclodextrins

- Cyclodextrins are cyclic oligosaccharides derived from starch containing six or more (α-1, 4) linked α-D-glucopyranose units.

- The glucopyranose unit being in the chair conformation makes the cyclodextrin molecule cone shaped with an interior cavity capable of non-covalently binding lipophilic drugs.

- The cyclodextrin's hydroxyl groups are oriented to the exterior of the cone while carbons and ether oxygens line the interior (Figure 4.15).

- The polarity of the cyclodextrin cavity is likened to a water–ethanol blend.

- The apparent solubility of the lipophilic drug in the cyclodextrin cavity is increased, increasing its dissolution rate and bioavailability.

- Cyclodextrin-complexed drugs may be more stable to hydrolysis, oxidation and photolysis, and demonstrate lower mucous membrane irritation as a result of lower concentration of free drug.

- Drugs are released from their cyclodextrin complexes when they are diluted by body fluids.

- Generally one drug molecule complexes with one cyclodextrin molecule, however, the complexation efficiency (ratio of complexed cyclodextrin to cyclodextrin concentration) varies from drug to drug and cyclodextrin to cyclodextrin.

Pharmaceutical example

Cyclodextrins are used in commercially available products such as the highly insoluble drug, itraconazole. Itraconazole oral solutions are solubilized by noncovalent binding with hydroxypropyl-beta-cyclodextrin. Its complex with beta–cyclodextrin increases its water solubility and solution stability.

Key Points

Drug derivatization

- The solubility of a drug molecule may be altered by covalently attaching a functional group with the desired solubility properties.

- The new functional group must be easily cleaved by body secretions or enzymes to regenerate the active drug.

- A drug molecule may be derivatized with a new functional group that enhances or detracts from the water solubility of the molecule.

  - Drug derivatization can be used to make drug molecules more water soluble and more readily absorbed or less water soluble to achieve sustained release.

  - Occasionally a drug molecule is derivatized so that it does not begin to dissolve in the mouth where its bitter taste would be detected.

  - Drugs that are unstable in water may be derivatized to reduce the amount of susceptible drug in solution and thus prolong the shelf life in liquid form.

- Functional groups that can be derivatized include hydroxyl groups, carboxylic acids and primary and secondary amines.

- Derivatized drug molecules are *prodrugs* because they are not active until the additional functional group is removed *in vivo*.

- Amides and esters are easily cleaved by body secretions or enzymes to regenerate the active drug.

- Proteases and esterases are ubiquitous enzymes, and if given by mouth strong acid in stomach can cleave those bonds to produce the original drug.

Summary of Key Points

- Strong acid or base or buffers may be used to prepare the ionized form of a weak acid or base.

- Cosolvents are water-miscible solvents that are less polar than water and may be mixed with water to make solution dosage forms.

- Drugs may be prepared as amorphous solids which have fewer attractive forces within the solid and therefore have better solubility and faster dissolution rates than their crystalline counterparts.

- Surfactants are amphiphilic molecules that assemble in spherical clusters called micelles with their polar ends towards water and their hydrophobic ends oriented inside the micelle. Hydrophobic drugs will move into the hydrophobic portion of the micelles to escape undesirable interactions with water.

- Cyclodextrins are cone shaped polysaccharides with a hydrophilic surface and hydrophobic core that can shield hydrophobic drugs from undesirable interactions with water.

- Water soluble functional groups that provide the molecule with improved water solubility may be attached covalently to drug hydroxyls, amines and carboxylic acids. These groups are cleaved off the drug in the body.

Questions and cases

1. Calculate the osmolarity of lidocaine hydrochloride injection for regional anesthesia. It has a lidocaine HCl concentration of 20 mg/mL (formula weight (FW) 288.82) and sodium chloride (FW 58.5) 6 mg/mL. Look up the drug structure and decide which van't Hoff factor you will use.

2. Calculate the milligrams of sodium chloride that should be added to 100 mg of vancomycin hydrochloride to make 10 mL of an isotonic solution.
What volume of water will make our drug weight isotonic?
How much NaCl is required to make 10 mL isotonic?

3. Which of the five methods for enhancing solubility presented in this chapter could be used to prepare a solution dosage form with each of the following drugs?

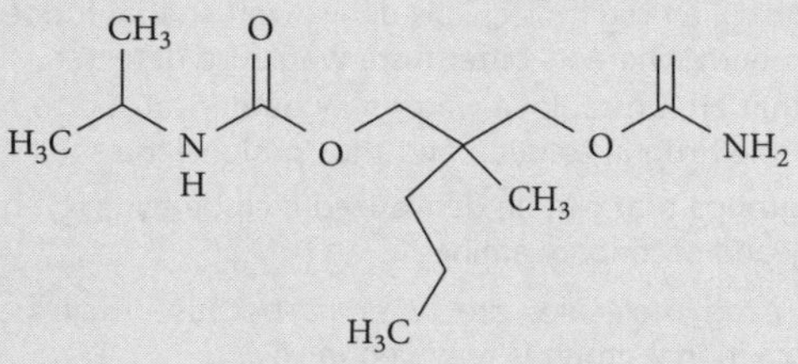

Carisoprodol

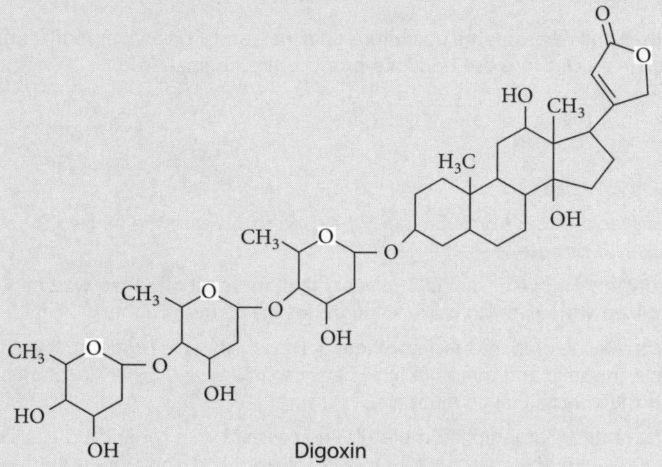

Digoxin

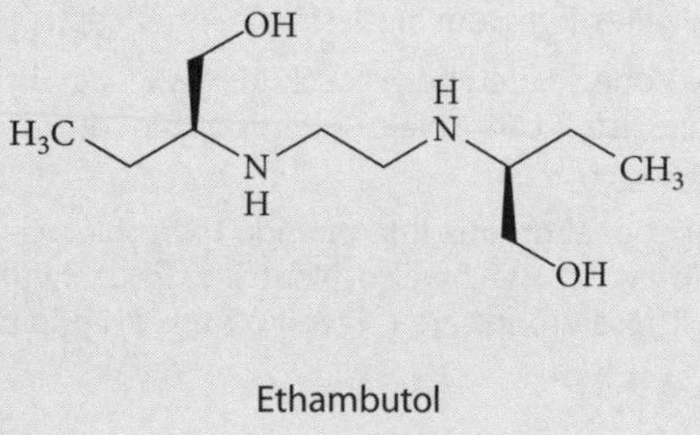

Ethambutol

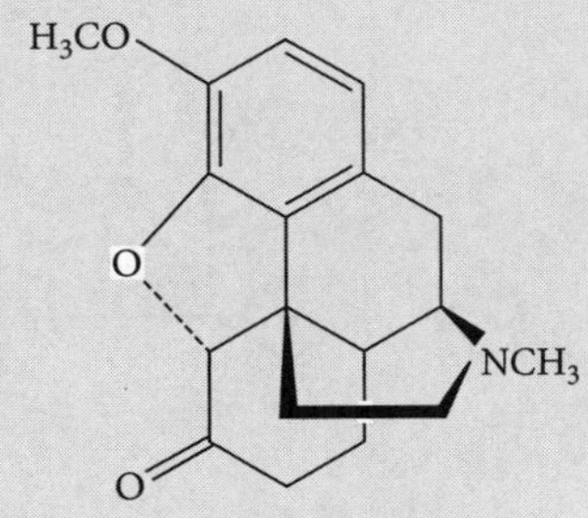

Hydrocodone

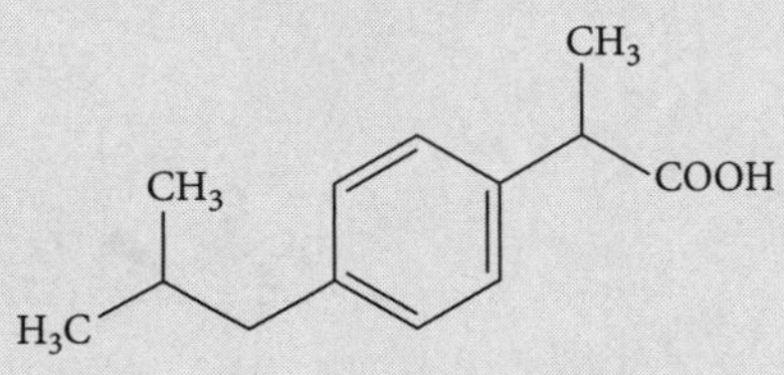

Ibuprofen

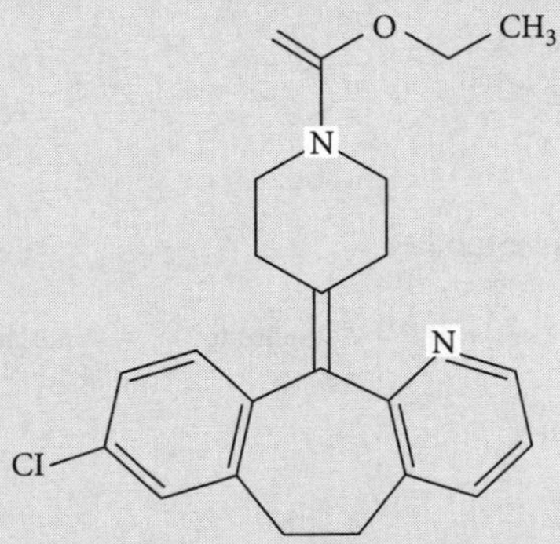

Loratidine

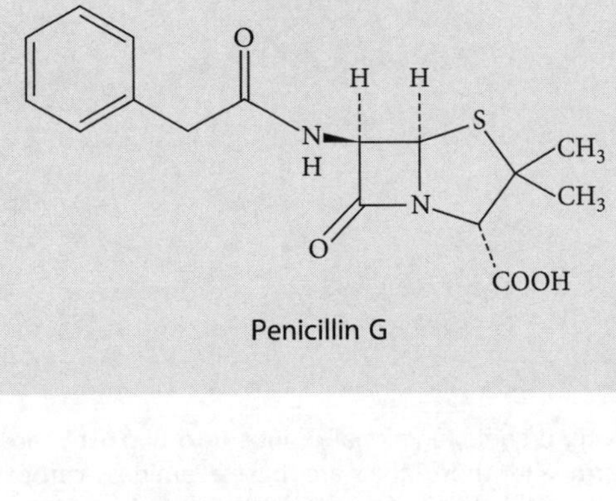

Penicillin G

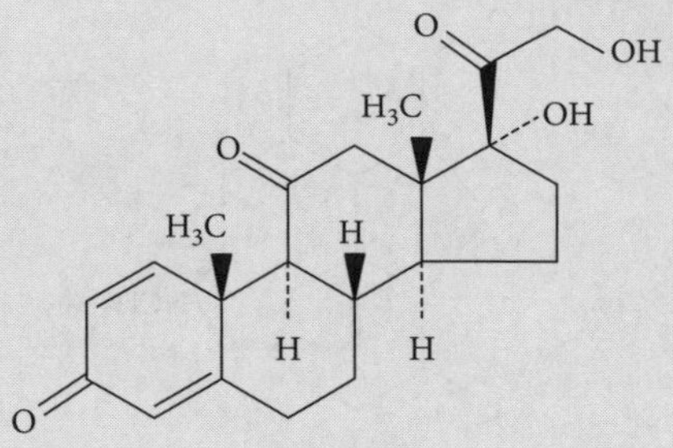

Prednisone

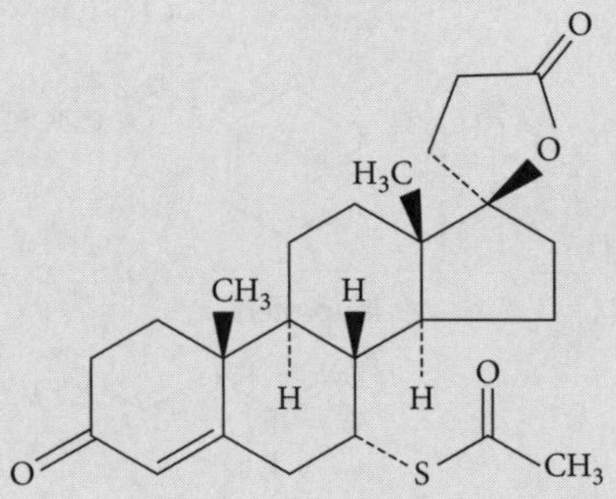

Spironolactone

Methods of manipulating solubility

Drug	pH	Cosolvents	Solubilizing agents	Amorphous solid	Derivatization
Carisoprodol					
Digoxin					
Ethambutol					
Hydrocodone					
Ibuprofen					
Loratidine					
Penicillin G					
Prednisone					
Spironolactone					

4. Put the following organic functional groups into two lists: those that are acids in uncharged form and those that are bases: amines, carboxylic acids, imides, phenols, sulfonamides, thiols. Use your lists to help you answer the following questions.

5. Amiodarone (FW 645) has a water solubility of 0.7 mg/mL and its pK_a is 6.56. It is available as a solution for injection containing 50 mg amiodarone hydrochloride (FW 681.8) per mL. Calculate the pH of precipitation. Can this drug be mixed with phenytoin injection that has a pH of 12 and remain in solution?

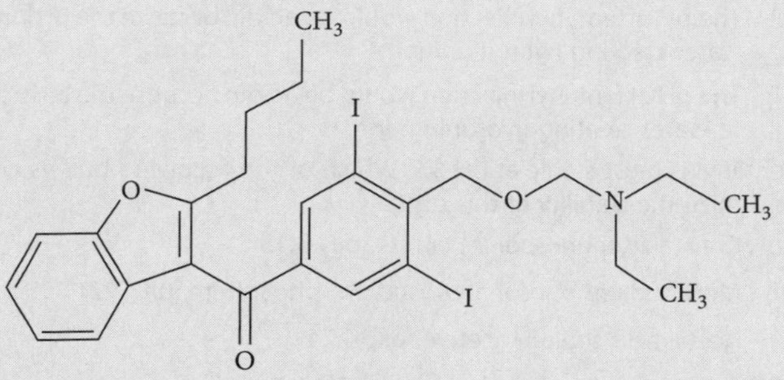

Amiodarone

6. Look up the structure and water solubility of naproxen (FW 230). You would like to make an oral solution of naproxen sodium (FW 252) containing 220 mg per 5 mL. What is the pH of precipitation for naproxen in this dosage form? From Table 4.4, select a commercially available oral liquid that would maintain this drug in solution. Drug examples using Henderson Hasselbach concepts

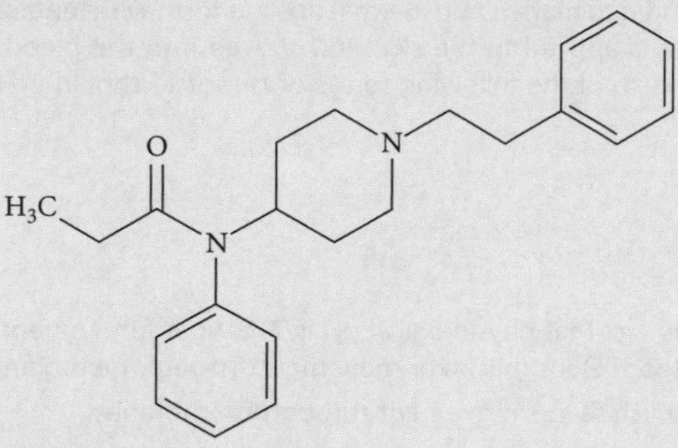

Fentanyl

Fentanyl has a pK_a of 8.4.

7. If we wanted to make an aqueous injection of this drug which of the following forms of the drug should be chosen?

(a) B:

(b) B:H⁺

(c) HA

(d) A⁻

8. What would be some possible counter ions that might be used with the above form of fentanyl?

(a) Chloride, sodium or potassium

(b) Sodium, potassium or calcium

(c) Chloride, sulfate or calcium

(d) Chloride, sulfate or citrate

9. Would the pH of the injection be acidic or basic and why?

(a) The pH of fentanyl injection would be basic because the B: picks up protons from water creating hydroxyl ion

(b) The pH of fentanyl injection would be basic because the B:H⁺ picks up protons from water creating hydroxyl ion

(c) The pH of fentanyl injection would be acidic because the B: donates protons to water creating hydronium ion

(d) The pH of fentanyl injection would be acidic because the B:H$^+$ donates protons to water creating hydronium ion

10. Fentanyl is most stable at pH 5.5. Which of the following buffers could be used to maximize the stability of this drug?

(a) Citric acid/monosodium citrate, pK_a 3.15

(b) Monosodium phosphate/disodium phosphate, pK_a 7.21

(c) Acetic acid/sodium acetate, pK_a 4.74

(d) benzoic acid/sodium benzoate, pK_a 4.2

11. Use of a citrate buffer with fentanyl citrate injection may:

(a) Increase solubility by decreasing the pH

(b) Decrease the solubility by decreasing the pH

(c) Decrease the solubility owing to the common ion effect

(d) Increase the solubility owing to the common ion effect

12. If we wanted to make a transdermal dosage form with fentanyl, that is a form of the drug that is applied to the skin and moves into the blood stream to the site of action, which of the following forms of the drug should be chosen?

(a) B:

(b) B:H$^+$

(c) HA

(d) A$^-$

13. Given the fact that physiological pH is 7.4, what form of fentanyl will predominate in the blood? Does this favor movement through membranes or water solubility?

(a) B: which favors movement through membranes

(b) B: which favors water solubility

(c) B:H$^+$ which favors movement through membranes

(d) B:H$^+$ which favors water solubility

(e) B: and B:H$^+$ are approximately equal at this pH

14. If a patient had had too much fentanyl and we wanted to increase the amount that he/she excreted in the urine, would we want the urine to be more basic or more acidic than the blood? (Hint: we want to trap the fentanyl in the urine so that it can't move back to the blood by passive diffusion.)

Dinoprostone

Dinoprostone has a pK_a of 4.6.

15. If we wanted to make an aqueous injection of this drug which of the following forms of the drug should be chosen?

(a) B:

(b) B:H$^+$

(c) HA

(d) A$^-$

16. What would be some possible counter ions that might be used with the above form of dinoprostone?

    (a) Chloride, sodium or potassium

    (b) Sodium, potassium or calcium

    (c) Chloride, sulfate or calcium

    (d) Chloride, sulfate or citrate

17. Would the pH of the injection be acidic or basic and why?

    (a) The pH of dinoprostone injection would be basic because the HA picks up protons from water creating hydroxyl ion

    (b) The pH of dinoprostone injection would be basic because the A$^-$ picks up protons from water creating hydroxyl ion

    (c) The pH of dinoprostone injection would be acidic because the HA donates protons to water creating hydronium ion

    (d) The pH of dinoprostone injection would be acidic because the A$^-$ donates protons to water creating hydronium ion

18. If we chose a buffer to maximize the solubility of this drug, which of the following buffers would be the best choice?

    (a) phosphoric acid/monosodium phosphate, pK_a 2.12

    (b) monosodium phosphate/disodium phosphate, pK_a 7.21

    (c) citric acid/monosodium citrate, pK_a 3.15

    (d) monosodium citrate/disodium citrate, pK_a 4.78

19. If we wanted to make a vaginal gel dosage form, that is a form of the drug that is applied to the mucous membranes and moves through a series of membranes to the site of action, which of the following forms of dinoprostone should be chosen?

    (a) B:

    (b) B:H$^+$

    (c) HA

    (d) A$^-$

20. If we measured the pH of the vaginal dosage form would it be acidic or basic?

    (a) The pH of dinoprostone gel would be basic because the HA picks up protons from water creating hydroxyl ion

    (b) The pH of dinoprostone gel would be basic because the A$^-$ picks up protons from water creating hydroxyl ion

    (c) The pH of dinoprostone gel would be acidic because the HA donates protons to water creating hydronium ion

    (d) The pH of dinoprostone gel would be acidic because the A$^-$ donates protons to water creating hydronium ion

21. Given the fact that vaginal pH is 4.6, what form of dinoprostone will predominate at the site of absorption? Does this favor movement through membranes or water solubility?

    (a) HA which favors movement through membranes

    (b) HA which favors water solubility

(c) A⁻ which favors movement through membranes

(d) A⁻ which favors water solubility

(e) HA and A⁻ are approximately equal at this pH

22. If a patient had received dinoprostone and was going to breast feed, assuming equilibrium would there be more dinoprostone in the blood (pH 7.4) or the breast milk (pH 7.0)?

23. A cosolvent problem:

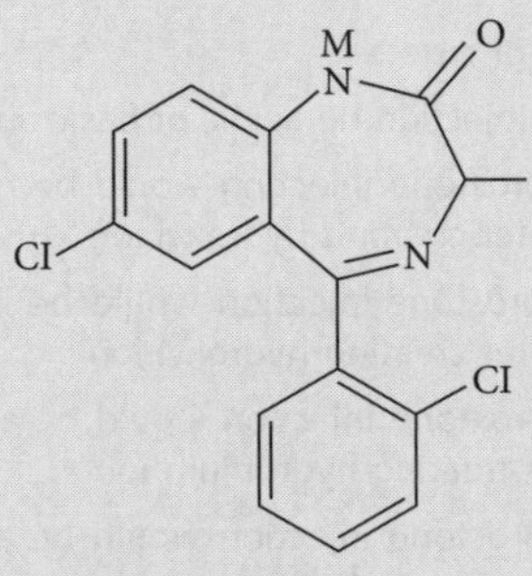

Lorazepam

The water solubility of lorazepam (FW 321) is 1 g/12 500 mL and the alcohol solubility is 1 g/86 mL. What concentration of alcohol would be required to dissolve lorazepam at a concentration of 1 mg/5 mL?

$\text{Log } S_{mixed} = \text{Log } S_{water} + \sigma fc$

	g/mL	Log (Y)	Fraction cosolvent (x)
Water solubility	1/12 500	−4.097	0
Alcohol solubility	1/86	−1.93	1
Desired strength	1/5000	−3.7	

Practice with derivatization:

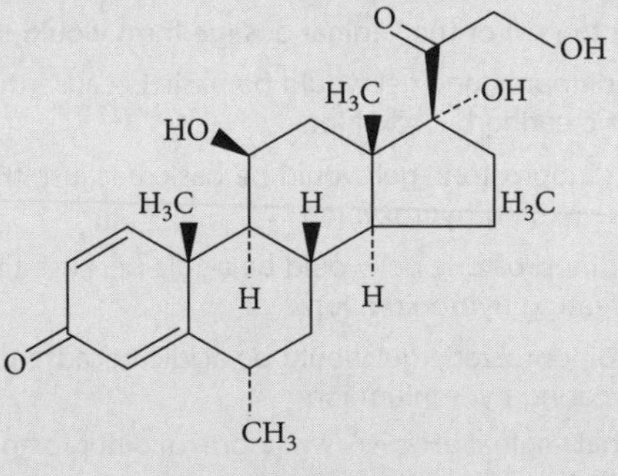

Methylprednisolone

24. What functional groups on this drug are derivatizable?

(a) The carboxylic acid or the two hydroxyls

(b) The three hydroxyls

(c) The carboxylic acid, the two hydroxyls or the ketone

(d) The three hydroxyls or the ketone

25. What derivative(s) could be attached to this functional group to enhance the water solubility of the drug?

(a) Glycine

(b) Phosphate

(c) Hemisuccinate

(d) All of the above

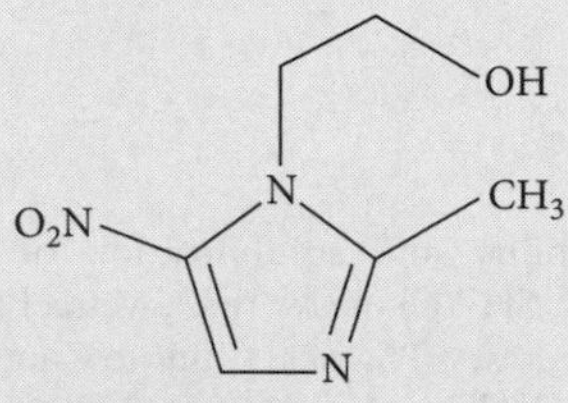

Metronidazole

26. What functional groups on this drug are derivatizable?

(a) The hydroxyl

(b) The amines

(c) The amines or the hydroxyl

(d) The amines or the nitro

27. What derivative(s) could be attached to this functional group to detract from the water solubility of the drug? (This is done to improve the taste.)

(a) Glycine

(b) Benzoate

(c) Phosphate

(d) Hemisuccinate

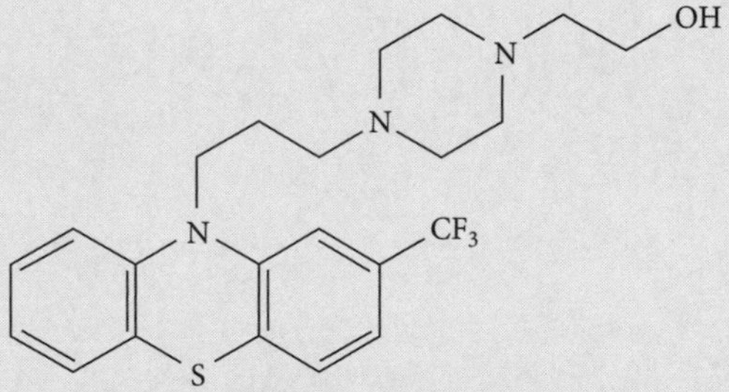

Fluphenazine

28. What functional groups on this drug are derivatizable?

(a) The hydroxyl

(b) The amines

(c) The amines or the hydroxyl

(d) The amines or the thioether

29. What derivative(s) could be attached to this functional group to detract from the water solubility of the drug? (This is done to prolong the absorption of the drug)

(a) Benzoic acid

(b) Decanoic acid

(c) Enanthic acid

(d) All of the above

Selected reading

1. SH Yalkowsky. Solubility and solubilization of nonelectrolytes, in Techniques of Solubilization of Drugs, SH Yalkowsky (ed.), Marcel Dekker, New York, 1981.
2. JE Thompson, LW Davidow. A Practical Guide to Contemporary Pharmacy Practice, 3rd edn. Lippincott, Williams & Wilkins, Philadephia, 2009.
3. TL Lemke. Review of Organic Functional Groups, 5th edn. Lippincott, Williams & Wilkins, Philadephia, 2012.
4. D Wishart. DrugBank (http://www.drugbank.ca/) University of Alberta, accessed daily.
5. ChemIDplus (http://chem.sis.nlm.nih.gov/chemidplus/) United States National Library of Medicine.
6. PubChem Compound (http://pubchem.ncbi.nlm.nih.gov/) from the United States National Library of Medicine.
7. L Felton (ed.) Remington: Essentials of Pharmaceutics. Pharmaceutical Press, London, 2012.

5

Chemical stability of drugs

Are you ready for this unit?
- Review organic functional groups and their reactivity to oxygen, light and water.
- Review the meaning of chirality, racemization and the order of a chemical reaction.

Learning objectives

Upon completion of this chapter, you should be able to:

- Identify the functional groups on a drug molecule that are susceptible to hydrolysis, oxidation, photolysis and racemization.
- Find reliable information on the pH of maximum stability for drugs available as injections.
- Find reliable information on the compatibilities of drugs with excipients including preservatives, antioxidants and buffers.
- Identify whether a drug degrades by first order or zero order kinetics from data given.
- Calculate the amount of drug remaining in a dosage form that degrades by first order or zero order kinetics.
- Given various storage conditions predict the effect on a drug's shelf life and recalculate shelf life when conditions are changed.
- Given a drug structure, design formulation, storage and packaging approaches to enhance the stability of hydrolyzable, oxidizable, photolyzable and racemizing drugs.

Introduction

There are three kinds of stability that we concern ourselves with as formulators and pharmacists: physical stability, microbial stability, and chemical stability. This chapter is focused on the chemical stability of drugs. Physical and microbial stability will be considered as we discuss each class of dosage form.

Definition

Chemical stability is concerned with maintaining the integrity of the chemical structure of the active ingredient.

Common pathways of drug degradation

For a compounding pharmacist developing a formulation or for pharmacists in other areas with the need to protect the stability of pharmaceuticals under their care, it is essential to be able to recognize the potential pathways of degradation.

Small drug molecules

Hydrolysis

Definition

Hydrolysis is the 'splitting' of drugs with water.

Functional groups that are susceptible to hydrolysis are mostly carboxylic acid derivatives that are 'split' into the carboxylic acid and the other reacting functional group. The susceptible groups include those shown in Figure 5.1.

Mechanism of hydrolysis

The relatively electron deficient carbonyl carbon is attacked by the oxygen of the water molecule to yield the carboxylic acid and, in the case of an ester, the alcohol (Figure 5.2).

The hydrolysis reaction may be catalyzed by hydroxyl ion (specific base catalysis) and hydronium ion (specific acid catalysis). Typically the rate of hydroxyl ion-catalyzed hydrolysis is greater than the rate of hydronium ion-catalyzed hydrolysis.

Key Points

- Hydrolysis cannot be detected by simple visual inspection in the pharmacy.
- Detection of hydrolysis requires that quantitative analytical instruments, such as ultraviolet spectroscopy, are preceded by a separation step to differentiate starting material from product.

Pharmaceutical example

Which of the drug molecules shown in Figure 5.3 are susceptible to hydrolysis? If susceptible, what functional group is involved?

Answers:

Susceptible: procaine (ester), cefuroxime (lactam, oxime), pilocarpine (lactone)
Not susceptible: ciprofloxacin

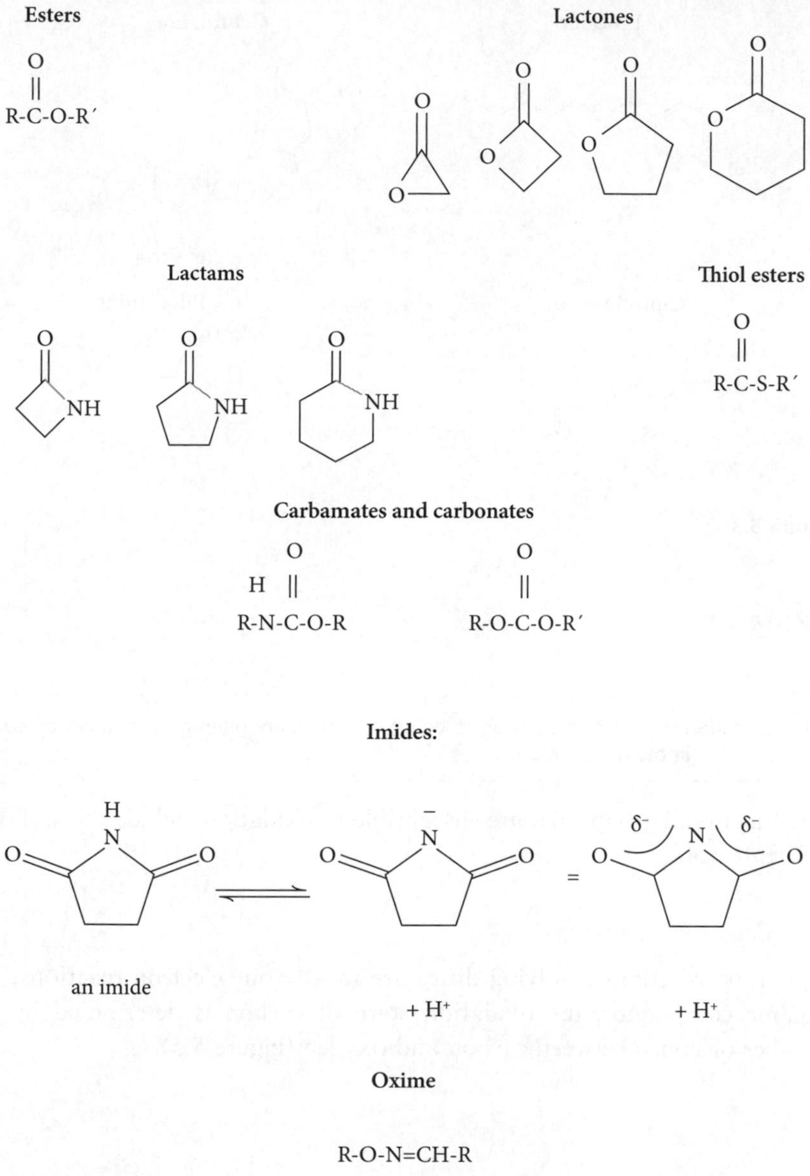

Esters

$$\begin{array}{c} O \\ \parallel \\ R\text{-}C\text{-}O\text{-}R' \end{array}$$

Lactones

Lactams

Thiol esters

$$\begin{array}{c} O \\ \parallel \\ R\text{-}C\text{-}S\text{-}R' \end{array}$$

Carbamates and carbonates

$$\begin{array}{c} H \quad \overset{O}{\parallel} \\ R\text{-}N\text{-}C\text{-}O\text{-}R \end{array} \qquad \begin{array}{c} O \\ \parallel \\ R\text{-}O\text{-}C\text{-}O\text{-}R' \end{array}$$

Imides:

an imide

+ H$^+$ + H$^+$

Oxime

R-O-N=CH-R

Figure 5.1 Functional groups susceptible to hydrolysis

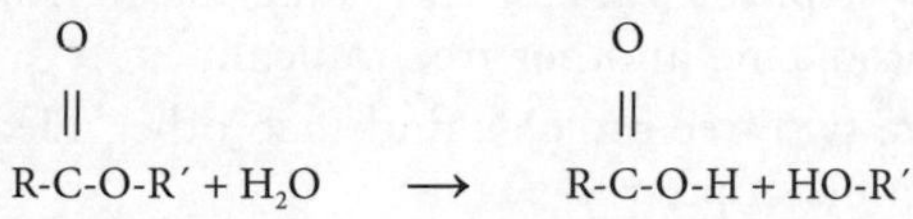

$$\begin{array}{c} O \\ \parallel \\ R\text{-}C\text{-}O\text{-}R' + H_2O \end{array} \quad \longrightarrow \quad \begin{array}{c} O \\ \parallel \\ R\text{-}C\text{-}O\text{-}H + HO\text{-}R' \end{array}$$

Figure 5.2 Mechanism of hydrolysis

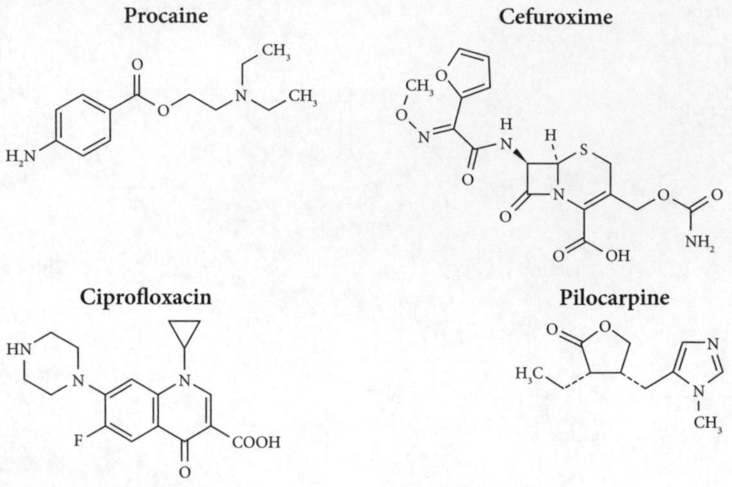

Procaine

Cefuroxime

Ciprofloxacin

Pilocarpine

Figure 5.3

Oxidation

Key Point

- Oxidation reactions, in addition to being common, are often accompanied by color changes in the drug solution.

The functional groups that are susceptible to oxidation include those shown in Figure 5.4.

Mechanism of oxidation

Oxidation reactions involving drugs are mostly one electron reactions. For organic compounds, the oxidation state of carbon is determined by the number of bonds between carbon and oxygen (Figure 5.5).

Definition

Oxidation is believed to be a chain reaction. The mechanism involves formation of a free radical, a form of the drug with an unpaired electron R˙

- Initiation step: formation of the free radical drug molecule.
- Propagation step: free radical takes an electron from another drug molecule, producing another free radical.
- Termination: two free radicals find each other. Electrons are now paired, reaction stops.

Alkenes (at the allylic position)

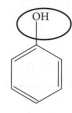

Aldehydes

Thiols

R-SH

Oxy substituents on aromatic rings

Enols such as ascorbic acid

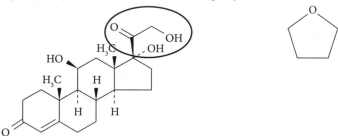

Alpha hydroxyketones such as the steroid drugs **Cyclic ethers such as tetrahydrofuran**

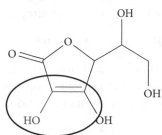

Figure 5.4 Functional groups susceptible to oxidation

$$CH_4 \rightarrow CH_3\text{-}OH \rightarrow H_2C{=}O \rightarrow HC\text{-}OH \rightarrow C{=}O$$

Reduced carbon   Oxidized carbon

Figure 5.5 Oxidation states of carbon

One of the potential initiators of oxidation is environmental oxygen. O_2 is a di-radical in its ground state and would like to fill its outer electron shell (Figure 5.6).

The O_2 extracts electrons from a drug molecule with a susceptible group:

$$R\text{--}H + O_2 \rightarrow R^{\cdot} + HO_2^{\cdot}$$

When oxygen is the initiator of the reaction, this is called auto-oxidation. There are two other initiators or catalysts that are also involved in drug oxidations:

$$\cdot \ddot{O} : \ddot{O}^\bullet + e^- \longrightarrow \ : \ddot{O} : \ddot{O}^{\bullet 1-} + e^- \longrightarrow \ : \ddot{O} : \ddot{O} :^{2-}$$

Figure 5.6   Oxygen as a di-radical

- Metal cations in trace quantities

$$Fe^{3+} + ArO^- \rightarrow Fe^{2+} + ArO^\cdot$$

- Shortwave visible or ultraviolet light

$$ArOH + h\upsilon \ \rightarrow \ Ar^\cdot + {}^\cdot OH$$

Unsaturation and aromaticity permit delocalization of the remaining unpaired electron and lower the energy of the radical therefore increasing the probability of its formation.

($h\upsilon$ = Planck's constant, $\hbar$, times light frequency, υ. Used here to depict the reaction of quantized light energy with a drug molecule.)

Key Point

- Initiator $\rightarrow$ R$^\cdot$
- R$^\cdot$ + O$_2$ $\rightarrow$ ROO$^\cdot$
- ROO$^\cdot$ + R–H $\rightarrow$ ROOH + R$^\cdot$
- ROO$^\cdot$ + R$^\cdot$ $\rightarrow$ ROOR

Key Points

- Many drug solutions will yellow or darken noticeably as a result of oxidation.
- Quantification of oxidation requires that quantitative analytical instruments, such as ultraviolet spectroscopy, are preceded by a separation step to differentiate starting material from product.

Pharmaceutical example

Which of the drug molecules in Figure 5.7 is susceptible to oxidation? If susceptible, what functional group is affected?

Answers:

Susceptible: epinephrine (phenols), captopril (thiol), and betamethasone (a-hydroxyketone)
Not susceptible: carisoprodol

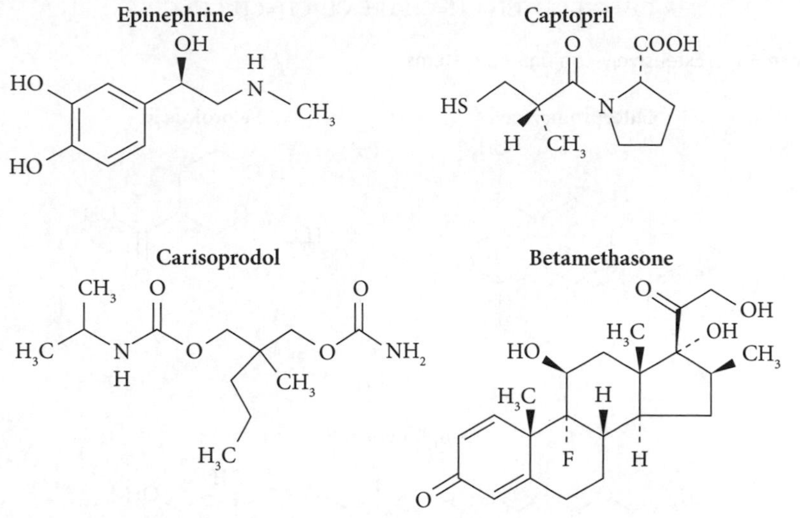

Epinephrine

Captopril

Carisoprodol

Betamethasone

Figure 5.7

Photolysis

> *Definition*
>
> Interaction of short wave visible (400–700 nm) and ultraviolet (200–400 nm) light with a drug molecule.

Mechanism of photolysis

- Provides sufficient energy to dissociate bonds in the molecule.
- Drug molecules that dissociate do so at fairly predictable locations e.g.

$$Ar^*R \rightarrow Ar^{\cdot} + R^{\cdot}$$

The functional groups that are affected by photolytic reactions are adjacent to extensive pi systems (Figure 5.8).

Types of photolysis reactions

- Oxidation: We have already seen that light can catalyze oxidations.
- Isomerization: Light can cause pi bonds to break and subsequent rotation around the single bond that will result in *trans-cis* conversions.
- Breaking of sigma bonds resulting in loss of side chains.
- Benzylic halogens are replaced by hydrogen from the solvent.

$$R\text{-}CH{=}CH\text{-}CH{=}CH\text{-}CH{=}CH\text{-}CH{=}CH\text{-}CH{=}CH\text{-}CH{=}CH\text{-}R$$

Figure 5.8 Extensively conjugated systems

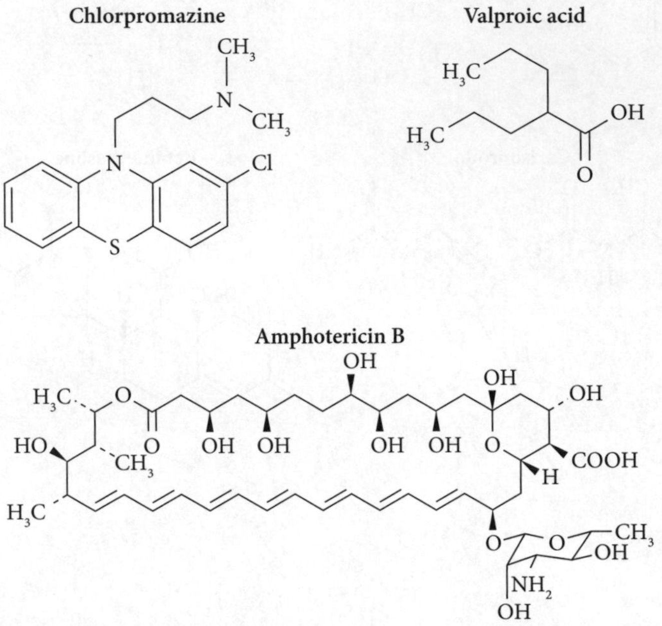

Figure 5.9

Key Points

- Many drug solutions will yellow or darken noticeably as a result of photo-oxidation.
- Detection and quantification of other types of photolytic reactions requires that analytical instruments, such as ultraviolet spectroscopy, are preceded by a separation step to differentiate starting material from product.

Pharmaceutical example

Which of the drug molecules in Figure 5.9 is susceptible to photolysis?
If susceptible, what functional group is affected?

Answers

Susceptible: chlorpromazine (benzylic amine or thioether), amphotericin B (extensively conjugated system)
Not susceptible: valproic acid (carboxylic acid does not oxidize further under environmental conditions)

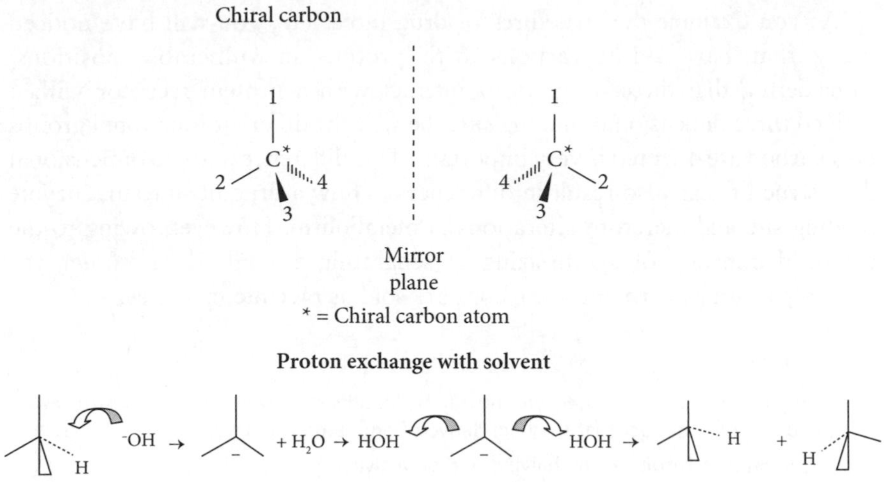

Chiral carbon

Mirror
plane
* = Chiral carbon atom

Proton exchange with solvent

Figure 5.10 Chiral carbon

Racemization

Pharmaceutical example

The racemization reaction can be considered a proton exchange with water meaning that the preparation of a suspension or a powder for reconstitution will reduce the rate of the reaction.

Drugs with chiral carbons can experience loss of optical activity (and loss of biological activity) in solution due to proton exchange with the solvent (Figure 5.10). For most drugs, the reaction only takes place at an appreciable rate in basic solution. However, epinephrine undergoes acid catalyzed racemization.

Susceptible functional groups are slightly acidic protons (Figure 5.11):

- drugs with protons on chiral carbons in the benzylic position *or*
- drugs with protons on chiral carbons alpha to a carbonyl group.

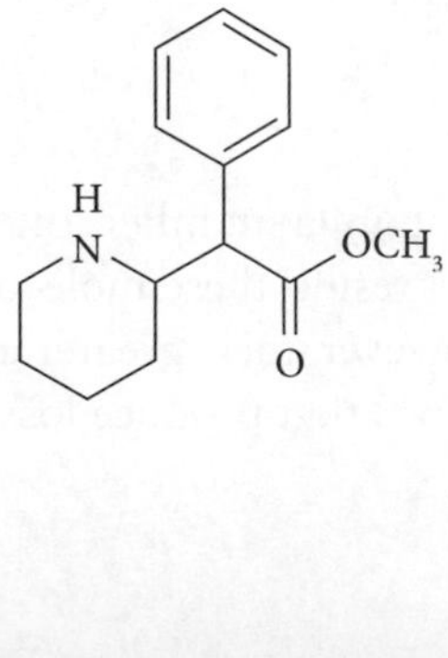

Figure 5.11 Methylphenidate

As you examine the structures of drug molecules, you will have noticed many that have chiral carbons with protons in vulnerable positions. Considering that these drugs *must* interact with a protein receptor with a defined three dimensional binding site, the way the different functional groups on a carbon are oriented is very important. This difference in three dimensional drug structure can also result in differences in how a drug fits into an enzyme binding site and therefore alterations in metabolism. However, owing to the technical difficulty of synthesizing or separating a single drug isomer, the majority of drugs with chiral carbons are sold as racemic mixtures.

Key Points

- Drugs sold as single isomers can usually be identified by their prefixes: *dextro, levo,* and *es,* or by the fact that they are derived from natural sources.
- Racemization results in the halving of drug activity.

Most vitamins contain chiral centers but because they are built by plant or animal enzymes, they will be single isomers. Thus despite the large number of drugs with chiral carbons, only a small number (those marketed as single isomers) are *deactivated* by the racemization pathway.

Key Points

- Detection of racemization requires that the isomers be separated before they are quantified.
- Separation of enantiomers can be accomplished with chiral columns or chiral derivatizing agents but in general it is difficult because the physical properties of enantiomers are identical.

Pharmaceutical example

Which of the drug molecules in Figure 5.12 is susceptible to racemization? If susceptible, what functional group is affected?

Answers:

Susceptible: hyoscyamine (proton on the benzylic carbon that is also alpha to a carbonyl), epinephrine (proton on the benzylic carbon)
Not susceptible: escitalopram has a chiral carbon but no proton

Protein drug degradation

Protein drug molecules

Protein drug molecules contain a number of the susceptible functional groups noted above, and as a result, these molecules can hydrolyze, oxidize, photolyze and racemize. However, their greater complexity means there are additional structural alterations that produce loss of pharmacologic activity.

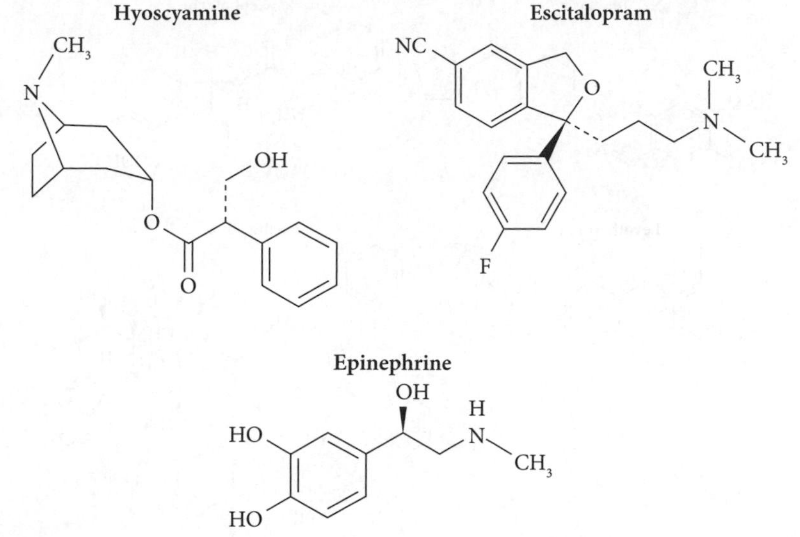

Figure 5.12

Definitions

- *Protein denaturation:* a change in the spatial arrangement of the polypeptide chain.
- *Aggregation:* a noncovalent assembly of protein molecules into multiples.
- *Fibrillation:* an aggregate consisting of protein strands.
- *Adsorption:* a monomolecular layer of protein molecules accumulated at a surface through unspecified attractive forces. This causes loss of protein molecules from the bulk solution and generally some unfolding occurs during the adsorption process.

Key Points

- Some proteins aggregate naturally while others aggregate only when they begin to unfold due to changes in ionic strength, pH, or shear stress caused by shaking, in which case, aggregation would be considered an instability.
- The surface most commonly involved in protein adsorption is glass.

DNA and RNA molecules

DNA and RNA molecules can also undergo hydrolysis and oxidation reactions that are dependent on pH, temperature and light.

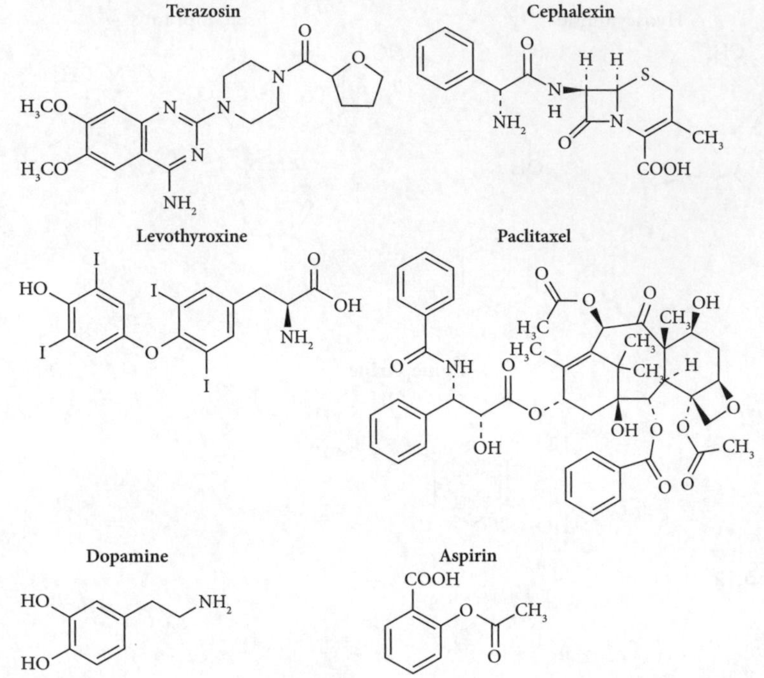

Figure 5.13

- RNA is less stable than DNA.
- As macromolecules the oligonucleotides are susceptible to conformation changes and aggregation.

Pharmaceutical example

The drugs in Figure 5.13 have stability problems in solution. What is the likely pathway of their degradation and what functional group is affected?

Answers:

Terazosin: photo-oxidation of the aromatic ethers
Cephalexin: hydrolysis of the lactam
Levothyroxine: photo-oxidation of the phenol or aromatic ether, racemization of the alpha proton
Paclitaxel: hydrolysis of the esters, possibly the amides
Dopamine: photo-oxidation of the phenol
Aspirin: hydrolysis of the ester, photolysis of the benzylic carbonyl

Rates of drug degradation reactions

The amide group and its cyclic version, the lactam, are both susceptible to hydrolysis, but the amide does not hydrolyze at an appreciable rate at room temperature, while drugs with lactam functional groups *must* be formulated without water.

It is important then, to determine the rate at which a particular drug degrades.

Rate of drug degradation

The rate of a chemical reaction is related to the number of collisions and therefore the concentration of reactants in solution. If we want to study how fast this occurs, we would measure the concentration of drug over time.

First order or pseudofirst order reactions

Consider the following data that was collected on a drug over a period of 6 days (Table 5.1).

We can see from Figure 5.14 that the concentration of drug declines over time. The shape of the relationship between percent drug remaining and time suggests exponential decline. If we transform the 'y' data into logs (Table 5.2).

Key Point

- The rate of the reaction is not constant as it decreases as the concentration of drug decreases.

Our equation is $\log [Drug]_t = \log [Drug]_0 - m \text{ (time)}$.

The slope m is negative; the variable y declines as x increases.

$$m = \frac{\Delta y}{\Delta x} = \frac{\log [drug]t_1 - \log [drug]t_2}{time_1 - time_2}$$

This equation tell us that the log of drug concentration changes linearly with time (Figure 5.15).

Table 5.1 Drug concentration over time	
Time (days)	[Drug] remaining (mg/mL)
0	100
1	75
2	53
3	40
4	30
5	21
6	17

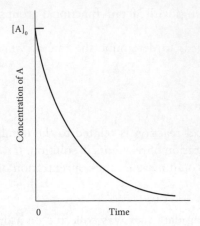

Figure 5.14 Decline of drug concentration over time. The rate (slope) decreases as the concentration of drug A decreases

Table 5.2 Log drug concentration over time		
Time (days)	[Drug] remaining (mg/mL)	Log [Drug]
0	100	2
1	75	1.875
2	53	1.724
3	40	1.602
4	30	1.477
5	21	1.322
6	17	1.230

What chemical species determines the rate of this reaction?

It is easier to see the answer to this question if we look at Figure 5.14. Mathematically the change in drug concentration over the change in time is represented by the slope of the line. If we drew tangents to the curve that represent the instantaneous rate we could see that the slope declines as the concentration of drug declines.

Key Point

- The order of a reaction is determined by the number of chemical species participating in the reaction.

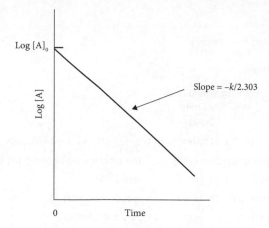

Figure 5.15 Log drug concentration over time

We can see that the concentration of only one species is included in the equation on page 119 and so this is a first order reaction.

Drug → Product

Key Point

- The first order equation is:

$$\frac{-d[Drug]}{dt} = k_1[Drug]$$

Do drugs decompose without reacting with anything?

Not very often. Nevertheless this equation describes many drug degradations. If we consider hydrolysis:

Drug + H$_2$O → Products

$$\frac{-d[Drug]}{dt} = k_2[Drug]\ [H_2O]$$

The concentration of water is very large compared with the drug concentration. When water molecules hydrolyze a few drug molecules the concentration of water remains constant. So we incorporate the constant water concentration into our rate constant, k:

$$k_1 = k_2[H_2O]$$

That simplifies the second order rate equation above to the *pseudofirst order* equation:

Key Points

The pseudofirst order equation is:

$$\frac{-d[\text{Drug}]}{dt} = k_1[\text{Drug}]$$

Using log 10, the working equation is

$$\log[\text{Drug}]_t = \log[\text{Drug}]_0 - \frac{k_1(\text{time})}{2.303}$$

where y is $\log[\text{Drug}]_t$, b is $\log[\text{Drug}]_0$, m is $\dfrac{k_1}{2.303}$ and k_1 is $-2.303(\text{slope})$ and x is time.

What is happening in solution is this:	The kinetics we observe will be:
Drug $\rightarrow$ Products	first order
Drug + H_2O $\rightarrow$ Products	pseudofirst order
Drug + O_2 $\rightarrow$ Products	pseudofirst order
Drug + light $\rightarrow$ Products	pseudofirst order

The rate of degradation of drug in solution declines as drug concentration declines.

Zero order kinetics

There is one other scheme that describes a common drug degradation situation (Figure 5.16).

As a molecule of drug becomes product, the reservoir of drug contributes to the concentration of drug in solution, and so the concentration of drug in solution does not change. This is the kinetic situation that we have in *suspensions*. We call this *zero order* because the rate is constant.

If we do our experiment with a drug in suspension we will get a concentration versus time curve that looks like the one in Figure 5.17.

Note that the untransformed data is linear. If we write the equation for this line:

y is drug concentration

b is drug concentration at time $= 0$

m is slope is negative $= \dfrac{\Delta y}{\Delta x} = \dfrac{[\text{Drug}]t_1 - [\text{Drug}]t_2}{\text{time}_1 - \text{time}_2}$

x is time.

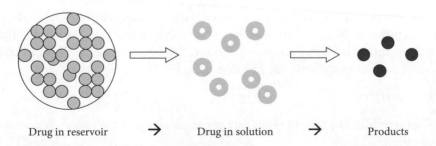

Drug in reservoir $\rightarrow$ Drug in solution $\rightarrow$ Products

Figure 5.16 Drug reservoir: zero order kinetics

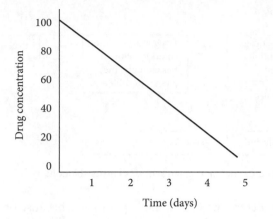

Figure 5.17 Zero order decline: remaining drug concentration versus time

The equation is:

$$[Drug]_t = [Drug]_0 - k_0(t)$$

The rate equation (differential form) is

$$\frac{-d[Drug]}{dt} = k_0 = slope \text{ of the drug concentration versus time plot}$$

For a zero order degradation, we can calculate the slope and find the rate of the reaction and it will be constant.

Key Points

- Because the concentration of drug in solution is constant, the rate of degradation in suspensions is constant.
- The working equation that describes the relationship is:

  $$[Drug]_t = [Drug]_0 - k_0(t)$$

Shelf life

> *Definition*
>
> Shelf life is the time period at which there will be 90% of the original concentration in the formulation. After this time period, the formulation is considered expired.

Would you expect a drug to last longer in a solution dosage form or a suspension dosage form (Figure 5.18)?

Perhaps we can answer this intuitively but we can also answer this question using shelf life calculations. Drug manufacturers do rate

In suspensions

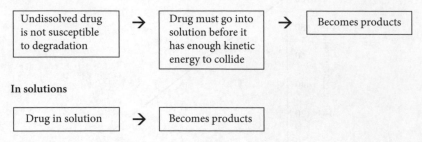

In solutions

Drug in solution → Becomes products

Figure 5.18 Drug degradation in suspensions versus solutions

experiments to determine k, a valuable parameter because of its relationship to shelf life. The relationship of k_1 to shelf life can be derived from the above equations substituting 0.9 and 1 for $[\text{Drug}]_t$ and $[\text{Drug}]_0$.

Key Points

- For a drug in solution (first order or pseudofirst order):

$$t_{90} = \frac{0.105}{k_1}$$

- For a drug in suspension (zero order kinetics):

$$t_{90} = \frac{0.1[\text{Drug}]_0}{k_0}$$

Pharmaceutical example

Aspirin (Figure 5.19) is a nonsteroidal anti-inflammatory drug that hydrolyzes in solution. The k_1 of this reaction at pH 2.5 is 5×10^{-7}/s (0.0018/hour). What is the shelf life?

$$t_{90} = \frac{0.105}{k_1} = \frac{0.105}{0.0018/\text{hour}} = 58.3 \text{ hours} = 2.4 \text{ days}$$

If we suspend aspirin rather than formulating it in solution, what is the shelf life if $[\text{Drug}]_0 = 0.65$ g/5 mL and the solubility of aspirin is 1 g/300 mL?

$$t_{90} = \frac{0.1\,[\text{Drug}]_0}{k_0}$$

We can calculate k_0 from k_1 and the drug solubility:

$$k_0 = k_1[\text{Drug}]_{\text{solution}} = (5 \times 10^{-7}/\text{s})(1 \text{ g}/300 \text{ mL}) = 1.65 \times 10^{-9} \text{ g/ml per second}$$

$$t_{90} = \frac{0.1(0.65 \text{ g/5 mL})}{5.5 \times 10^{-10}\text{g/mL per second}} = 7\,878\,787.8 \text{ s} = 91 \text{ days}$$

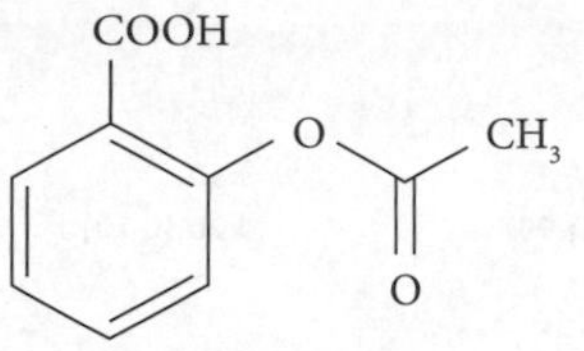

Figure 5.19 Aspirin

Factors that affect shelf life

pH

There are several reference books that provide information about the effect of pH and pharmaceutical necessities on the stability of drugs (see selected reading 1–10). Another useful guide is to find the pH of an injectable or oral solution of the drug since the manufacturer will choose conditions that provide a long shelf life.

Temperature

Increase in temperature will increase the number of collisions between molecules and therefore the rate of reaction.

The Arrhenius plot: The relationship of reaction rate to temperature was described by Arrhenius as follows:

$$\log k = \log A - \frac{E_a}{2.303RT}$$

where k is the rate constant of any order, A is a constant for the reaction (actually the hypothetical rate at infinite temperature), E_a is activation energy in calories per mole; R is the gas constant (1.987 cal/K per mole) and T is temperature in kelvins ($0°C = 273$ K).

The pharmaceutical industry uses the Arrhenius plot, a graph of log k versus the reciprocal of temperature to determine the shelf life for drugs that are relatively stable at room temperature.

- The drug is tested at higher temperatures, a plot constructed and the room temperature rate constant extrapolated from the plot.
- Room temperature shelf life can be calculated from the extrapolated rate constant, k.
- E_a, the energy barrier that the reactants have to climb to become products, ranges between 12.2 and 24.5 kcal/mol for drug degradations.

Practical applications: If we have information about the stability of a drug at one temperature, given the linear relationship between log rate constant and reciprocal temperature, we can calculate a ratio of rate constants that can be used to estimate shelf life.

Activation energy (E_a) (calories/mol)	k-70 k20 (Q-90)	k30 k20 (Q10)	k120 k20 (Q100)
12.2	9.24×10^{-5}	2	210
19.4	3.85×10^{-7}	3	4805
24.5	7.92×10^{-9}	4	44 626

Table 5.3 Ratio of rate constants at different temperature spreads

Here is the Arrhenius equation rearranged to facilitate these calculations:

$$\log \frac{k_2}{k_1} = \frac{E_a}{(2.303R)} \frac{(T_2 - T_1)}{(T_2 \, T_1)}$$

If we were interested in determining how much faster a drug degradation would proceed at autoclave temperature (120°C) than room temperature (20°C), assuming an energy of activation of 12.2 kcal/mol, our equation would be:

$$\log \frac{k_2}{k_1} = \frac{12\,200}{(2.303)(1.987)} \frac{(393 - 293)}{(393 \times 293)} = (2666)(0.00086844) = 2.315$$

antilog (2.315) = 210 times faster

The results of Arrhenius calculations for pharmaceutically important temperature spreads for energies of activation between the limits of 12.2 and 24.5 kcal/mol are provided in Table 5.3.

Q_{10} *method*: Simonelli and Dresback presented a convenient method for approximating the change in rates using the ratio of the rate constants at a 10°C spread which is referred to as the Q_{10} method.

- Q_{10} is the ratio of two rate constants for a drug measured at two temperatures that are 10°C apart.
- This relationship can be used to estimate the change in the rate and therefore a new shelf life at any new temperature using the equations below.

The ratio of the rate constants for any temperature spread, ΔT, is:

$$\frac{k_2}{k_1} = Q_{\Delta T} = Q_{10}^{\Delta T/10}$$

And the new shelf life is:

$$T_{90T2} = \frac{T_{90T1}}{Q_{\Delta T}}$$

Table 5.4 Pharmaceutically important temperatures		
Description	Temperature (°C)	Temperature (K)
Ultracold freezer	−70	203
Freezer	−20	253
Freezing point of water	0	273
Refrigerator	5	278
Room temperature	20–25	293–298
Boiling point of water	100	373
Autoclave temperature	120	393

Pharmaceutical example

Let's say a newly reconstituted product that is susceptible to hydrolysis is labeled to be stable for 24 hours in a refrigerator (see Table 5.4 for important temperatures). What is the estimated shelf life at room temperature?

$T_1 = 5°C$ $T_2 = 25°C$ $\Delta T = +20°C$ (sign is important!)

First we calculate the ratio of the rate constants for an increase in 20°C. If we do not have E_a, one approach would be to do two estimates using both the intermediate (19.4) and the more conservative estimate (24.5) kcal/mol of E_a.
For E_a of 19:

$$\frac{k_2}{k_1} = Q_{\Delta T} = Q_{10}^{\Delta T/10} = 3^{20/10} = 3^2 = 9$$

And the new shelf life would be:

$$T_{90T2} = \frac{T_{90T1}}{Q_{\Delta T}} = \frac{24 \text{ hours}}{9} = 2.7 \text{ hours}$$

For E_a of 24.5:

$$\frac{k_2}{k_1} = Q_{\Delta T} = Q_{10}^{\Delta T/10} = 4^{20/10} = 4^2 = 16$$

And the new shelf life would be:

$$T_{90T2} = \frac{T_{90T1}}{Q_{\Delta T}} = \frac{24 \text{ hours}}{16} = 1.5 \text{ hours}$$

The estimated shelf life of the drug if it is left on the counter instead of stored in the refrigerator is between 1.5 and 2.7 hours.

Pharmaceutical example

You are a community pharmacist and have just reconstituted a Ceclor (cefaclor) suspension for a patient with otitis media (Figure 5.20). The product has a shelf life of 14 days at refrigerator temperature (5°C). The patient's mother wants to know if it would hurt the drug to keep it in the car for an hour while she does some shopping. You estimate the temperature inside of the car on this particular day to be 35°C (about 90°F). Estimate how much time will be lost from the shelf life if the suspension is stored inside the car for 1 hour.

If we use E_a of 19, what is the ratio of rate constants for this change in temperature?

$$T_1 = 5°C \quad T_2 = 35°C \qquad \Delta T = +30°C$$

$$\frac{k_2}{k_1} = Q_{\Delta T} = Q_{10}^{\Delta T/10} = 3^{30/10} = 3^3 = 27$$

How much time is lost from the shelf life?

Now we need to do something different. The product will not sit for all of its shelf life at 35°C. It will be at that temperature for 1 hour. We know that the rate at 35°C is 27 times the rate at 5°C. This means that sitting for 1 hour at 35°C is equivalent to sitting for (1 hour × 27) 27 hours at 5°C.

So 27 hours have been lost from the original shelf life and if we were to write a new expiration date on the container for the patient it would be:

$$(14 \text{ days}) \frac{24 \text{ hours}}{1 \text{ day}} = 336 \text{ hours} - 27 \text{ hours} = 309 \text{ hours}/24 \text{ hours} = 12.9 \text{ days}$$

Pharmaceutical example

Asparaginase is an antineoplastic drug that has a shelf life of 8 hours at refrigerator temperature (5°C) after reconstitution. The patient will use only half of the vial and will require another dose the following day. Estimate the shelf life of the reconstituted drug when stored in the freezer (−20°C). Because asparaginase is a large protein with many functional groups, and data on its routes of degradation are not readily available, we will assume an activation energy of 19 kJ/mol.

$$T_1 = 5°C \quad T_2 = -20°C \qquad \Delta T = -25°C \text{ (sign is important!)}$$

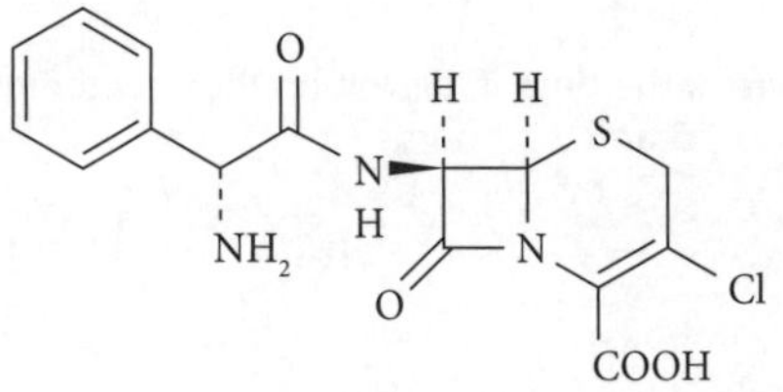

Figure 5.20 Cefaclor structure.

For E_a of 19.:

$$\frac{k_2}{k_1} = Q_{\Delta T} = Q_{10}^{\Delta T/10} = 3^{-25/10} = 3^{-2.5} = 0.06415$$

And the new shelf life would be:

$$T_{90T2} = \frac{T_{90T1}}{Q_\Delta T} = \frac{8 \text{ hours}}{0.06415} = 125 \text{ hours}$$

Key Points

- if the pharmacist would like to store a product at a temperature other than that recommended in the labeling a new shelf life can be calculated from the equation:

$$T_{90T2} = \frac{T_{90T1}}{Q_{\Delta T}}$$

- if a product has been stored at a different temperature for a portion of its shelf life, a shelf life adjustment can be calculated from

$$T_{90T2} = T_{90T1} - (\text{storage time} \times Q_{\Delta T})$$

Managing chemical stability problems

Pharmacists must be able to recognize drugs with stability problems and have strategies to prevent those problems. The approaches that may be used to prevent chemical stability problems fall into three main categories:

- Formulation changes
- Storage conditions
- Packaging

Formulation changes

Hydrolyzable drugs

- Reduce or eliminate water from the formulation.
- Control of pH with buffers, HCl or NaOH: There is generally one pH at which each drug is most stable. Buffers must be selected with attention to their pK_a, appropriateness for the route and compatibility with the other ingredients in the formulation.

Oxidizable drugs

- Replace the air in the container's head space with nitrogen and boil water prior to its use to remove dissolved O_2.

Table 5.5 Characteristics of antioxidant compounds

Antioxidant	Concentration (%)	Formulation constraints
EDTA	0.002–0.05	Water soluble
Citric acid	0.3–2	Weak chelator compared with EDTA
Preferentially oxidized		
Sodium bisulfite	0.05–1.0	More stable at pH 1–5, may cause airway reactions in asthmatics
Sodium metabisulfite	0.025–0.1	More stable at pH 1–5, may cause airway reactions in asthmatics
Sodium sulfite	0.01–0.2	More stable at pH 7–10, may cause airway reactions in asthmatics
Sodium thiosulfate	0.1–0.5	More stable at pH 7–10
Ascorbic acid	0.02–0.5	Most stable between pH 3 and 6
Chain terminators		
Thioglycerol	0.1–0.5	Unstable at alkaline pH, strong odor
BHA/BHT	0.005–0.02	Lipid soluble
Propyl gallate	0.05–0.1	Lipid soluble
Tocopherol	0.05–0.075	Lipid soluble

EDTA = ethylenediaminetetraacetic acid; BHA = butylated hydroxyanisole; BHT = butylated hydroxytoluene.
Adapted from MJ Akers. J Parent Sci Tech 1982: 36: 222–228.

- Control pH: Most drugs that oxidize are more stable at acidic pH, in protonated form.
- The addition of chelating agents stabilizes oxidizable drugs by trapping metals responsible for initiation step: Fe^{+++}, Cu^{++}, Co^{+++}, Ni^{++}, Mn^{+++}.
- The addition of antioxidants (Table 5.5): A drug reference should be consulted to discover incompatibilities when choosing antioxidant ingredients:
  - preferentially oxidized compounds (O_2 scavengers)
  - chain terminators donate an H radical and form a radical that is stable and incapable of continuing propagation.

Drugs that racemize

- Reduce or eliminate water from the formulation.
- The rate of racemization may be reduced through control of pH with buffers, HCl or NaOH. The racemization of epinephrine is acid catalyzed but sufficiently slow at pH 3.5 (a compromise between the oxidation and

Table 5.6 USP definitions of storage conditions

Term	Specifications (°C)	Specifications (°F)
Cold	≤8°C	≤46°F
Refrigerator	2–8°C	36–46°F
Freezer	−10°C to −20°C	14°F to −4°F
Cool	8–15°C	46–59°F
Room temperature	20–25°C	68–77°F

Reproduced with permission from selected reading 9.

racemization) to allow an adequate shelf life. Most other racemizations are catalyzed by hydroxyl ion.

Storage conditions (Table 5.6)

- Dry environment for hydrolyzable drugs: Do not store in dry form in the bathroom or in the refrigerator. Ideal humidity 40–60%.
- Low temperature: Stability of drugs in liquid form can be increased by decreasing temperature in most cases.

Key Point

- Note that elixirs may precipitate at refrigerator temperatures.

- Oxidizable drugs generally are more stable at lower temperatures because the activity of metal ions and hydroxyl ion is reduced at lower temperatures.

Key Point

- Remember that the solubility of O_2 increases at colder temperatures.

- It takes about 3 hours for 500 mL to reach room temperature after refrigeration. Infusion fluids in lines take about 20 minutes to reach room temperature.

Key Point

- Be aware that freezing protein drugs can cause them to denature (unfold and not refold correctly).

- Thawing of frozen minibags (50 mL) can take 90 minutes, while thawing 3-liter bags can take 8–10 hours.

Packaging

Packaging should be selected after consideration of:

- permeability
- leaching
- adsorption.

Key Point

Packaging can be chosen to provide protection from oxygen, water vapor and light.

Definition

Permeability is defined as the ability of volatile substances such as water vapor, oxygen or drug vapors to move through the packaging material. Only a few drugs are volatile enough to penetrate packaging. For example, nitroglycerin should be packaged in glass so it is not lost through the packaging over time.

Key Point

- Amber containers effectively retard the photolysis of drugs. Enclosing colorless glass in cardboard may be used to occlude light if it is important to be able to inspect the contents of the package (parenterals).

Definition

Leaching is the loss of materials from the packaging into the drug solution.

What other considerations could influence the choice of packaging?

- Type I glass is primarily composed of silicon dioxide and boric oxide, and this composition leaches relatively low levels of metal oxides (alkali) into waters and drug solutions.
- Other glass types have higher levels of leachable metal oxides and thus must be used with greater caution particularly for drug solutions with pH >7.
- Plastics have a number of additives that are potential leachables into drug solutions including antioxidants, lubricants and plasticizers. The plasticizer, diethylhexylphthalate (DEHP), added to polyvinyl chloride to produce flexibility is leached in to drug solutions containing lipids, surfactants or cosolvents and is a concern because it is classified as a human carcinogen.

Table 5.7 Materials used for drug container packaging

Material	Leaching	Leaching extent	Adsorption extent	Extent permeable water	Extent permeable O$_2$	Sterilization
Glass, borosilicate	Alkali (metal oxides)	1	2	0	0	Dry heat, autoclave
Glass, soda lime	Metal oxides	5	2	0	0	Dry heat
High density polyethylene	antioxidants	1	2	3	3	Irradiation, gas
Low density polyethylene	Plasticizers, antioxidants	2	2	5	5	Irradiation, EtO gas
Polyvinyl chloride	Plasticizers, stabilizers	4	2	5	2	Irradiation, EtO gas
Polypropylene	Antioxidants, lubricants	2	1	5	3	EtO gas, autoclave
Rubber, natural	Metal salts, lubricants	5	3	3	3	Autoclave, irradiation
Rubber, butyl	Metal salts, lubricants	3	2	1	1	Autoclave, irradiation
Rubber, silicon	Minimal	2	1	5	5	Autoclave, irradiation

Extent 0–5: 0= none; 5= extensive. EtO = ethylene oxide.

Adapted from DC Liebe. Packaging of pharmaceutical dosage forms. In Modern Pharmaceutics, 3rd edn, New York: Marcell Dekker, 1995 and E Vadas. Stability of pharmaceutical products. In AR Gennaro (ed.). Remington: The Science and Practice of Pharmacy, 20th edn, Philadephia, PA: Lippincott, Williams & Wilkins, 2000.

- Rubber materials in stoppers and syringe plungers can leach metals, 2-mercaptobenzothiazole and nitrosamines.

> ### Definition
>
> *Adsorption* is when active or inactive components of the drug solution may adhere to the surface of packaging materials reducing their concentration in solution.

Please see Table 5.7 for qualities of materials used for packaging.

Key Points

- Most drugs are polyfunctional, organic molecules that are formulated into products containing multiple pharmaceutical excipients.
- Our discussion has not included all of the potential pathways of drug degradation in dosage forms, the potential of excipients and drugs to react, nor the degradation of drug molecules by second or higher order reactions. Nor are most pharmacies equipped with the separation or analytical capability to determine whether a drug has degraded in a dosage form. However, much can be learned about a drug's stability in solution from reviewing its functional group chemistry. The pharmacist can use this information to:
  - protect hydrolysable drugs from water
  - protect photolyzable drugs from light
  - stabilize easily oxidized drugs with appropriate buffers or antioxidants
  - protect fragile drugs from environmental conditions.
- If a patient's response to a medication is not what would be anticipated, the pharmacist may consider drug instability among the different possible causes of the unexpected response.

Questions and cases

1. The following drugs have stability problems in solution. State the likely pathway of their degradation and circle the functional group involved.

Cloxacillin

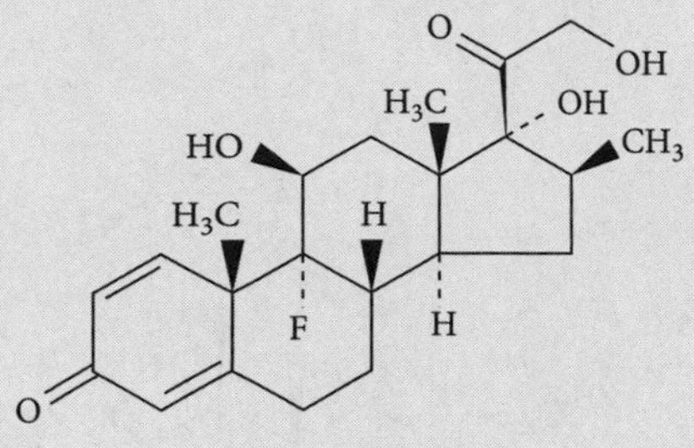

Betamethasone

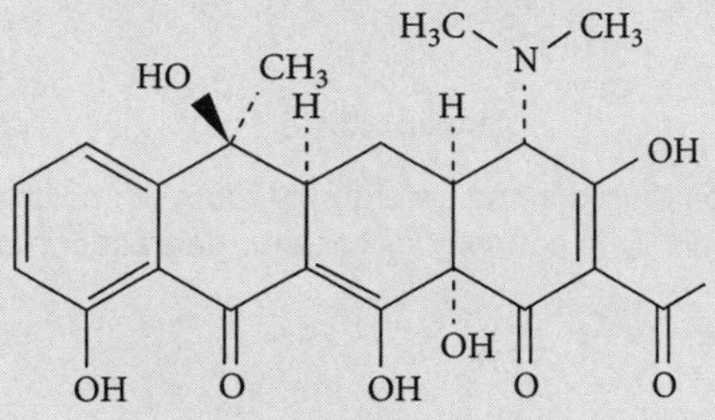

Tetracycline

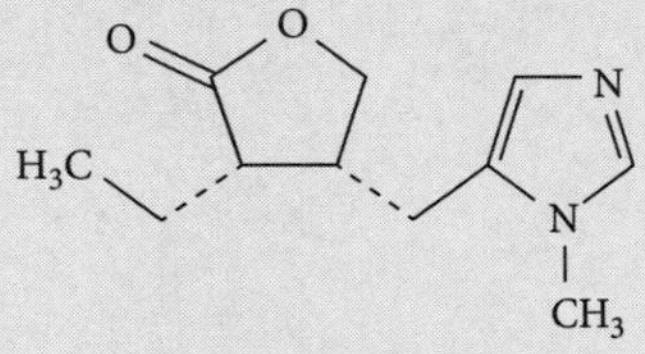

Pilocarpine

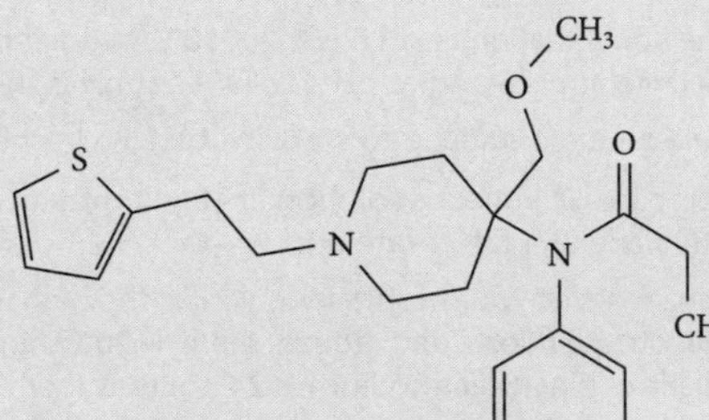

Sufentanil

Simvastatin

2. Cocaine HCl solution is used in the emergency room to anesthetize wounds before suturing. What is the likely pathway of cocaine degradation and what functional group is involved?

Cocaine

3. The k_1 for cocaine's degradation at pH 8 is 1.2×10^{-4}/s and at pH 4.5 is 3.4×10^{-10}. Calculate the shelf life at pH 8 and at pH 4.5. At what pH is the drug most stable?

4. What buffer would be most suitable to stabilize cocaine in solution and why?

5. If cocaine were prepared in aqueous solution at 40 mg/mL and at 20 mg/mL, which solution would degrade at a faster rate and why?

6. The director of pharmacy at your hospital would like the night shift pharmacists to reconstitute antibiotic solutions and freeze them ($-20°C$) for use during busier times. If the shelf life of piperacillin sodium is 24 hours at room temperature ($20°C$), what is the shelf life of the drug when it is frozen? Use an E_a of 12 kcal/mol.

7. You are a pharmacist with the Indian Health Service in Alaska. You make a rifampin suspension for a child who travels about 6 hours by car to the clinic. The suspension is stable for 45 days at refrigerator temperature ($5°C$). If the temperature in the car is $25°C$, estimate how much time will be lost from the shelf life on the trip home from the clinic? Use an E_a of 19 kcal/mol.

8. A compounded, liquid preparation of procainamide HCl is stable for 7 days at room temperature ($25°C$). What would the shelf life be if the patient stored it in the refrigerator? ($5°C$) Use E_a of 19 kcal/mol.

9. You are in charge of the formulation of a parenteral solution of ketorolac, a nonsteroidal antiinflammatory drug. Preformulation data is as follows: water solubility 1 g/300 mL, alcohol solubility 1 g/2 mL, pK_a 3.54, pH of maximum stability = 4.0, k_1 at $25°C$ 1.65×10^{-9}/s. The product's concentration should be 15 mg/mL.

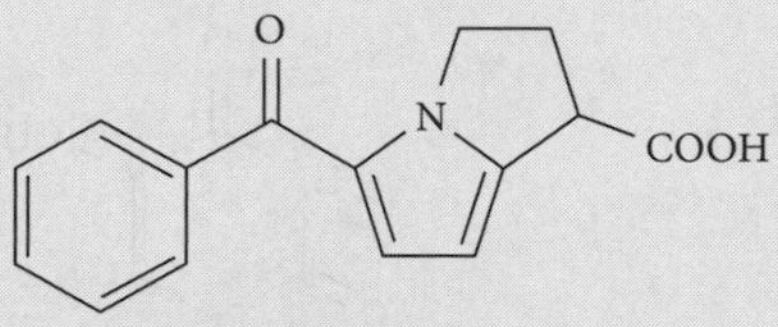

Ketorolac

Calculate the shelf life of ketorolac.

10. Your pharmacy makes isoniazid suspension for children with tuberculosis. The suspension has a shelf life of 21 days at refrigerator temperature (5°C). The patient's mother wants to know if it would hurt the drug to keep it in the car for an hour while she does some shopping. You estimate the temperature inside of the car on this particular day to be 35°C (about 90°F). Estimate how much time will be lost from the shelf life if the suspension is stored inside the car for 1 hour. Assume an E_a of 24.5.

11. What is the likely pathway of venlafaxine's degradation? Circle the functional group involved.

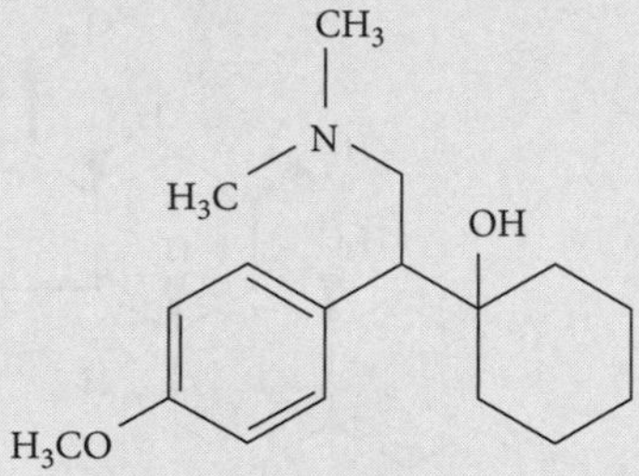

Venlafaxine

Briefly describe four formulations, storage and/or packaging strategies you would recommend to be used in preparing a liquid formulation of venlafaxine.

12. What is the likely pathway of cephalexin's degradation? Circle the functional group involved.

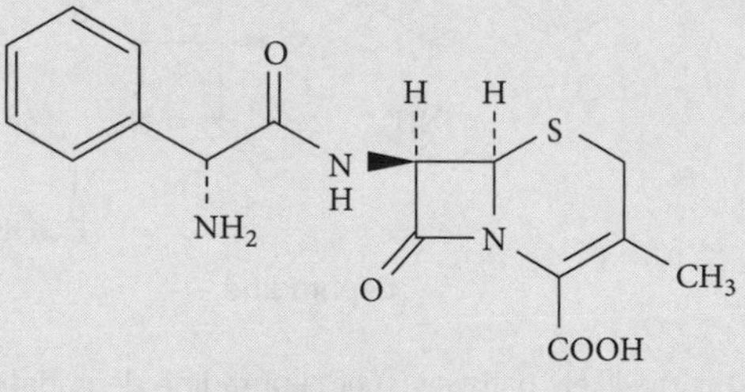

Cephalexin

Briefly describe four formulations, storage and/or packaging strategies you would recommend to be used in preparing a liquid formulation of cephalexin.

13. What is the likely pathway of stavudin's degradation? Circle the functional group involved.

Stavudin

Briefly describe four formulations, storage and/or packaging strategies you would recommend to be used in preparing a liquid formulation of stavudin.

14. You are compounding a betamethasone liquid for inhalation. Given the potential instability that you identify, briefly describe four formulations, storage and/or packaging strategies you would recommend to be used in preparing the liquid betamethasone formulation.

Betamethasone

15. Cephradine is a cephalosporin antibiotic.

Cephradine

(a) What is the likely pathway(s) of cephradine degradation and what functional group(s) is (are) involved?

(b) What formulation, storage and packaging strategies would you recommend to be used in preparing an easy-to-swallow dosage form of cephradine?

16. Dimercaprol is an antidote for heavy metal poisonings.

Dimercaprol

(a) What is the likely pathway(s) of dimercaprol degradation and what functional group(s) is (are) involved?

(b) What formulation, storage and packaging strategies would you recommend to be used in preparing an injection dosage form of dimercaprol?

17. Methylphenidate is a stimulant drug used to treat attention deficit disorder.

Methylphenidate

(a) What is the likely pathway(s) of methylphenidate degradation and what functional group(s) is (are) involved?

(b) Methylphenidate is most stable at pH 3.5. What buffer pair would you choose to stabilize a liquid formulation of methylphenidate? What storage and packaging strategies would you recommend to be used in preparing an easy-to-swallow dosage form of methylphenidate?

Selected reading

1. American Hospital Formulary Service: Drug Information, yearly.
2. KA Connors, GL Amidon, VJ Stella. Chemical Stability of Pharmaceuticals, 2nd edn. John Wiley & Sons, New York, 1986.
3. R Jew, W Soo-hoo, S Erush. Extemporaneous Formulations, 2nd edn. ASHP, 2010.
4. LA Trissel. Handbook on Injectable Drugs, 16th edn, ASHP, 2010.
5. RC Rowe, PJ Sheskey, ME Quinn. Handbook of Pharmaceutical Excipients, 6th edn. London: Pharmaceutical Press, 2009.
6. King JC, Catania PN. King Guide to Parenteral Admixtures, King Guide Publications, 2006.
7. Martindale: The Complete Drug Reference, Pharmaceutical Press, http://www.medicines complete.com/mc/martindale/current/ Remington: The Science and Practice of Pharmacy, 21st edn. Mack Publishing, 2005.
8. LA Trissel. Trissel's Stability of Compounded Formulations, 4th edn, APhA, 2008.
9. United States Pharmacopeia, 34th edn, and National Formulary, 29th edn. Baltimore, MD: United Book Press, 2011.
10. American Journal of Health-System Pharmacy.
11. International Journal of Pharmaceutical Compounding.

12. Journal of Pharmaceutical Science and Technology.
13. Journal of Pharmaceutical Sciences.
14. AP Simonelli, DS Dresback. Principles of formulation of parenteral dosage forms. In Perspectives in Clinical Pharmacy, 1st edn, DE Franke and HAK Whitney (eds). Hamilton, IL: Drug Intelligence, 1972.
15. MD Dibiase, MK Kottke. Stability of Polypeptides and Proteins, in JT Carstensen, CT Rhodes (eds), Drug Stability, Principles and Practices, 3rd edn. New York: Marcel Dekker, 2000.

6

Drug travel from dosage form to receptor

Learning objectives

Upon completion of this chapter, you should be able to:

- Define log distribution coefficient, transcellular, paracellular, transepithelial electrical resistance, transporter, endocytosis, transcytosis, pinocytosis, phagocytosis, permeability, permeability coefficient and enzyme induction.
- Describe whether each of the following will increase or decrease how quickly a drug arrives at the site of action, how much arrives and how long it remains at the site:
  - physicochemical factors: water solubility, log P, number of hydrogen bond donors, number of hydrogen bond acceptors, pH/pK_a, molecular weight
  - biological properties of the drug: affinity for transporters, protein or tissue binding, elimination pathways, accessibility and selectivity of the target
  - dosage form factors: design of the dosage form, drug release dissolution and diffusion rates
  - patient factors: use of the dosage form or drug delivery system, disease states, pharmacogenetics.
- Predict whether each of the following characteristics of the routes of administration will increase or decrease how quickly a drug arrives at the site of action, how much arrives and how long it remains at the site:
  - the permeability of the absorptive surface from layers and type of epithelial membrane, transporters present and type of capillary
  - given a tissue, whether it is well perfused, intermediate perfused, or poorly perfused
  - the consequences of substrate drug encountering metabolizing enzymes before it is distributed to its target.
- Outline the steps of drug travel from administration to elimination.

Introduction

In this chapter we consider the processes that move drugs from the dosage form to the portion of the body where the drug's receptor is located, and the factors that influence or limit them. Unless a drug dosage form is an intravenous solution, the drug must be released from the dosage form and dissolve in the body fluids near the site of absorption. The drug must penetrate one or more membranes as it moves into and out of the blood or lymph for distribution. The drug will distribute with the blood flow and will penetrate into other tissues as its chemistry and membrane permeability allow. If the dose of the drug was properly chosen, it will produce the desired effect at the site of action. If the dose was too low, the patient will experience therapeutic failure and if too high, adverse effects. Blood flow will carry the drug to the liver where enzymes may break it into a more polar species, and to the kidneys where the drug or its metabolite will be eliminated. When the amount of drug leaving the body has exceeded the amount of drug coming into the body sufficiently long to drop the concentration at the site of action below the therapeutic threshold, the patient will no longer experience the effects of the drug. Ultimately we seek to understand this aspect of pharmaceutical science in order to predict how fast and how much drug is presented to the site of action. This understanding will enable the pharmacist to answer our patients' questions about when a medication will begin to improve their condition or when a nurse should evaluate the effect of a dose.

Pharmaceutical scientists have developed efficient approaches to identify drug candidates that have the appropriate biopharmaceutical characteristics to warrant the investment required to bring a drug to market. These same approaches can be used by the practicing pharmacist to make predictions about the rate and extent of travel of drugs in our patients. First the four categories of factors that affect drug travel are reviewed and then their influence is examined as a drug is followed through its liberation, absorption, distribution, metabolism and elimination. The factors that determine how much drug and how quickly a drug is available at the site of action fall into four categories:

- physicochemical factors
- biological factors
- dosage form factors
- patient related factors.

Physicochemical factors

Drug absorption is a balancing act between aqueous solubility and membrane permeability. In order for a drug to be absorbed, it must have sufficient water solubility to be dissolved in body fluids and sufficient lipid solubility to cross biological membranes.

Biological membranes are essentially lipid structures and act as barriers to highly polar species. It is not surprising that the vast majority of drugs (95%) are weak organic acids or bases with a lipid soluble form that dissociates to form an ionized, water soluble form.

The physicochemical factors that affect drug travel include:

- lipophilicity, which can be estimated with log P
- the degree of ionization, which is determined by the pH of the body fluid at the membrane and the pK_a of the drug

- the size of the drug, which is estimated with molecular weight
- drug solubility in the body fluid at the site of administration.

Key Points

- Lipophilicity, degree of ionization and size of the molecule are all determinants of membrane permeability.
- However, dissolution in biological fluids is where all drugs must start before they can move across the membranes that separate them from their site of action.

Water solubility

Recall that solubility is the maximum amount of drug that remains in solution in a given volume of solvent.

Key Points

- Biological fluids that act as solvents for drugs in the body are aqueous, thus the number of functional groups that a drug offers for hydrogen or ionic bonding will determine its ability to dissolve in these fluids.
- Solubility is the driving force for diffusion across a membrane; that is, the highest concentration a drug can achieve in the biological fluid provides the high point of the concentration gradient. Because of this, dissolution rate can be the rate limiting step for absorption of low solubility drugs.
- For very low solubility drugs, all of the dose might not dissolve in the amount of time the dosage forms spends at the site of absorption, thus affecting absorption extent as well.

Pharmaceutical example

Poorly soluble itraconazole is incompletely absorbed from its solid dosage form (40–55%) due to incomplete dissolution during small intestine transit. The extent of its absorption is improved by formulation as an oral solution using cyclodextrin as a solubilizing agent. This formulation has an oral bioavailability of 72% on an empty stomach.

Log P

The log of the partition coefficient of the drug between octanol and water is used widely in the pharmaceutical sciences to predict permeability of a drug through a biological membrane.

Key Points

- More specifically as log of partition coefficient increases, the ability of a drug to diffuse passively through a biological membrane increases.
- The optimal log of partition coefficient range for a particular absorption barrier differs very little in the different routes of administration. (See Appendix Table A.1).

Log of partition coefficient, both measured and calculated, and permeability coefficients in Caco-2 cells have been tabulated for many drugs at DrugBank (http://www.drugbank.ca/).

Log *D*

For drugs that are weak acids or bases we can more properly understand their lipophilicity using the *log distribution coefficient*, which quantifies the pH dependent partitioning behavior of these drugs. For example, two lipid lowering drugs, atorvastatin (log P 4.23) and simvastatin (log P 4.42), appear to have similar lipophilicity based on their log P. But when log D is calculated we find that simvastatin (Log $D^{pH\ 7.0}$ 4.41) is much more permeable than atorvastatin, (log $D^{pH\ 7.0}$ 1.54) at a pH close to physiologic. Log $D^{7.4}$ for many available drugs can be found in selected reading 5.

$$\text{Log } D^{pH\ 7.4} = \log \frac{[\text{Unionized} + \text{Ionized}]_{oct}}{[\text{Unionized} + \text{Ionized}]_{water}}$$

pH/pK_a

Key Points

- The solubility and dissolution rates of weak acids and bases are dependent on the pH of the biological fluid they encounter on administration.
- The degree of ionization is also important to membrane permeability. The pH partition hypothesis postulates that ionizable compounds penetrate membranes in the nonionized form.

In general, acidic drugs showed higher permeability at pH 6.5 or lower, whereas bases showed higher permeability at pH 7.4. While pHs more basic than 7.4 are not usually encountered *in vivo*, some dosage forms may provide a higher pH to enhance diffusion across the membrane for a basic drug. It is important to recognize that there are absorption barriers where this pH partition principle will not apply. For instance in the stomach where pH is very low and the concentration of membrane soluble (uncharged) acidic drugs is very high, there is in fact very little drug absorption across the stomach wall due to its thick muscular walls and low surface area. In addition there is some evidence that drugs can be absorbed in charged form although at a slower rate than the uncharged form.

Molecular weight

Key Points

- The larger the drug molecule the more difficult it becomes for the molecule to move across epithelial barriers.
- The molecular weight preference is different for different routes of administration (see Appendix Table A.1), and can be expected to be revised upward as new penetration enhancers and technologies such as ultrasound, iontophoresis and microneedles are developed.

Biological factors

These factors determine how much and how quickly a drug is available at the site of action:

- membrane structure and transport
- characteristics of the route of administration
  - enzymatic activity
  - blood flow.

Membrane structure and transport mechanisms

Epithelial cell membranes consist of a lipid bilayer with associated proteins that provide transport, ligand binding and cell attachment (Figure 6.1). Unlike cells in many tissues (consider connective tissues, for example), epithelial cells attach to adjacent cells and their foundation of structural protein creating the barrier properties that we associate with epithelial membranes lining and covering body structures. The term 'cell membrane'

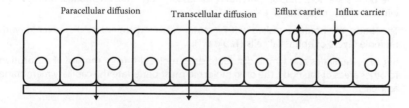

Apical side of the epithelial cells faces the organ

Basal side of the cells faces the collagenous foundation

Figure 6.1   Epithelial membrane

Table 6.1 Types of epithelial cells

Type	Description	Location	Permeability
Simple squamous	Single layer of flattened, interlocking cells	Lining blood vessels, alveoli	Excellent
Simple columnar	Single layer of rectangular cells	Stomach, small intestine, upper respiratory tract	Very good
Transitional	Multiple layers of different shapes, suited for distention and contraction	Ureters, urinary bladder	Poor
Stratified squamous, nonkeratinized	Multiple layers of flattened, nucleated cells, suited for abrasion	Buccal and sublingual, lower pharynx, esophagus, vagina, conjunctiva, cornea	Intermediate to very good
Stratified squamous, keratinized	Multiple layers of flattened, nonnucleated cells filled with keratin, suited for abrasion and waterproofing	Hard palate, gums, nasal vestibule, skin	Poor

Adapted from JE Hall. Guyton and Hall Textbook of Medical Physiology, 12th edn, Philadelphia: Saunders, 2011 and N Washington, C Washington, CG Wilson. Physiological Pharmaceutics, 2nd edn, London: Taylor & Francis, 2001.

will be applied to the lipid bilayer surrounding the epithelial cell and the phrase 'epithelial membrane' to the organized arrangement of epithelial cell layers and collagen that cover and line body structures.

Key Points

- A drug molecule that encounters an epithelial membrane moves first through the apical cell membrane, then through the aqueous intracellular compartment, out through the cell's basal membrane, perhaps traversing two or more cell layers before encountering the connective tissue where capillaries and a lymph vessel are found.
- The body's surfaces that are intended for absorption (intestine, lungs) are covered with one layer of epithelial cells.
- The body membranes tasked with greater barrier properties are covered with multiple layers of epithelial cells (Table 6.1).
- Because capillaries and lymph distribute drug molecules from the site of absorption to other areas of the body, the drug molecule must penetrate the capillary membrane as well.
  - The capillary wall consists of a single layer of highly permeable endothelial cells that, in most locations, create a lesser barrier to drug diffusion than epithelial membranes.
  - The permeability of capillary membranes varies from tissue to tissue (Table 6.2).

Table 6.2 Capillary permeability in different tissues

Capillary location	Capillary description	Pore size restricts flow of lipid insoluble macromolecules with diameters larger than
Retinal, brain–spinal cord, enteric nervous system, lymphoid tissue	Continuous with tight junctions	<1 nm
Skin, muscle, cortical bone, adipose tissue, lung, intestinal mesentery, developing ovarian follicle	Continuous with loose junctions	Approx. 5 nm
Connective tissue, kidney, intestinal mucosa	Fenestrated	6–15 nm
Hepatic	Sinusoidal, open fenestrae	Approx. 180 nm
Myeloid bone marrow	Sinusoidal, non-fenestrated except when blood cells are transported across endothelial membrane	Approx. 5 nm
Splenic terminal capillaries	Sinusoidal, with intermittent basement membrane	≤5 μm

Information from H Sarin. J Angiogenesis Res 2010; 2: 1–19.

Movement of drugs across membranes

Dissolved solutes (such as drugs) permeate epithelial membranes by:

- passive diffusion
  - transcellular diffusion across the cell membrane
  - paracellular diffusion between cells
- carrier mediated transport
- cytosis.

Passive diffusion

Key Points

- Most drug molecules move across epithelial membranes via transcellular permeation or diffusion across the lipid bilayer. The ability of drugs to diffuse via the transcellular route is favored for drugs with higher lipid solubility, lower molecular weight and no charge.
- The movement of water and hydrophilic, low molecular weight solutes occurs through the distension and constriction of the tight junctions between cells. These channels, referred to as the paracellular route, open and close in response to various physiological stimuli.
- Transcellular and paracellular transport occur down a concentration gradient.

The absorptive membranes that drugs encounter in different tissues will differ in their permeability particularly through the paracellular route. For example the paracellular channels of the stratum corneum are filled with lipids that dramatically limit drug diffusion, while most capillary membranes admit solutes as large as inulin (FW 5000) through their intercellular clefts. The *transepithelial electrical resistance* (TEER, Ω cm^2) is a measure of the charge flow across the membrane and thus provides you with an appreciation of the relative paracellular permeability of different epithelia: the lower the transepithelial electrical resistance, the higher the permeability (see Appendix Table A.1).

Drug transporters

Carrier mediated transport describes the movement of drug molecules bound to a saturable and somewhat structurally specific protein site.

Key Points

See Table 6.3 and Figure 6.2.

- Protein carriers called *transporters* have evolved at some absorbing membranes such as the small intestine and at organs of elimination such as the kidneys and liver.
- The apical cell surface exposed to the lumen of the organ has different transport proteins than the basal surface attached to the collagen foundation.
- Some carrier mediated transport occurs down a concentration gradient.
- Some carrier mediated transport requires energy to move drug across the membrane and thus can occur against the concentration gradient.
- Drug molecules that resemble nutrients (amino acid drugs), may be assisted across absorptive membranes or reintroduced into the body from excretory fluids such as urine by the influx transporters.
- Drugs which resemble toxins (like anticancer drugs) will be escorted out of the cells of the small intestine back into the lumen or pushed into excretory fluids by the efflux transporters.
- It is important to recognize that carrier mediated transport is saturable and this property invites the possibility of drug–drug interactions at specific transporters with broad substrate specificity. Depending on whether the transporter is efflux or influx, inhibition may increase or decrease levels of the substrate drug at the receptor.

Endocytosis and transcytosis

Definitions

- *Endocytosis* is a membrane transport mechanism in which a macromolecule or particle is engulfed by a segment of the cell plasma membrane which pinches off to form a vesicle inside the cell.

Table 6.3 Transporters in epithelial membranes

Transporter	Type	Locations
ATP binding cassette (ABC) transporters		
MDR1, p-glycoprotein, multidrug resistant protein	Active, efflux	Intestine (into lumen), liver (into bile), kidney (into urine), brain (into blood), retina (into blood)
MRP2 (multidrug resistant protein 2)	Active, efflux	Intestine (into lumen), liver (into bile), kidney (into urine)
BCRP (breast cancer resistant protein)	Active, efflux	Breast cancer cells, placenta (into blood), Intestine (into lumen), liver (into bile), kidney (into urine)
Solute carrier family		
OATP1B1 (Organic anion transport polypeptide)	Passive, influx	Liver (into hepatocyte)
OATP2B1	Active, influx	Intestine (into enterocyte)
OCT1 (Organic cation transporter 1)	Passive, influx	Liver (into hepatocyte)
OCT2 (Organic cation transporter 2)	Passive, influx	Kidney (blood into kidney cell)
OAT3 (Organic anion transporter 3)	Passive, influx	Kidney (blood into kidney cell)
MATE 1 (Multidrug and toxin extruder)	Passive, efflux	Liver (into bile), Kidney (into urine)
LAT1 (Large amino acid transporter 1)	Passive, influx	Brain (blood to brain), retina (blood to retina)
CAT1 (Cationic amino acid transporter 1)	Passive, influx	Brain (blood to brain), retina (blood to retina)
PEPT1 (Peptide cotransporter)	Active, influx	Intestine (lumen to cell)
PEPT2 (Peptide cotransporter)	Active, influx	kidney (urine to cell)

The term *pinocytosis* is used when the material engulfed is a macromolecule such as a protein. Pinocytosis occurs continually in the cell membranes of most cells.

- *Phagocytosis* is unique to tissue macrophages and neutrophils, which envelop large particles such as bacteria, whole cells or tissue debris.
- *Transcytosis* is a similar process in the endothelial cells of capillaries in which the cell membrane engulfs a macromolecule and transports it through the cell to the other membrane in a small vesicle.

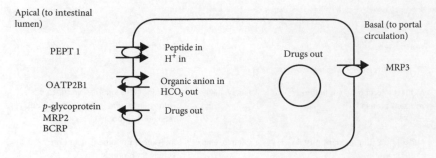

Figure 6.2 Selected transporters in the enterocyte (small intestine)

Receptor mediated cytosis

When bound by the appropriate ligand, a cell surface receptor will initiate the formation of vacuoles that transport the ligand into the cell. There are a number of endogenous substances that are transported into cells via receptor mediated endocytosis: insulin, epidermal growth factor, luteinizing hormone, thyroid hormone and immunoglobulins. This transport mechanism could be used to target a drug to the epithelial cells containing the specific receptor by linking a drug to the ligand (an immunoglobulin, for example).

Permeability of absorptive membranes

The sum of all of the above processes that carry a drug through a biological membrane can be considered the membrane permeability. *Permeability coefficient* is defined as the distance traveled by a drug through the membrane per unit time (P_{eff}, cm/s, cm/h).

This biological property is measured in a variety of tissue cultures and animal models and is expressed in centimeters per second. The cell-based systems may differ from their membranes of origin in terms of transporters and paracellular dimensions, however they are widely used. Table 6.1 may provide the pharmacist with a perspective on the relative permeability of the absorptive membranes encountered via different routes of administration.

Blood flow, tissue perfusion, and other modes of distribution

The rate of drug penetration across a membrane is the net result of all of the above processes: diffusion across the membrane, diffusion between cells, carrier mediated transport and cytosis.

- Most drugs move across membranes by passive diffusion, active transport being important only for a few drugs.
- When a drug penetrates an absorbing membrane into a capillary, blood moves the drug from the site of administration to other perfused areas including the site of action and the sites of elimination.

Table 6.4 Blood flow to body organs			
Fluid compartment	Tissues	Flow rate (mL/minute per 100 g tissue)	Time to complete drug distribution
Well perfused viscera	Brain, lungs, heart, liver, kidneys, adrenal, thyroid, GI tract	50–400	<2 minutes
Intermediate perfusion: lean tissue	Muscle, skin, red marrow, non-fat subcutaneous tissue	5–12	60–100 minutes
Poorly perfused: fat	Adipose tissue, fatty marrow, bone cortex	1–3	300 minutes

- The slow step of this transport can occur at the epithelial membrane or at the blood flowing by the area of administration.

Blood flow limited transport is illustrated by intramuscular injections where the only membrane encountered by the drug is the capillary membrane, which does not produce a barrier for the movement of most drugs. The muscle group with the highest blood flow provides the fastest rate of drug absorption.

The organs and tissues in the body can be grouped according to their perfusion primacy. In other words the blood flowing into the major arteries from the heart goes first to an extensively perfused group of vital organs – the lungs, heart, kidneys, brain and liver. Next the blood flows to an intermediately perfused group, which could also be referred to as the lean tissue – skeletal muscle, skin and nonfatty subcutaneous tissue. Finally blood reaches the poorly perfused adipose tissue, fatty marrow and bone cortex. Drugs traveling with this blood reach these organs and tissues in this same order and extent; thus the effect of a drug on the heart will be observed before the effect on skeletal muscle (Table 6.4).

Tissue and plasma protein binding

- A number of drugs have been noted to bind reversibly to protein molecules as they travel with the blood.
- Because these protein molecules are large, drug that is bound does not move outside of the vascular system, and for a drug that is extensively (>90%) bound, such binding can reduce the access of a drug to its receptor and to the organs that eliminate it.
- Free or unbound drug is able to diffuse or be transported across membranes to produce an effect.
- The amount of free drug available to cross membranes may be affected by displacement by other drugs and the amount of protein available to bind the drug.

- Because the changes in drug response resulting from changes in protein binding can be complex, it is recommended that the pharmacist monitors susceptible patients for both drug efficacy and drug adverse effects.

Drugs may enter or bind to non-fluid portions of the body, thus reducing the amount of drug in the blood and the amount of drug that is delivered to the receptor and organs that will eliminate it. For example, very lipid soluble drugs may concentrate in the adipose tissue resulting in very low concentrations in plasma. Drugs with structures that chelate multivalent cations tend to collect in the bone. Tissue and intracellular drug binding is not easily characterized but the extent of drug preference for tissues outside of the vasculature is measured by a pharmacokinetic parameter called *volume of distribution*. Very polar drugs will have smaller volumes of distribution that are equal to or slightly larger than the plasma volume. Drugs with affinity for non-fluid body tissues will have large volumes of distribution.

Drug metabolism (Table 6.5)

- Drug metabolizing enzymes can be found in many tissues though the most significant populations are in the liver, kidneys, lung, intestine and skin.
- The metabolism of a drug may deactivate or activate it. Drugs that are administered in inactive form with the expectation that enzymes will activate them *in vivo* are called *prodrugs*.

Table 6.5 Drug metabolism

Metabolic phase	Enzymes	Location
Phase I oxidation	CYP3A4	Liver, lung, small intestine
Phase I oxidation	CYP2C19	Small intestine, liver
Phase I oxidation	CYP1A2, CYP2D6, CYP2C9	Liver, lung
Phase I reduction	Azoreductase, nitroreductase	Liver, colonic bacterial enzymes
Phase I hydrolysis	Amidases (peptidases)	Ubiquitous
Phase I hydrolysis	Esterases	Ubiquitous
Phase II	Uridine diphosphate glucuronosyl-transferase	Liver, lung, nose, intestine, kidney, skin, brain
Phase II	Sulfotransferases	Liver, small intestine, brain, kidneys, platelets
Phase II	N-acetyl-transferases	Liver, spleen, lung, intestine

Information from C Ionescu, N Caira. Drug Metabolism: Current concepts. Dordrecht: Springer, 2005.

- Metabolic enzyme proteins catalyze reactions that transform fat soluble drugs into more water soluble forms or add large moieties such as amino acids to make an amphiphilic metabolite.
  - Phase I reactions introduce or expose a functional group on a molecule that makes the molecule more water soluble. Phase I reactions include oxidation, reduction and hydrolysis. The cytochrome P450 enzymes are the major proteins involved in phase I oxidations.
  - Phase II reactions produce a covalent bond between a functional group on the drug or metabolite and an endogenous compound such as glucuronic acid, sulfate, glutathione, or an amino acid, creating a highly polar molecule with a charged group.
- Changes in drug structure facilitate the final elimination of the drug metabolite through urine or feces.
- A significant feature to note about drug metabolizing enzymes in the context of the different routes of administration is whether they are encountered by the drug before or after the drug is able to distribute to its site of action.
- Although the respiratory tract and skin have appreciable populations of drug metabolizing enzymes, the only well characterized 'first pass effect' is in the gastrointestinal tract. The amount of drug lost to enzymatic activity must be factored into the dose chosen for each route.
- Like other saturable processes, drug metabolizing enzymes can be inhibited by other substrate drugs or non-competitive inhibitors resulting in a reduction of the fraction of the drug that is deactivated (or activated). This is particularly true because one enzyme may have broad substrate specificity, meaning that it can metabolize a large number of structurally different drugs.
- Inhibition of a drug's metabolism via one enzymatic reaction may not result in accumulation of the unchanged drug if alternate routes of metabolism are available since a particular drug is often a substrate for more than one type of enzyme.
- Repeated administration of certain drugs and chemicals increases the levels of certain drug metabolizing enzymes, a process referred to as *enzyme induction*. This is an adaptive response; it takes time to develop. The inducer causes the cellular machinery to synthesize greater amounts of an enzyme. Thus the liver has the capacity to respond to certain exogenous chemicals to clear them at a faster rate by making more enzymes to catalyze the chemical's breakdown.
- If patients on a drug that is a substrate for an inducible enzyme begin to take an inducer, over time they may find that the substrate drug no longer produces a therapeutic effect.

Management of drug interactions

- Metabolic drug–drug interactions may be prevented by avoiding interacting medications, however, this is not always foreseen and not always possible.
- Many interacting drug combinations can be managed by monitoring for either therapeutic failure (inducers) or substrate drug toxicity (inhibitors and when discontinuing an inducer drug). In some cases, patients cannot monitor themselves, and must be referred to the physician for blood testing or other procedures.
- In other cases, the pharmacist should recommend replacing the interacting drug with a noninteracting drug of the same class or an appropriate drug from another class.
- Individual patients can differ significantly in their expression of phase I and phase II enzymes resulting in a variety of unexpected responses to drugs.

Key Points

- Biological membranes with multiple layers and tight paracellular channels are less permeable than those with single layers and loose paracellular channels.
- Capillary membranes consist of a single layer of epithelium with relatively large channels between cells.
- Drugs move through biological membranes across the lipid bilayer (transcellular diffusion), between cells (paracellular diffusion), on transporter proteins or via cytosis.
- The most common pathway of drug movement through membranes is via transcellular diffusion.
- Transporters can move drugs across absorptive membranes towards their targets or out of absorptive membranes away from their targets.
- Drug metabolizing enzymes that are encountered before the drug is distributed to its target can reduce the amount of drug reaching its active site.
- Highly perfused tissues include the organs; muscle and skin receive intermediate perfusion, and adipose tissue is poorly perfused.

Dosage form factors

The design of the dosage form has considerable effect on the extent and rate of drug movement to the site of action. In fact pharmaceutical scientists have addressed each of the factors described above with a drug delivery system designed to overcome problems with a drug's biological fate. This section reviews the processes by which drugs are liberated from dosage forms and explores how dosage form design can be used to influence challenges to our ability to deliver a drug to the site of action.

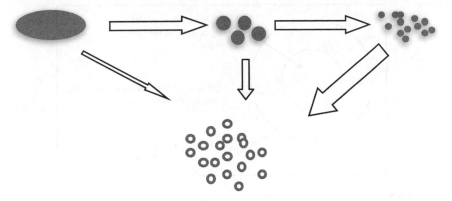

Figure 6.3 Drug dissolution from oral solid dosage form

Release from the dosage form

Broadly speaking drugs are released from dosage forms by one of two processes: dissolution or diffusion. Drugs that are presented to the body as aqueous solutions do not require 'release' and are ready to move across membranes.

Release of drug by dissolution

Dosage forms that present the drug in solid or incompletely dissolved form, such as tablets, capsules and suspensions must dissolve to release drug for absorption. During the process of dissolution the solid molecules must exchange interactions with other solid molecules for interactions with aqueous body fluids. Two factors that are commonly used to facilitate this process are small particle size and high water solubility.

> ### Definition
>
> *High water solubility* is the ability to dissolve the highest dosage strength of a drug in 250 mL water (Figures 6.3 and 6.4).

Release of drug by diffusion

When we apply a drug to the skin, whether we do this for local treatment or systemic treatment, the drug will need to diffuse out of the drug delivery system to reach the skin. Other dosage forms that release drug by diffusion include the diffusion-controlled, extended release tablets and capsules. Diffusion is the movement of a dissolved drug from an area of high concentration to an area of low concentration by random molecular motion.

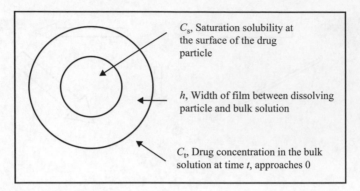

Figure 6.4 Dissolving particle

C_s, Saturation solubility at the surface of the drug particle

h, Width of film between dissolving particle and bulk solution

C_t, Drug concentration in the bulk solution at time t, approaches 0

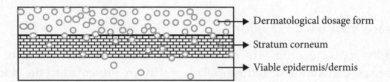

Dermatological dosage form

Stratum corneum

Viable epidermis/dermis

Figure 6.5 Drug diffusion from a topical dosage form

It is essential to the process of diffusion that the diffusing drug be dissolved in the diffusion medium. This means that in order to diffuse out of a topical cream or ointment the drug must be in solution in the dosage form. And for drugs in solid dosage forms such as extended release tablets, aqueous body fluids must enter the dosage form and dissolve the drug before it can diffuse out of the tablet (Figure 6.5).

Dosage form design

Dosage forms are designed to overcome a number of challenges including the ability to measure the dose, palatability, comfort and stability of the drug for the period of treatment. One of the most interesting challenges has been to design dosage forms to deliver drugs more efficiently to the site of action. In addition to addressing problems with drug solubility, membrane permeability and drug metabolism, pharmaceutical scientists have developed dosage forms that are able to deliver a drug more selectively to the site of action and permit better control over the rate of release of drugs. Table 6.6 summarizes these achievements. The optimized drug product delivers the right amount of drug to the right place for the appropriate length of time.

Key point

- A dosage form can be designed to overcome suboptimal physicochemical or biological properties of a drug or route of administration.

Table 6.6 Design of dosage form to overcome drug travel challenges

Factor	Dosage form design	Example
Poor water solubility	Use of surfactants in tablets	Lipitor, Septra DS
	Use of cosolvents to make a solution dosage form	Lorazepam, diazepam, phenytoin, phenobarbital
	Preparation of a soft gelatin capsule	Lanoxin, Prometrium
	Preparation of liposomal formulation	Amphotericin liposomal preparations
Poor permeability	Use of pH modifiers to increase concentration of unionized form	Lidocaine patch
	Use of absorption promoters	Butrans
	Carrier made from drug and part of an immunoglobulin capable of entering epithelial cells via receptor mediated cytosis	Transceptor drug delivery technology
Retention of the drug at the site of absorption	Slow eroding ocular inserts	OcusertR for sustained delivery of ophthalmic drugs
	Mucoadhesive dosage forms that stick at the absorbing membrane	Striant mucoadhesive buccal tablets
Drug instability at extreme pHs	Preparation of enteric coated dosage forms	Omeprazole and other proton pump inhibitors
Rapid drug metabolism	Preparation of extended release tablets and capsules	Morphine sulfate extended release tablets can be administered once or twice daily rather than six times daily
	Linking polyethylene glycol to protein molecules to reduce their elimination by amidases and the mononuclear phagocytic system	Pegfilgrastim (Neulasta) solution for injection, pegaptanib (Macugen) solution for injection, pegaspargase (Oncaspar) solution for injection
Extensive drug metabolism	Preparation of sublingual dosage forms	Nitroglycerin tablets
	Formulation with enzyme inhibitors	Levodopa/carbidopa
Selective site delivery	Respiratory drugs are delivered by aerosol inhalation	Proventil HFA, Ventolin HFA, Advair Diskus
	Enteric coats for delivery to colon	Pentasa, Asacol
	Targeted delivery systems	Herceptin, Rituxan, Erbitux
Controlled rate of delivery	Infusion pumps	Insulin pumps
	Reservoir transdermal patches	Transderm Scop, Transderm Nitro, Estraderm, Androderm
	Osmotic oral dosage forms	OROS, GITS and COER technology

Table 6.7 Issues relating to patient use of dosage forms

Dosage form	Issue
Tablet or capsule	Takes without water resulting in sticking to the esophagus and subsequent mucosal injury
Extended release tablet or capsule	Use as an as needed medication not helpful because of slow release properties
	Crushing results in conversion to an immediate release form
Enteric coated tablets	Crushing exposes the drug to stomach acid or the stomach lining to the drug
Buccal or sublingual tablets	Swallowing results in extensive loss of the drug to first pass metabolism
Suspensions	Failure to shake well results in non-uniform doses
Eye drops	Failure to wait 5 minutes between drops to the same eye results in loss of the second dose
Nose drops	Head back administration results in loss of drug to the nasopharynx
Suppositories	Must be unwrapped
Metered dose inhalers	Fails to empty lungs; mis-times activation of device with inhalation; fails to hold breath resulting in less medication reaching the lungs
	Fails to rinse mouth after inhaled corticosteroids resulting in oral thrush infections
Dry powder inhaler capsules	Swallows capsule rather than inhaling with dry powder inhaler
Transdermal patches	Placement must be rotated
	Reservoir devices should not be cut

Patient factors

There are three patient related factors that can influence the rate and extent of drug travel to the site of action. They are:

- use of the dosage form or drug delivery system
- disease states
- pharmacogenetics.

It is the pharmacist's responsibility to ensure that the patient knows how to use a dosage form appropriately. The acquisition of this knowledge is one of the goals of your education in pharmaceutics. Table 6.7 summarizes selected examples of issues relating to the appropriate use of dosage forms.

A number of disease states impact absorption, distribution, metabolism and elimination of drugs through effects on membrane permeability, stomach emptying, intestinal transit, drug binding to proteins and ability to metabolize or eliminate drugs in the liver or kidneys. It is useful for the pharmacist to recognize that these disease states can contribute in predictable ways to the drug response. Table 6.8 provides examples of disease states that affect the rate or extent of drug travel from the site of administration to the site of action.

The proteins responsible for drug transport and metabolism are susceptible to genetic variation that results in reduced drug transport and increased or decreased metabolism. Metabolic variations have been categorized as ultrarapid metabolizers (UM), extensive metabolizers (EM generally considered to be the normal enzyme), intermediate metabolizers (IM) and poor metabolizers (PM). Table 6.9 presents the predicted effects of these different polymorphisms on drug action. Genetic variations in transporter proteins have been identified but most are not yet well characterized. Two variants of the organic anion transporter polypeptide that transports statin drugs into hepatocytes have been identified and associated with statin induced myalgia.

Key Points

- Patients' abilities to use a dosage form properly will determine whether they receive maximum benefit from the drug therapy.
- Patients with disease states that impact the absorption, distribution, metabolism and elimination of drugs should be monitored for safety and efficacy.
- Genetic variation in the activity of drug transporting or metabolizing proteins can create differences in the amount of drug reaching the target and subsequent patient response.

Following a drug from administration to elimination

Figure 6.6 shows a summary of drug travel.

To pull together the discussion of factors affecting drug travel from the dosage form to the site of action, we will follow an oral tablet containing metoprolol tartrate from administration to elimination. Metoprolol has a log P of 1.9, a molecular weight of 267 and the tartrate salt is very soluble in water (1 g in less than 1 mL water). It is an aliphatic secondary amine with a pK_a of 9.68.

Administration

The patient takes the metoprolol tablet by mouth with a full glass of water (as suggested by the pharmacist) before breakfast. The tablet is washed through the esophagus and into the stomach where it is mixed with stomach fluids and water and encounters a pH of 2.0.

Table 6.8 Effect of disease processes on drug travel from dosage form to receptor

Disease state	Effect
Absorption	
Dysphagia (myasthenia gravis, elderly)	Difficulty delivering drug to site of absorption
Achalasia (inability to empty esophagus)	Difficulty delivering drug to site of absorption
Asthma/emphysema increased airway resistance	Difficulty delivering inhaled drug to site of absorption
Achlorhydria; pH of gastric fluids ≥6.5	Reduced absorption of atazanavir, ketoconazole, digoxin, calcium, vitamin B12, iron, vitamin C
Liver disease	Slow gastric emptying; slow oral drug absorption
Vomiting	Loss of medication from the route of administration
Distribution	
Obesity changes the lean to fat tissue ratio	Dosing metric will depend on drug lipophilicity – use total body weight for lipophilic drugs and lean body mass for hydrophilic drugs
Circulatory shock reduces cardiac output	Reduces drug distribution to peripheral tissues
Multiple sclerosis, AIDS/AIDS related dementia, Alzheimer's disease, encephalitis/meningitis, Hypertension, seizure disorder	Increased permeability of blood–brain barrier
Liver disease, reduced protein synthesis	Decreased albumin; more unbound drug
Inflammatory diseases, stress, malignancy	Increased alpha-1 glycoprotein; more bound drug
Chronic kidney disease, loss of protein through kidney	Decreased albumin; more unbound drug
Metabolism	
Liver disease reduced phase I enzyme activity, phase II enzymes less affected	Decreased metabolism of morphine, nifedipine, lovastatin, metoprolol, and others
Elimination	
Acute renal failure	Sudden, reversible reduction in renal elimination of drugs
Chronic renal failure progressive loss of functioning nephrons	Reduction in renal elimination of drugs: amikacin, cephalexin, digoxin, gentamycin, hydrochlorothiazide, and others
Biliary obstruction	Reduces drug and metabolite elimination in the feces

Table 6.9 Predicted effects of genetic variations in metabolic enzymes

Metabolizer phenotype	Effect on drug	Effect on patient	Effect on prodrug	Effect on patient
Ultrarapid (UM)	Rapid deactivation of the drug	Reduced therapeutic effect	Rapid activation of the drug	Increased therapeutic or adverse effect
Extensive (EM)	Extensive deactivation of drug, considered the normal phenotype	Expected therapeutic effect	Extensive activation of the drug	Expected therapeutic or adverse effect
Intermediate (IM)	Lower than normal deactivation of the drug	Increased therapeutic or adverse effect	Lower activation of the drug	Reduced therapeutic or adverse effect
Poor (PM)	Loss of drug deactivation via this enzyme	Increased therapeutic or adverse effect	Loss of drug activation via this enzyme	Reduced therapeutic or adverse effect

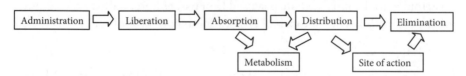

Figure 6.6   Overview of drug travel

Release from the dosage form

Metoprolol tartrate is available as an immediate release tablet (the extended release formulation is prepared with the succinate salt), which disintegrates readily in the stomach contents. Metoprolol is 99.999% ionized at this pH so it dissolves rapidly.

Absorption

After approximately 60 minutes the dissolved drug has mostly moved into the small intestine where the pH is 6.5. At this pH the ionized form still predominates (99.93%), however, with a log P of 1.9 and molecular weight of 267, metoprolol easily permeates by passive diffusion and it is rapidly and completely absorbed into the capillary system of the small intestine within 2.5 to 3 hours.

Distribution

The capillary system of the small intestine drains into the portal vein that goes to the liver. The capillaries of the liver are very permeable allowing drug rapidly into hepatocytes.

First pass metabolism

Metoprolol is distributed to the hepatocytes before it flows to the heart where it can be delivered to its site of action. Within the hepatocyte about 50% of the metoprolol is inactivated by CYP2D6 enzymes to water soluble metabolites.

Distribution

Upon leaving the liver in the hepatic veins, the remaining active metoprolol and metabolites move into the inferior vena cava and into the heart where they are pumped into the arteries. Because metoprolol is both soluble and permeable and only about 11–12% protein bound, it is widely distributed into tissues, starting with the well perfused viscera where it will encounter its site of action at the beta-receptor in the heart. The active metoprolol and metabolites are also rapidly delivered to the kidneys. Metoprolol is then distributed to the intermediate perfused lean tissues and finally to the adipose of the patient's body. Metoprolol crosses the blood–brain barrier.

Metabolism

The arteries carry the active metoprolol back to the liver where CYP2D6 deactivates another 50% of the metoprolol passing through.

Elimination

The arteries carry the active metoprolol and its water soluble metabolites to the kidneys where they are filtered through the glomeruli into the urine. The water soluble metabolites remain in the urine while the lipid soluble form of metoprolol may be reabsorbed into the blood by passive diffusion. Metoprolol is not known to be actively secreted from the blood towards the urine by transporters as some drugs are. Only about 5% of the active metoprolol is eliminated in the urine. When the amount of metoprolol metabolized in the liver or eliminated in the urine exceeds the amount of drug coming into the body sufficiently long to drop the concentration at the heart below the therapeutic threshold, the patient will no longer experience the reduced cardiac output produced by the drug.

Questions and cases

1. The pH of the stomach is between 1.0 and 3.5 and the pH of the small intestine is 6.5–7.6. In which of these portions of the gastrointestinal tract will atenolol, pK_a 9.6, exist in greater amounts in its membrane soluble form?

Atenolol

   (a) The stomach, pH 1.0–3.5

   (b) The small intestine pH 6.5–7.6

   (c) There is no membrane soluble form at either of these pH ranges

2. Amlodipine is a popular calcium channel blocker used in the treatment of hypertension and angina. Log P 1.9, water solubility 75.3 mg/L, molecular weight (MW) 408.9. Which of the following would you predict about its oral absorption/membrane permeability?

Amlodipine

   (a) Absorption will be good because log P <5, MW <500, there are fewer than 5 hydrogen bond donors and 10 hydrogen bond acceptors

   (b) Absorption will be poor because there are more than 5 hydrogen bond donors

   (c) Absorption will be poor because there are more than 10 hydrogen bond acceptors

3. Phenytoin has a formula weight of 252, log P 2.2 and is 90–95% plasma protein bound. Its site of action is in the brain. Which of the following would you expect of phenytoin's distribution?

   (a) Unbound drug will cross the blood–brain barrier via the paracellular route

   (b) Unbound drug will move into the brain via passive diffusion

   (c) Bound drug will move out of the blood stream via facilitated diffusion

   (d) Bound drug will move out of the blood stream via active transport

4. You have a patient with a new prescription for cephalexin 500 mg po every 12 hours for a skin and soft tissue (subcutaneous) infection. Cephalexin is mostly eliminated as unchanged drug by the kidneys. Which of the following describes the movement of this drug from the site of administration to the sites of action and elimination?

   (a) Cephalexin reaches its site of elimination before it reaches the heart

(b) Cephalexin reaches its site of elimination before it reaches the site of the infection

(c) Cephalexin reaches the site of infection before it reaches the heart

(d) Cephalexin reaches the site of infection before it reaches the site of elimination

5. Felodipine has a log P of 4.8. You would predict that felodipine:

(a) Easily crosses the blood–brain barrier

(b) Has good water solubility

(c) Is a substrate for the organic anion transporter

(d) Is not metabolized by cytochrome P450 enzymes

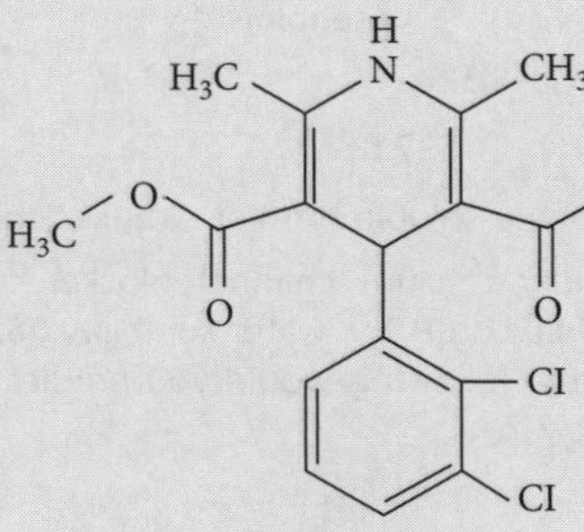

Felodipine

6. MK is on felodipine ER tablets 10 mg once daily. Felodipine is a calcium channel blocker that is also a p-glycoprotein substrate. What happens to the amount of felodipine absorbed if MK takes felodipine with grapefruit juice?

(a) The amount of felodipine in the blood increases

(b) The amount of felodipine in the blood decreases

(c) The amount of felodipine in the intestine increases

(d) The amount of felodipine reaching the vascular smooth muscle cells decreases

7. Glycopyrrolate and atropine are two drugs with anticholinergic activity. Which drug is more likely to distribute to the brain and cause central nervous system side effects?

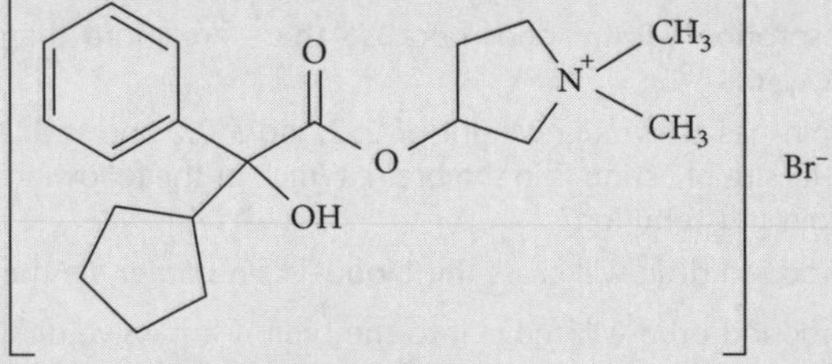

Glycopyrrolate

Atropine

(a) Glycopyrrolate is more likely to distribute to the CNS because it is more lipophilic

(b) Glycopyrrolate is more likely to distribute to the CNS because it is more hydrophilic

(c) Atropine is more likely to distribute to the CNS because it is more lipophilic

(d) Atropine is more likely to distribute to the CNS because it is more hydrophilic

8. Rifampin is a potent inducer of Cyp3A4. Nifedipine is a substrate of Cyp3A4. What happens to the amount of nifedipine reaching drug targets in blood vessels if the patient is started on rifampin?

(a) The amount of nifedipine increases because the number of metabolic enzymes increases

(b) The amount of nifedipine decreases because the number of metabolic enzymes increases

(c) The amount of nifedipine increases because the number of metabolic enzymes decreases

(d) The amount of nifedipine decreases because the number of metabolic enzymes decreases

9. Amoxicillin and methotrexate are both secreted by the organic anion transporter in the kidney. What happens to the amount of methotrexate in the blood if the patient is started on amoxicillin?

(a) The amount of methotrexate increases because amoxicillin increases its secretion

(b) The amount of methotrexate increases because amoxicillin decreases its secretion

(c) The amount of methotrexate decreases because amoxicillin increases its secretion

(d) The amount of methotrexate decreases because amoxicillin decreases its secretion

10. The pharmacist advises a patient with a prescription for pain medication to drink a full glass of water with the tablet. Which of the following dissolution variables would be affected?

(a) Water would increase the width of the boundary layer (h)

(b) Water would decrease the area exposed to solvent (A)

(c) Water would reduce the viscosity of the dissolution medium (D)

(d) Water would increase the saturation solubility of the drug (C_s)

11. The use of the salt form, hydrocodone tartrate, instead of the free base, hydrocodone, in the tablet for pain has which of the following effects on drug dissolution variables?

(a) The salt form would increase the width of the boundary layer (h)

(b) The salt form would decrease the area exposed to solvent (A)

(c) The salt form would reduce the viscosity of the dissolution medium (D)

(d) The salt form would increase the saturation solubility of the drug (C_s)

12. The manufacturer of the hydrocodone immediate release tablet has dispersed the drug in a solid material that wicks water into the dosage form causing it to swell. Which of the following dissolution variables would be affected?

(a) The ingredient would increase the width of the boundary layer (h)

(b) The ingredient would decrease the area exposed to solvent (A)

(c) The ingredient would reduce the viscosity of the dissolution medium (D)

(d) The ingredient would increase the saturation solubility of the drug (C_s)

13. The pharmacist learns that the patient with the hydrocodone prescription has a condition called diabetic gastroparesis that reduces the motility of his stomach. Which of the following dissolution variables would be affected?

(a) The condition would increase the width of the boundary layer (h)

(b) The condition would decrease the area exposed to solvent (A)

(c) The condition would reduce the viscosity of the dissolution medium (D)

(d) The condition would increase the saturation solubility of the drug (C_s)

14. Drugs bound to plasma proteins:

(a) Cannot cross membranes

(b) Are transported across membranes on specialized carrier proteins

(c) Are filtered by the kidneys and rapidly excreted

(d) Are covalently bound

15. As the dose of a drug that is absorbed via passive diffusion goes up, the:

(a) Percentage of drug absorbed will remain constant because diffusion is not saturable

(b) Percentage of drug absorbed will go down because diffusion is not saturable

(c) Percentage of drug absorbed will go down because diffusion is saturable

(d) Percentage of drug absorbed will go up because diffusion is saturable

16. As the dose of a drug that is absorbed via carrier mediated influx goes up, the:

(a) Percentage of drug absorbed will remain constant because carrier mediated influx is not saturable

(b) Percentage of drug absorbed will go down because carrier mediated influx is not saturable

(c) Percentage of drug absorbed will go down because carrier mediated influx is saturable

(d) Percentage of drug absorbed will go up because carrier mediated influx is saturable

17. As the dose of a drug that is absorbed via carrier mediated efflux goes up, the:

(a) Percentage of drug absorbed will go up because carrier mediated efflux is not saturable

(b) Percentage of drug absorbed will go down because carrier mediated efflux is not saturable

(c) Percentage of drug absorbed will go down because carrier mediated efflux is saturable

(d) Percentage of drug absorbed will go up because carrier mediated efflux is saturable

Cases

These cases require that the student refers to other drug references.

18. Find the log P and $CaCO_2$ permeability coefficient for each of the following drugs from the DrugBank website http://www.drugbank.ca/. Graph permeability on the y axis and log P on the x axis. What is the relationship of permeability to lipophility? To what route of administration is $CaCO_2$ *in vitro* data relevant? Are there any molecules that do not fit with the rest of the data? Are any of them substrates for transporters?

Drug	Log CaCO$_2$ permeability coefficient	Log partition coefficient
Amoxicillin		
Acyclovir		
Clonidine		
Diazepam		
Enalapril		
Felodipine		
Hydrochlorothiazide		
Ibuprofen		
Labetalol		
Methotrexate		
Metoprolol		
Naproxen		
Phenytoin		
Ranitidine		
Verapamil		
Warfarin		

19. There are several polymorphs of the organic cation transporter (OCT) 1 that moves metformin from blood into the hepatocyte towards its hepatic site of action. (Metformin is eliminated in the urine, not by metabolim). These polymorphic variations decrease the activity of the transporter compared with the normal protein. Diagram the transport of metformin on OCT1 into the liver in the normal and polymorphic cases. You have a diabetic patient who has the reduced function OCT1 polymorph who is starting on the usual initial dose of metformin 500 mg twice daily.

You will monitor the patient's drug therapy with daily blood glucose and quarterly hemoglobin A1C test. What results would you expect from this patient's glucose and hemoglobin A1C tests compared with a diabetic patient with the normal transporter? Based on the predicted results will you need to increase or decrease that patient's metformin dose?

20. Use a drug reference to read about the effect of rifampin on CYP1A2. You have a patient who has been on rifampin for about 6 months for the treatment of tuberculosis. She comes into the pharmacy with a new prescription for warfarin for 'a heart problem'. The doctor starts her at 5 mg warfarin daily and, eventually, raises the dose to 5 mg Monday, Wednesday and Friday and 7.5 mg Sunday, Tuesday, Thursday and Saturday. Two months later, she comes into the pharmacy to refill her usual warfarin prescriptions and happily announces that she is done with the rifampin therapy. What do you need to advise the patient and physician about this sequence of drugs?

21. You are the consultant pharmacist at a long term care center. One of the residents, BE, is on furosemide 20 mg q am, atenolol 25 mg qd (for hypertension), valproic acid 125 mg bid (for dementia), and lovastatin 10 mg with evening meal (for hyperlipidemia). Her recent lab tests indicate that her albumin level is reduced. What effect will this have on the movement of these drugs from dosage form to receptor or to eliminating organ? Would you predict that the therapeutic effect will increase or decrease? How would you monitor a patient on each of these drugs that is greater than 90% protein bound?

22. Mr Lawson takes for hypertension: Dilacor XR 240 mg caps qd and hydrochlorothiazide 25 mg caps qd. On Thursday he had surgery on his jaw which has caused soreness and difficulty swallowing. His wife opens his medications and puts them in apple sauce for him. On Saturday Mrs Lawson comes into the pharmacy to pick up post-surgical pain medications and tells the pharmacist that Mr Lawson has fallen twice in the last two days. He becomes extremely dizzy when he stands up to move around the house. Explain his sudden onset of dizziness? What can the pharmacist recommend to assist Mr Lawson?

Selected reading

1. MN Martinez, GL Amidon. A mechanistic approach to understanding the factors affecting drug absorption: a review of fundamentals. J Clin Pharmacol 2002; 42: 620. http://jcp.sage pub.com/content/42/6/620.
2. CA Lipinski, F Lombardo, BW Dominy, et al. Experimental and computational approaches to estimate solubility and permeability in drug discovery and development settings, Adv Drug Del Rev 2001; 46: 3–26.
3. O Zolk, MF Fromm. Transporter-mediated drug uptake and efflux: Important determinants of adverse drug reactions, Clin Pharmacol Ther 2011, 89: 798–805.
4. H Sarin. Physiologic upper limits of pore size of different blood capillary types and another perspective on the dual pore theory of microvascular permeability, J Angiogenesis Research 2010; 2: 1–19.
5. LZ Benet, F Broccatelli, TI Oprea. BDDCS applied to over 900 drugs. AAPS J 2011; 13: 519–547.
6. SA Scott, Personalizing medicine with clinical pharmacogenetics, Genet Med 2011; 13: 987–995.
7. MK DeGorter, CQ Xia, JJ Yang, RB Kim, Drug transporters in drug efficacy and toxicity. Annu Rev Pharmacol Toxicol. 2012; 52: 249–273.

7

Bioavailability, bioequivalence and the Biopharmaceutical Classification System

Learning objectives

Upon completion of this chapter, you should be able to:

● Define reference listed drug; define and differentiate bioavailability, pharmaceutical equivalence, bioequivalence and therapeutic equivalence.

● Define the following pharmacokinetic variables and state whether each provides information about absorption rate or extent.

— C_{max}, T_{max}, area under the curve.

● Given comparative pharmacokinetic parameters, determine whether two products are bioequivalent.

● Given Orange Book data, determine whether two products are therapeutically equivalent.

● Explain the difference between brand name drugs and their generic counterparts in lay terms.

● Define high solubility and high permeability according to the Biopharmaceutical Classification System and, given a drug's water solubility and membrane permeability, determine its Biopharmaceutical Classification and predict which of these parameters limits the drug's absorption rate.

Introduction

We have seen that many factors affect the rate and extent of drug travel from its dosage form to the site of action. When a new, patented drug product is approved for sale in the United States, its manufacturer has provided detailed evidence that the drug product is safe and effective for a specified purpose or condition. After the period of patent exclusivity has ended, generic drug products may be marketed that purport to be effective for the same clinical indications as the innovator product based on the fact that they contain the same drug in the same dose in the same dosage form. In recognition of the complexity of the processes that bring drugs to their receptors, the Food and Drug Administration (FDA) requires manufacturers proposing to introduce a product onto the market with an Abbreviated New Drug Application (e.g.

new generic drug products) to provide evidence that the drug product is indeed equivalent to an established innovator product. As pharmacists who choose the products we will provide to our patients, it is essential that we understand how these important quality determinations are made and how to interpret information provided by the FDA about the equivalence of drug products. This chapter briefly reviews the general approach that is used to study the bioequivalence of drug products. The study of oral drug products has received the most attention but we will also consider other product categories that have been the subject of FDA Guidances to the pharmaceutical industry. This chapter also considers the properties that pharmaceutical scientists use to predict the biological fate of drug molecules under development. Drug candidates are screened using these properties to focus the industry's efforts on those molecules that are most likely to produce pharmacological activity via the intended route of administration. The Biopharmaceutical Classification System is presently used to waive the requirement for bioequivalence evaluations in human subjects for drugs that are known to have high water solubility and high membrane permeability and therefore are unlikely to present problems with the rate or extent of drug absorption.

Pharmaceutical equivalence + bioequivalence = therapeutic equivalence

Key Points

- The innovator drug product or branded product is referred to in FDA databases as the *reference listed drug*.
- Bioavailability data describes the rate and extent of drug absorption from an innovator product and sets the standard by which generic products are compared.
- Two drug products that have the same active ingredient, dosage form, strength, route of administration and conditions of use are termed *pharmaceutically equivalent*.
- Two drug products that have substantially similar rate and extent of absorption of the active ingredient are called *bioequivalent*.
- A generic drug product that is both pharmaceutically equivalent and bioequivalent is deemed *therapeutically equivalent* and may be substituted for the innovator product as permitted by state laws.

Definition

Two drug products that have the same active ingredient, dosage form, strength, route of administration and conditions of use are termed *pharmaceutically equivalent*.

- To receive approval for a generic drug product, the manufacturer generally must establish that its product is pharmaceutically equivalent *and* that the active ingredient of the proposed drug product is absorbed into the blood stream to the same extent and at the same rate as the innovator product.

— The innovator drug product is referred to in FDA guidances and databases as the *reference listed drug*.

> *Definition*
>
> The term used to describe substantially similar rate and extent of absorption of the active ingredient is *bioequivalence*.

- A generic drug product that contains the same active ingredient in the same dose in the same dosage form, and demonstrates bioequivalence, will be deemed *therapeutically equivalent* and may be substituted for the innovator product as permitted by state laws.

The distinction between bioequivalence and therapeutic equivalence is significant in that a tablet and a capsule dosage form containing the same active ingredient can (sometimes do) provide the same extent and rate of drug absorption and thus could be considered bioequivalent but these two products are not *therapeutically equivalent* because they are different dosage forms.

Bioavailability

When a branded or innovator drug product is approved, the manufacturer, in addition to safety and efficacy studies, has provided bioavailability data that detail the process by which a drug is released from the dosage form and moves to the site of action.

- Bioavailability data provide an estimate of the fraction of the drug absorbed, the rate of absorption and the drug's subsequent distribution and elimination.
- Bioavailability studies may be required for new formulations (change in inactive ingredients) of drug products already approved to ensure that the change in ingredients will not significantly affect the performance of the drug in patients.
- This information provides the benchmark for bioequivalence studies that must be performed when generic products are developed that will compete with innovator products.
- The bioavailability of systemic drug products is generally studied in healthy human volunteers.

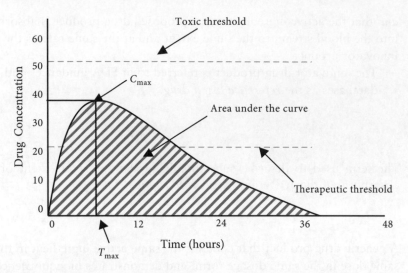

Figure 7.1 Pharmacokinetic parameters used to evaluate the performance of dosage forms *in vivo*

In a study of a new oral tablet, healthy volunteers receive one dose of the drug by intravenous injection and plasma levels of the drug are measured over time. This allows the assessment of drug exposure when there is no absorption step and no first pass metabolism. When the drug from the intravenous dose has been eliminated from the body, each of the subjects takes the drug tablet by mouth, and the data collected from the two separate routes are compared.

- Three parameters are used to quantify and compare rate and extent of drug absorption (Figure 7.1):
  - peak concentration (C_{max})
  - time to peak (T_{max})
  - area under the drug concentration versus time curve (AUC).
- Peak concentration and area under the drug concentration versus time curve provide information about the extent of exposure.
- Time to peak provides information about the rate of drug absorption.

A ratio of the area under the curve provided by the tablet to the area under the curve provided by the intravenous injection can be considered the absolute availability or the amount of intact drug available systemically from the tablet.

$$\text{Absolute availability} = F = \frac{[\text{AUC}]_{po}/\text{Dose}_{po}}{[\text{AUC}]_{iv}/\text{Dose}_{iv}}$$

An absolute availability of 1 would mean that 100% of the dose was absorbed into the systemic circulation as intact drug. The absolute bioavailability of most orally administered drugs is between 0.15 and 1

(15% and 100%). Many prodrugs such as the statins have absolute bioavailabilities of less than 0.05 (<5%) because the active form is intentionally different than the administered form. Note that absolute availability is different from a related parameter, fraction absorbed, in that a drug may be 100% absorbed through the small intestine but a significant fraction is lost to hepatic metabolism before it reaches the systemic circulation. Please see a pharmacokinetic text for more detailed information about these parameters.

Dissolution rate data

Accompanying the bioavailability data for an oral drug product will be dissolution rate profiles for all strengths of the dosage form. These studies are done *in vitro* in at least three dissolution media (pH 1.2, 4.5 and 6.8 buffer). The amount of drug released from the dosage form is measured over time to determine if drug dissolution plays a role in limiting the absorption rate of the drug.

Bioequivalence of systemically available drug products

Bioequivalence studies are conducted to compare the performance of a proposed generic product that is expected to produce systemic effects, with a referenced listed drug product of the same drug and dosage form. These requirements would apply to any product that will be used to provide therapeutic effects at an active site different from the site of absorption: oral tablets, capsules and suspensions, transdermal products, nasal or rectal dosage forms. Oral solutions generally do not require bioequivalence studies because their behavior *in vivo* is not expected to differ from product to product. Healthy, fasted volunteers are randomized into two groups: one receives the reference (innovator) dosage form, the other the test (generic) dosage form in the first period of the study. After the first dose has been eliminated the subjects receive the other dosage form in the second period of the trial. Blood samples are taken at established intervals and analyzed for drug and/or metabolites.

Data are analyzed to determine the C_{max}, T_{max} and AUC for each dosage form. To be considered therapeutically equivalent to a reference product, there must be no statistically significant difference between the test product parameters (C_{max}, T_{max}, AUC) and the reference product parameters. Specifically this means that a ratio is constructed of each of the drug products' parameter means:

$$\frac{\text{Mean } C_{max_{test}}}{\text{Mean } C_{max_{ref}}} \quad \frac{\text{Mean } T_{max_{test}}}{\text{Mean } T_{max_{ref}}} \quad \frac{\text{Mean AUC}_{test}}{\text{Mean AUC}_{ref}}$$

The 90% confidence interval about these ratios must be within ±20%. When log transformed data are used, the 90% confidence interval is set at 80–125%.

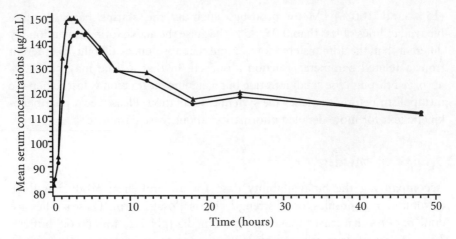

Figure 7.2 Drug concentration versus time for two levothyroxine drug products, Unithroid (▲) and generic (●). (Data from US Food and Drug Administration Center for Drug Evaluation and Research, http//www.fda.gov/cder/orange/default.htm)

Table 7.1 Rate and extent of absorption of levothyroxine from two dosage forms

Parameter	Mylan (mean ± SD)	Jerome Stevens (mean ± SD)
AUC (ng · hour/mL)	5734 (±12.8)	5824 (±13.9)
C_{max} (ng/mL)	155 (±15.6)	161 (±15.2)
T_{max} (hours)	3.4 (±48.3)	2.5 (±52.4)

AUC = area under the drug concentration versus time curve.
Data from http://www.fda.gov/ForHealthProfessionals/Drugs/default.htm.

The observed average differences in the test parameters between reference and generic products are generally between 3% and 4% (Figure 7.2 and Table 7.1).

Bioequivalence of products that produce local effects

The FDA has several guidances for industry scientists on the comparison of generic products to reference listed locally acting products that do not produce measurable concentrations of drug in an accessible biological fluid. The methods recommended generally include *in vivo* studies to demonstrate equivalent pharmacological effects and/or *in vitro* tests to establish comparable product release of drug. The FDA also has prepared documents to provide guidance for the development of biosimilar generic protein and peptide products.

Using the FDA Orange Book

The FDA evaluates the bioequivalence data provided by generic manufacturers to determine if they can be considered therapeutically equivalent and therefore interchangeable with reference listed drug (brand name) products. The results of their evaluations are posted at http://www.fda.gov/cder/orange/default.htm. Approved generic products are given one of several two letter codes that describe the acceptability of the data (A codes are bioequivalent, B codes have not been determined to be bioequivalent) as well as the basis for the judgment (*in vitro* data, *in vivo* data, identical formulation). These codes are more explicitly defined in Table 7.2.

When you navigate to the FDA website the information for healthcare professions is at http://www.fda.gov/Drugs/ResourcesForYou/HealthProfessionals/default.htm. From this page there are two links that are of interest to our discussion of therapeutically equivalent drug products:

- Drugs@FDA current and archival drug labels, drug approval histories and bioequivalence data
- Approved Drug Products with Therapeutic Equivalence Evaluations (Orange Book) database of all FDA approved prescription drugs, including new and generic drugs with bioequivalence codes

Therapeutic equivalents

Let's first go to Drugs@FDA and search 'Bupropion'. Select 'Wellbutrin SR' from the table displayed and then 'Therapeutic Equivalents'. The table shown contains all of the FDA approved generic versions of Wellbutrin SR that may be used twice daily. All generic equivalents of Wellbutrin SR are coded AB1.

Approval documents

- Click on the 'Back to Search Results' button right below Drugs@FDA.
- Select 'Bupropion Hydrochloride' and then 'Bupropion Hydrochloride (ANDA #075913)' by IMPAX labs, you will find a page of Drug Details on the IMPAX products.
- Select 'Approval History, Letters, Reviews and Related Documents'. You will see a Table of Reviews and Letters.
- Select the Review dated 1/28/2004. This review is 260 pages long but we want to slide the view bar down to page 109. You will see Pharmacokinetic Parameters Tabulated for their study of the Bupropion ER 150 mg strength (Table 7.3).

Using these documents pharmacists can inform themselves about the data used to make decisions about generic drug products.

Now let's look at the other part of the FDA Orange Book site for health professionals:

Table 7.2 Therapeutic equivalence codes for FDA approved drug products

Code	Dosage form	Interpretation
A codes		
AA	Any	Products not presenting bioequivalence problems for which *in vitro* data have been evaluated
AB, AB1	Any	Products judged bioequivalent based on *in vivo* data
AB2, AB3	Submitted	
AN	Aerosolized	Judged bioequivalent based on *in vitro* methods
AO	Oil solution	Judged bioequivalent because they are identical in active ingredient, concentration and oil used as a vehicle
AP	Parenteral aqueous solutions	Judged bioequivalent because they are identical in active ingredient and labeled concentration
AT	Topical products	Considered bioequivalent based on manufacturing data. *In vivo* studies not required for these products
B codes		
BC	Extended release	Without data to demonstrate bioequivalence to RLD products
BD	Any	Active ingredients and dosage forms with documented bioequivalence problems
BE	Delayed release, oral	Without data to demonstrate bioequivalence to RLD (enteric coated) products
BN	Aerosolized	Without data to demonstrate bioequivalence to RLD products
BP	Any	Active ingredients and dosage forms with potential bioequivalence problems
BR	Suppository, enema	Without data to demonstrate bioequivalence to RLD products
BS	Any	Product with drug standard (analytical data) deficiencies
BT	Topical	Without data to demonstrate bioequivalence to RLD products
BX	Any	Drug products with insufficient data

RLD = reference listed drug.
Adapted from http://www.fda.gov/ForHealthProfessionals/Drugs/default.htm.

- Close the Approval document and navigate back to http://www.fda.gov/ForHealthProfessionals/Drugs/default.htm
- Select Approved Drug Products with Therapeutic Equivalence Evaluations (Orange Book) This is a database of all FDA approved prescription drugs, including new and generic drugs with therapeutic equivalence codes.
- Select 'Orange Book Search' (middle of the page) and 'Search by active ingredient'.

Table 7.3 Pharmacokinetic parameters for bupropion ER by IMAX laboratories

Parameter	Test	Reference	T/R	90% CI (log)
$AUC_{(0-t)}$ ng · hour/mL	776.5 ± 28	765.7 ± 26	1.01	94.9–107.4
AUC_{inf} (ng · hour/mL)	806.4 ± 27	795.6 ± 26	1.01	95.2–107.2
C_{max} (ng/mL)	88.8 ± 30	89.5 ± 27	0.99	
T_{max} (hours)	3.4	3.2		
k_{el} (/hour)	0.05	0.06		
$T_{1/2}$ (hours)	15.09	14.97		

AUC = area under the drug concentration versus time curve; C_{max} = peak concentration; T_{max} = time to peak concentration; k_{el} = elimination rate; $T_{1/2}$= half-life.
Reproduced with permission from http://www.fda.gov/ForHealthProfessionals/Drugs/default.htm

- Search again for 'bupropion'.
- It brings up a table of all of the Bupropion HCl products: Application No., Therapeutic Equivalence code, reference listed drug, Active ingredient, Dosage form, route, strength, Proprietary name and Manufacturer.

Questions to answer about Bupropion products:

- Bupropion HCl ER 200 mg tablets by Watson Labs are rated … ?
- What does AB mean?
- Bupropion HCl ER 200 mg tablets by Watson Labs are bioequivalent to … ?
- Bupropion HCL ER 150 mg tablets by Watson (Application A077715) are *not* bioequivalent to … ?
- To provide a generic equivalent of Zyban 150 mg ER tablets, the pharmacist should dispense … ?

More Orange Book searches can be found in the Questions and Cases section of this chapter.

Biopharmaceutical Classification System

Drug manufacturers expend considerable effort in the development of drug products that are safe and effective. The large number of drug candidates synthesized by medicinal chemists is screened for pharmacological and pharmaceutical properties to focus the industry's resources on those new chemical entities most likely to be active drugs.

- The two properties of biopharmaceutical interest, that is, the properties that predict the ability of the drug candidate to move from dosage form to receptor, are water solubility and membrane permeability.

Table 7.4 Biopharmaceutical Classification System

Class	Solubility	Permeability	Considerations for dosage form development
I	High	High	Absorption not limited by dissolution or permeability; stomach emptying (rate of presentation to the absorbing membrane) may be rate limiting
II	Low	High	Dissolution of drug is rate limiting; absorption can be enhanced by formulation to maximize dissolution rate
III	High	Low	Permeability is rate limiting for drug absorption
IV	Low	Low	Difficult to formulate to provide good absorption

- These two properties, which can be measured *in vitro*, are used to classify drugs into four categories (Table 7.4).
- The theoretical basis for the Biopharmaceutical Classification System (BCS) is a simplified version of Fick's first law, applied to the intestinal membrane:

Rate of flux $= J = P(C)$

where J is the rate of flux in $g/cm^2/s$, P is the permeability in cm/s and C is the concentration of drug at the membrane surface.

Solubility limits the driving force for diffusion across a membrane, since it represents the highest point of the concentration gradient, in other words, concentration = solubility. According to this equation, the absorption rate may be limited either by the water solubility of the drug or by the permeability of the drug.

- Drugs in Class I, that is, those drugs with high water solubility and high membrane permeability, predictably demonstrate good *in vivo* bioavailability.

Key Point

- The *in vivo* absorption of high solubility, high permeability drugs correlates well with *in vitro* dissolution tests.

- Because of the strong correlation of *in vitro* properties with *in vivo* performance, the FDA permits generic drug companies to submit *in vitro* evidence in lieu of testing bioequivalence in human volunteers for drugs with the following properties:
  - *High solubility*, defined as the highest dose strength of the drug, is soluble in ≤250 mL water over a pH range of 1–7.5 at 37°C.
  - High permeability defined as the extent of absorption in humans is determined to be ≥90% of an administered dose based on mass balance or in comparison to an intravenous reference dose.

— Rapid dissolution defined as ≥85% of the labeled amount of drug dissolves within 30 minutes in a volume of ≤900 mL dissolution medium at 37°C.
— Similar dissolution profile to brand product.
— Excipients chosen for the dosage form have been used previously in FDA approved immediate release solid dosage forms.

In addition the FDA specifies that the following classes of drugs *cannot* receive a BCS biowaiver:

- Narrow therapeutic range drug products which include products containing drugs that are subject to therapeutic drug concentration or pharmacodynamic monitoring. Examples of narrow therapeutic range drugs include digoxin, lithium, phenytoin, theophylline and warfarin.
- Products designed to be absorbed in the oral cavity, such as sublingual or buccal tablets.

Water solubility revisited

The definition of high water solubility in the BCS takes into account not only the amount of a drug that dissolves in a specified volume of solvent but also the amount of drug required to produce an effect (highest dose strength available). Thus a drug such as digoxin which is considered very slightly soluble by the USP classification system (0.986 mg/mL at 37°C), is considered highly soluble by the BCS because its highest dose strength is 0.25 mg.

$$\frac{(0.986 \text{ mg digoxin})}{(\text{mL})} \, (250 \text{ mL})$$

= 246.5 mg digoxin should dissolve in 250 mL

Considering the small amount of digoxin required to produce the therapeutic effect, it is a high solubility drug. In all more than 40% of the Top 200 drugs are classified as practically insoluble by the traditional descriptive solubility categories. However, taking into consideration the ability to dissolve the highest dose strength in 250 mL, 55% could be classified as high solubility drugs.

The solubility of drugs at different pH's or at temperatures other than 25°C can be difficult to find in standard references. The effect of pH on solubility can be conservatively estimated by using the solubility of the uncharged form of drug. The water solubility at 37°C has been tabulated for 900 drugs by Benet et al (selected reading 4) or the water solubility at 25°C may be used to calculate a conservative estimate of the solubility of the highest dose strength in 250 mL.

Permeability and human intestinal absorption

> ### Definition
>
> The FDA defines *high permeability* as the extent of absorption in humans greater than or equal to 90%.

There are several approaches that the FDA considers acceptable for the measurement of permeability for a BCS waiver, and there are other surrogate measures of this property that are used by the pharmaceutical industry to predict the permeability of a new chemical entity that may only be available in very small quantities.

It is important to note that while human intestinal permeability is an excellent predictor of fraction of dose absorbed and bioavailability, each of these parameters has a distinct meaning (Figure 7.3). Intestinal permeability is the distance traveled by drug per unit time through the intestinal membrane. Fraction of dose absorbed refers to the portion of the dose given that is absorbed into the intestinal capillaries. Bioavailability is the fraction or percent of the drug available in its active form to the systemic circulation considering losses in the gastrointestinal tract, first pass hepatic metabolism and solubility limited absorption. This distinction means there are many drugs that are highly permeable, able to extensively penetrate the intestinal membrane, but are much less than 90% bioavailable because of extensive metabolism on first pass through the liver.

Permeability data for available drugs is not widely published (Table 7.5). Some scientists will use measured or calculated log partition coefficient (log P or C log P) as a 'surrogate' to estimate the permeability of drugs when data in humans are not available or when during drug development, quantities of the drug are limited. Using metoprolol (log P 1.88, log P_{app} −4.59) as the drug that defines the lowest permeability in the high permeability class, log P

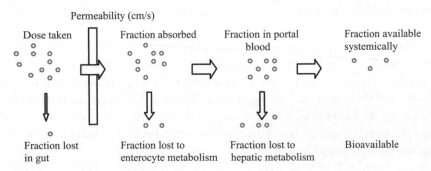

Figure 7.3 Permeability, fraction absorbed and bioavailability. (Based on a PharmPK discussion on March 29, 2010)

Table 7.5 Permeability, fraction absorbed, and bioavailability

Drug	Permeability (log P_{app}, cm/s)	Fraction absorbed (f_a, %)	Bioavailability (F, %)	BCS class
Amoxicillin	−5.83	93	74–92	I
Aspirin	−5.06	84	65–71	III
Atenolol	−6.44	50	45–55	III
Caffeine	−4.41	100	100	I
Chlorothiazide	−6.72	36–61	8–20	IV
Cimetidine	−5.89	64	60–70	III
Diazepam	−4.32	100	85–100	I
Felodipine	−4.64	88	20	II
Hydrochlorothiazide	−6.06	65–72	61	III
Ibuprofen	−4.28	95	80	II
Metoprolol	−4.59	95	50	I
Propranolol	−4.58	99	26–46	I
Theophylline	−5.4	100	80–100	I
Verapamil	−4.70	100	20–35	I

correctly predicted BCS permeability class about two thirds of the time. For drugs that are weak acids and bases log D would be the preferred surrogate. Measured log P and log $D^{7.4}$ values have been tabulated in several articles discussing the BCS (see selected reading 4). The Biopharmaceutical Drug Disposition Classification System (see below) considers percent of dose hepatically metabolized as a surrogate of permeability since the drug is not metabolized by the liver unless it gets absorbed.

Biopharmaceutical Drug Disposition Classification System

The BCS is a scientific framework that predicts the *in vivo* absorption of drugs from water solubility and intestinal permeability data. The Biopharmaceutical Drug Disposition Classification System has extended the prediction of *in vivo* drug behavior to include their routes of elimination, distribution and potential for drug interactions. Developed for industry scientists to predict the disposition of new chemical entities, the BDDCS may also be useful to pharmacists as a framework for understanding the factors that govern drug absorption, distribution, metabolism and elimination. This classification system replaces the high permeability criteria of the BCS with ≥70% hepatic metabolism, a criterion that is more conveniently measured than human intestinal permeability. There is considerable overlap between drugs with sufficient lipophilicity to be highly permeable and drugs that are metabolized to more polar molecules.

Questions and cases

- Go to the FDA Orange Book to learn about the bioequivalence of oral tablets and capsules: http://www.fda.gov/Drugs/ResourcesForYou/HealthProfessionals/default.htm
- Select Approved Drug Products with Therapeutic Equivalence Evaluations (Orange Book)

Questions about bupropion products from the Orange Book website:

1. Bupropion HCl ER 200 mg tablets by Watson Labs are rated:
    (a) AB
    (b) AB_1
    (c) AB_2
    (d) AB_3
2. What does AB mean?
    (a) The product does not present bioequivalence problems for which *in vitro* data have been evaluated
    (b) The product is judged bioequivalent based on *in vivo* data submitted
    (c) The product has documented bioequivalence problems
    (d) The product has potential bioequivalence problems
3. Bupropion HCl ER 200 mg tablets by Watson Labs are bioequivalent to:
    (a) Wellbutrin SR
    (b) Wellbutrin XL
    (c) Wellbutrin
    (d) Zyban
4. Bupropion HCl ER 150 mg tablets by Watson (Application A077715) are *not* bioequivalent to:
    (a) Bupropion HCl ER 150 mg tablets by Mylan (A090942)
    (b) Bupropion HCl ER 150 mg tablets by IMPAX (A077415)
    (c) Zyban 150 mg ER tablets by GSK
    (d) Wellbutrin XL 150 mg tablets by Valeant
5. To provide a generic equivalent of Zyban 150 mg ER tablets, the pharmacist should dispense:
    (a) Bupropion HCl 150 mg ER tabs by Sun Pharma (A078866)
    (b) Bupropion HCl 150 mg ER tabs by Watson (A079095)
    (c) Bupropion HCl 150 mg ER tabs by Watson (A077715)
    (d) Bupropion HCl 150 mg ER tabs by Watson (A079094)

Drugs, Orange Book Search, by active ingredient 'medroxyprogesterone':

6. What product is the reference listed drug for medroxyprogesterone acetate oral tablets?

(a) PremPro

(b) Depo-Provera

(c) Provera

(d) PremPhase

7. Medroxyprogesterone acetate by USL Pharma is rated BP. What does this mean?

(a) The product does not present bioequivalence problems for which *in vitro* data have been evaluated

(b) The product is judged bioequivalent based on *in vivo* data submitted

(c) The product has documented bioequivalence problems

(d) The product has potential bioequivalence problems

Go to Drugs, Drugs@FDA, 'Coumadin', Therapeutic Equivalents:

8. All of the warfarin products in these tables have been evaluated as therapeutic equivalents (bioequivalent) to Coumadin. What is the difference between Coumadin and the warfarin sodium product by Barr? How would you explain this to a patient?

(a) Coumadin contains warfarin potassium and the Barr product is warfarin sodium

(b) Coumadin is a tablet and the Barr product is a capsule

(c) Coumadin contains different inactive ingredients than the Barr warfarin product but is otherwise equivalent

(d) Coumadin tablets release drug at a different rate and to a different extent than the Barr tablets

Drugs, Orange Book Search, by active ingredient 'lithium carbonate':

9. Can lithium carbonate 300 mg capsules by Roxane be used interchangeably with Lithobid 300 mg tablets by Noven Therapeutics? Explain.

(a) Yes, they both contain the same amount of the same chemical entity

(b) Yes, they are both AB rated

(c) No, one is a capsule and one is a tablet

(d) No, one is immediate release and one is extended release

Drugs, Orange Book Search, by active ingredient 'metoprolol':

10. Can metoprolol 50 mg tablets by IPCA Labs (A078459) be used interchangeably with Toprol XL 50 mg tablets by Astra Zeneca? Explain.

(a) Yes, they are both AB rated

(b) Yes, they are both 50 mg tablets

(c) No, one is a succinate and the other is a tartrate

(d) No, one is extended release and the other is immediate release

Drugs, Orange Book Search, by active ingredient 'diltiazem':

11. What product(s) can be used interchangeably with Cardizem CD?

(a) Dilt-CD and Tiazac

   (b)   Dilt-CD and diltiazem HCl by Valeant (A075116)

   (c)   Tiazac and Diltzac

   (d)   Dilacor XR and Cartia XT

Case example
Drugs, Orange Book Search, by active ingredient 'levothyroxine':

12. If you are the pharmacist at a hospital where patients may be admitted on one of several different brands of levothyroxine, which product would you choose for the formulary? Explain.

Use of the Biopharmaceutical Classification System (BCS)

13. For the following drugs, determine whether each would be considered water soluble by the USP definition or highly soluble by the BCS classification and by the USP criteria (≥33 mg/mL).

Drug	Highest strength oral dosage form	Water solubility	Water soluble by USP criteria	Water soluble by BCS criteria
Prednisone	50 mg	312 mg/L		
Levonorgestrel	1.5 mg	2.05 mg/L		
Griseofulvin	500 mg	8.64 mg/L		
Simvastatin	80 mg	0.76 mg/L		
Digoxin	0.25 mg	0.0648 mg/mL		

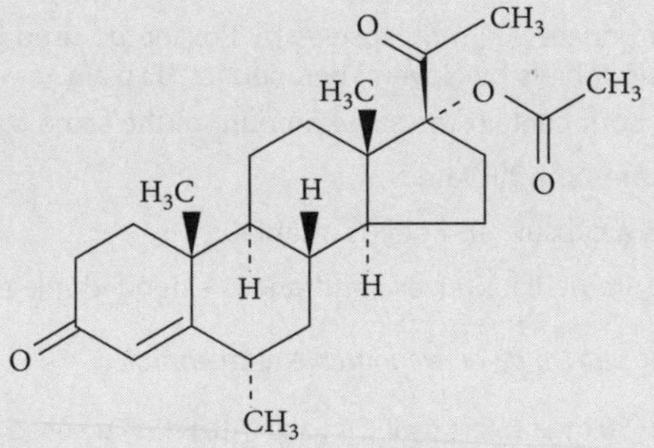

Medroxyprogesterone

14. Medroxyprogesterone acetate:

- Synthetic oral progestin
- Available as 5 and 10 mg tablets
- Solubility of 22.2 mg/L water
- Dissolution rate of ≥50% of labeled amount in 45 minutes
- Bioavailability orally (high first pass) 10%

- Log P 3.5

- MW 385.5

- Generic medroxyprogesterone acetate tablet contains lactose, magnesium stearate, methylcelluose, microcrystalline cellulose, D&C red no. 30, D&C Yellow no 10, and talc.

- Provera by UpJohn contains calcium stearate, corn starch, lactose, mineral oil, sorbic acid, sucrose, and talc.

Does the highest dose of this drug dissolve in 250 mL water? Will the solubility be different at pH 1.2, 4.5 or 6.5?

Is this drug likely to have good permeability? What properties should you evaluate to determine this if you do not have human intestinal permeability data?
This drug is 10% bioavailable orally. How do you reconcile this with your prediction of the drug's permeability? Explain how this is different from percent absorbed orally.

In what BCS classification would you place this drug? Which of its properties is likely to limit its absorption?

Should the FDA require *in vivo* bioequivalence data for this drug?

Acyclovir

15. Acyclovir
   - Antiviral drug
   - Water solubility 2.5 mg/mL
   - Available as 200 mg capsule, 400 and 800 mg tablets
   - Log P −1.56, log D 7.4−1.76
   - pK_{a1} 2.27 (amine), pK_{a2} 9.25 (enol)
   - Log P_{app} −6.15 cm/s
   - Bioavailability 10−20%
   - Fraction absorbed 0.2
   - MW 225

Does the highest dose of this drug dissolve in 250 mL water? Will the solubility be different at pH 1.2, 4.5 or 6.5?

Is this drug likely to have good permeability?

Permeability may be assessed using bioavailability, fraction absorbed, log P, log D, P_{eff} from human or animal experiments, and P_{app}. Rank these measures of permeability in terms of their accuracy (ability to predict the true value).

In what BCS classification would you place this drug? Which of its properties is likely to limit its absorption?

$$
\begin{array}{c}
\text{CH}_3 \\
|\\
\end{array}
$$

Metformin

In what BDDCS classification would you place this drug? How is it likely to be eliminated?

16. Metformin
    - Glucose lowering agent
    - Log P −0.5, log D 7.4–5.41
    - pK_a 12.4
    - Water solubility 500 mg/mL (HCl salt)
    - Highest dose available: 1000 mg tablet
    - Fraction absorbed 0.53
    - Bioavailability 52%
    - MW 129

Does the highest dose of this drug dissolve in 250 mL water? Will the solubility be different at pH 1.2, 4.5 or 6.5?

Is this drug likely to have good permeability?

In what BCS classification would you place this drug? Which of its properties is likely to limit its absorption?

In what BDDCS classification would you place this drug? How is it likely to be eliminated?

17. The pharmacy buys a generic drug tablet from a new manufacturer that has an area under the curve (AUC) that is twice the AUC of the product previously stocked. Which of the following would you expect to observe in your patients who use the new product?
    (a) Symptoms will be relieved more rapidly
    (b) Symptoms will be relieved more slowly
    (c) Side effects may occur due to higher drug levels
    (d) Therapeutic failure may occur due to lower drug levels

18. The pharmacokinetic parameter that tells the pharmacist how quickly a drug is absorbed is:
    (a) C_{max}
    (b) T_{max}
    (c) Area under the curve
    (d) Half life

Selected reading

1. US Food and Drug Administration Center for Drug Evaluation and Research, http://www.fda.gov/ForHealthProfessionals/Drugs/default.htm.
2. GL Amidon, H Lennernas, VP Shah et al. A theoretical basis for a Biopharmaceutics Drug Classification: the correction of in vitro drug product dissolution and in vivo bioavailability, Pharm Res 1995; 12: 413–420.
3. H Lennernas. Human intestinal permeability, J Pharm Sci 1998; 87:4: 403–410.
4. LZ Benet, F Broccatelli, TI Oprea. BDDCS applied to over 900 drugs. AAPS J 2011; 13: 519–547.
5. CY Wu, LZ Benet. Predicting drug disposition via application of BCS: transport/absorption/ elimination interplay and development of a biopharmaceutic drug disposition classification system. Pharm Res 2005; 22: 11–23.
6. MVS Varma, S Khandavilli, Y Ashokraj et al. Biopharmaceutic Classification System: A scientific framework for pharmacokinetic optimization in drug research, Curr Drug Metabol 2004; 5: 375–388.

8

Parenteral drug delivery

Are you ready for this unit?

- Review isotonicity, osmotic pressure, cosolvents, surfactants and amorphous solids.
- Review suspensions and emulsions.
- Review the instabilities of organic functional groups.
- Review the organic functional groups that are weak acids or bases and whether a drug's solubility will be reduced by raising or lowering the pH of the solution.

Learning objectives

Upon completion of this chapter, you should be able to:

- Define phlebitis, hemolysis, extravasation, sterility, lyophilization, pyrogen, ISO 5 air quality, large volume parenteral, small volume parenteral, cleanroom, horizontal laminar flow hood, aseptic technique and buffer zone.
- Identify the parenteral routes that require isotonicity and explain the relevance of isotonicity to parenteral product safety and comfort.
- Distinguish between the parenteral routes and identify the appropriate route(s) of administration for the following dosage forms: aqueous solutions, cosolvent solutions, surfactant solubilized solutions, powders for reconstitution (solutions and suspensions), oily solutions, suspensions and emulsions.
- List the categories of ingredients that are added to parenteral products, and identify how the ingredient contributes to the manageable size of the dosage form, its palatability or comfort, its stability (chemical, microbial, physical), the convenience of its use and the release of drug from the dosage form.
- Identify cosolvents and surfactants used in parenteral products to maintain the solubility of drugs and state the appropriate handling of these products to prevent precipitation.
- Given a formulation identify what class of parenteral dosage form it represents and whether it is immediate or extended release.
- State the purpose of the methods used to process sterile products: distillation, air filtration, lyophilization and sterilization.
- Given a proposed sterile product preparation determine whether the procedure is low, medium or high risk by USP definitions.
- Locate information on the testing and training required for personnel involved in the preparation of compounded sterile products.
- Describe the testing of finished compounded sterile products.
- Given two drugs' structures and formulations, predict potential physical, chemical and packaging incompatibilities when combined in the same IV bag, syringe or Y site.

- Identify antimicrobial preservatives used in parenteral products and state the appropriate handling and beyond-use date for multidose vials.
- Advise patients and other healthcare professionals on proper injection procedures, medication storage and expiration dating.

Introduction

The term parenteral means the delivery of drugs by routes other than enteral, i.e. other than the gastrointestinal tract. In practice we use the term to describe drugs that are injected, a delivery method that bypasses the body's protective chemistry in the gastrointestinal tract and the low permeability barrier of the skin. The preparation and handling of parenteral products requires extraordinary care to maintain the potency, microbial and particulate standards assigned to these drugs.

Characteristics of parenteral administration routes

The most common routes of parenteral administration are the intravenous, intramuscular and subcutaneous routes. We will also briefly examine the intradermal, intra-arterial, intrathecal, epidural and intra-articular routes in terms of volume and comfort constraints, dosage forms used and typical time of onset (see Table 8.1 and Appendix Table A.1).

Intravenous route

Intravenous injections are placed in a vein, usually a smaller, peripheral vein if the drug therapy is short term and non-irritating, and a larger, central vein for hypertonic drugs or long term drug therapy.

- Peripheral veins include any vein that is not in the chest or abdomen but most generally means the cephalic or median cubital.
- Central veins must be accessed by tunneling under the skin or surgically placing sterile tubing into a smaller vein and threading it into the superior vena cava.
- The intravenous route produces rapid distribution and onset of action since there is no absorption step.
- The effects of drugs administered intravenously are more predictable than other routes, allowing doses to be fine-tuned according to the patient's response.

Table 8.1 Dosage forms for parenteral use

Route	Volume constraints	Comfort constraints	Dosage forms	Onset
Intravenous – into a vein	1–1000 mL	pH 3–11	Aqueous solutions	Immediate
Bolus		Osmolarity 215–900 mosmol/L	Emulsions	10–20 minutes
Infusion		Particle free	Liposomes	
Intramuscular	0.2–2.5 mL	Preferably isotonic, pH 3–9	Aqueous solutions	15–30 minutes
Deltoid			Emulsions	
Vastus lateralis			Oily solutions	24–48 hours
Gluteus			Suspensions	2–4 hours
Subcutaneous	0.2–2 mL	Preferably isotonic, pH 3–7		
Upper arm			Aqueous solutions	15 to 60 minutes
Abdomen			Suspensions	1–4 hours
Lateral thigh			Implants	1–4 hours
Intradermal – between dermis and epidermis	0.05–0.1 mL	Isotonic	Aqueous solutions, suspensions	Immediate
Intra-arterial – into an artery	2–20 mL	Preferably isotonic; particle free	Aqueous solutions	Immediate
Epidural – epidural space	6–30 mL	Isotonic, preservative free; pH 7.4	Aqueous solutions, suspensions	15–30 minutes
Intrathecal – subarachnoid space	1–4 mL	Isotonic; preservative free; pH 7.4	Aqueous solutions, suspensions	15–30 minutes
Intra-articular – into a joint	2–20 mL	Isotonic	Aqueous solutions, suspensions	3–4 hours

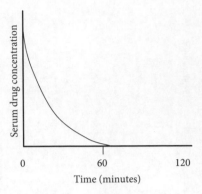

Figure 8.1 Serum concentration versus time curve for an intravenous bolus injection. The drug solution is placed in a syringe and administered over a short period of time (minutes) directly into a vein or intravenous tubing that goes into a vein. Note that the IV push injection has a very rapid peak and then typically declines exponentially

- Drugs administered via this route must traverse the comparatively permeable endothelial membrane of the capillary system in the drug's target tissue but have no absorption step because they are placed directly into the vascular system.
- It is technically difficult to administer medications by the intravenous route and continued competency requires frequent training in the use of new devices.
- There are several hazards associated with parenteral therapy:
  - Caution must be taken to prevent introduction of air into intravenous products and administration devices because the injection of a few milliliters of air into a vessel can obstruct flow and seriously harm the patient.
  - Certain devices such as catheters and in-dwelling needles are prone to clot formation that, if dislodged, can be harmful to the patient.
  - A blood vessel may be irritated by hypertonic solutions or extreme pH resulting in an inflammatory condition called *phlebitis*.
  - *Hemolysis* may occur when hypotonic solutions or membrane active drugs such as amphotericin are administered.
  - If the drug solution is allowed to leak around the vessel where it is not rapidly carried away, a problem called *extravasation*, it can cause serious damage, particularly if the drug is cytotoxic.
- There are three methods of injection into a vein:
  - The direct intravenous injection, also known as IV bolus or IV push, provides the most rapid drug levels, administering a relatively small volume of drug solution over less than 5 minutes (Figure 8.1).
  - The intermittent infusion administers 50 to 200 mL over 20 to 60 minutes, at evenly spaced intervals if the dose is to be repeated (Figure 8.2).

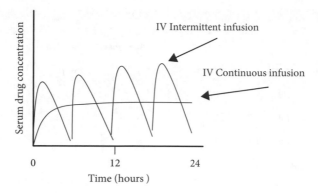

Figure 8.2 Intravenous infusions

- Continuous infusions are 250 to 1000 mL parenterals that are slowly infused into a vein over 8 to 24 hours. This method of infusion produces constant blood levels of the drug, reduces irritation to veins because drug solutions are more dilute and usually isotonic, but requires frequent monitoring by nursing staff and cannot be used on fluid restricted patients (see Figure 8.2).

Because particles larger than 5 μm will be trapped by the pulmonary capillaries, drugs administered intravenously must be aqueous solutions or oil-in-water emulsions. The USP sets standards for the number of subvisible particles large and small volume parenteral solutions may contain. The diameter of droplets in IV emulsions is maintained between 0.2 and 0.5 μm. The body fluids have an osmotic pressure of approximately 300 mosmol/L. The American Society for Parenteral and Enteral Nutrition guidelines recommend a maximum osmolarity of 900 mosmol/L for peripherally infused parenteral nutrition. However, phlebitis is common in patients receiving solutions via peripheral vessels with an osmolarity in excess of 600 mosmol/L. Small volumes can be hypotonic or hypertonic without causing significant irritation, but use of central veins is preferred for large volumes of hypertonic solution, such as parenteral nutrition solutions.

Intramuscular injection

- Intramuscular injections are administered into three of the larger muscle groups, taking care to avoid any vessels or nerves that may be nearby.
  - the deltoid in the upper arm
  - the lateral thigh muscles (vastus lateralis)
  - the gluteal muscles of the buttock.
- The drug mixes with or dissolves in the interstitial fluid and is absorbed into the blood supply of the muscle before it begins to distribute.

Table 8.2 Guidelines for maximum amounts (mL) to be injected intramuscularly

Muscle group	Age (years)				
	0–1½	1½–3	3–6	6–15	>15
Deltoid	NR	NR	0.5	0.5	1–2
Gluteal	NR	NR	1.5	1.5–2	2.5–5
Vastus Lat.	0.5–1	1	1.5	1.5–2	2–2.5

NR = not recommended.
Adapted from LV Howry, RM Bindler, Y Tso. Pediatric Medications, Philadelphia, PA: Lippincott, 1981 and A Gray, J Wright, V Goodey, et al. Injectable Drugs Guide, London: Pharmaceutical Press, 2011.

- Intramuscular injections are technically easier to administer than intravenous and the formulation can be varied to modify onset and duration.
- A number of drugs have been prepared with oils or as aqueous suspensions in what are called depot dosage forms.
- The volume of drug solution that may be injected in any one muscle is limited to 5 mL and is lower for children (Table 8.2).

Subcutaneous injection

- Subcutaneous injections are placed into the tissue just under the skin, between the dermis and the muscle. The drug mixes with or dissolves in the interstitial fluid and is absorbed via the subcutaneous blood supply.
- The subcutaneous tissue is largely adipose tissue and therefore somewhat poorly perfused.
- The areas most commonly used for subcutaneous injections are the thigh, the upper arm and the abdomen.
- Subcutaneous injections are technically easier than intravenous or intramuscular injections and this route is chosen most often for parenteral drugs that will be used at home.
- Like intramuscular injections, the formulation can be varied for rapid onset (aqueous solutions) or longer duration (suspensions and implants). Suspensions produce continued absorption from the injection site (depot dosage forms) and can maintain blood levels for hours to days.

Intradermal administration

- Intradermal injections are administered into the skin just beneath the epidermis. The procedure requires a very small volume (0.1 mL) and if used for diagnostic purposes, the solution must be isotonic.
- The route is used primarily for local effects, vaccines and administering antigens for allergic reaction testing.

Intra-arterial administration

- An intra-arterial injection is placed into an artery. This route can be used for regional delivery of drugs when it is desirable to limit the distribution of the therapy to one organ or body area (cancer chemotherapy).
- Administration of an intra-arterial injection is technically difficult, prone to embolism, arterial occlusion and local drug toxicity.
- The formulations administered in this fashion include solutions and emulsions.

Injection into the central nervous system

- The epidural or intrathecal routes are used for the treatment of central nervous system infections, for inducing spinal anesthesia and for relieving chronic pain.
  - An epidural injection is placed above the dura mater.
  - An intrathecal or spinal injection is placed into the subarachnoid space.
- Injection into the central nervous system requires careful attention to formulation constraints.
  - It must be isotonic, pH 7.4 with no preservatives and, as always, sterile and pyrogen free.

> *Definition*
>
> *Pyrogens* are bacterial debris that elicit an immune response (fever, chills) if injected.

Intra-articular administration

- An intra-articular injection is placed into the joint space, and is used for treatment of joint pain and inflammation.
- To reduce the possibility of increased inflammation of the tissue, intra-articular dosage forms must be isotonic as well as sterile and pyrogen free.

Drug properties

- Good water solubility is essential for the preparation of immediate release, aqueous solutions for injection.
- Although drugs administered parenterally will need to move through capillary membranes to reach target areas, these membranes are

permeable to relatively large molecular weight molecules, such as proteins, as well as drug molecules with a broad range of polarity.

- Parenteral routes may be used for drugs that would be degraded by enzymes or by extremes of pH if administered by the oral route.

Parenteral dosage forms

Key Points

- Parenteral products must be sterile and pyrogen free.
- Aqueous solution dosage forms may be administered by any route.
- Suspensions may be administered by any route but intravenously. Oily solutions are only administered intramuscularly.
- Cosolvent and surfactant solutions may precipitate when mixed with other injections or if injected intravenously too rapidly.
- The release of drugs from parenteral dosage forms can be prolonged by the preparation of suspension or oily solution dosage forms.
- Devices are essential for the convenient use of parenteral products.

Design of parenteral dosage forms

1. One dose in a manageable size unit
   - Good water solubility is essential for the preparation of a drug dose in a manageable volume for intramuscular or subcutaneous administration. The intravenous route permits drug doses in a large range of volumes.
2. Comfort
   - Tonicity agents enhance the comfort of injected solutions and may be required for dosage forms administered parenterally.
   - Because of the rapid dilution of drug solutions administered intravenously, this route tolerates a wider range of pH and osmolarity than any other.
3. Microbial stability

Definition

Sterility is defined as the complete absence of viable microorganisms.

- Parenteral products must be sterile to prevent the introduction of microbes into the blood stream or interstitium.
- Parenteral products must be pyrogen free.
- Drug products that will be used to withdraw more than one dose must be preserved to prevent growth of microorganisms that may be introduced upon entry of the drug container.

4. Chemical stability
   - Drugs with chemical stability problems may be adjusted to an appropriate pH using buffers or strong acid or base.
   - Many unstable drugs are prepared as powders for reconstitution. The term reconstitution is used because these dosage forms are prepared as solutions, sterilized by filtration and freeze dried to remove the water to provide for extended shelf life. The protectants are ingredients used to prevent freeze concentration of drugs during lyophilization. Human serum albumin may be added to provide protection against adsorptive loss of proteins on packaging and administration materials.

5. Convenience and ease of use
   - Parenteral devices provide control over the release of parenterally administered drugs into the body, and provide caregivers and patients with greater convenience in the use of drugs that must be given by these routes (Tables 8.3 and 8.4).

6. Release of drug
   - The release of drugs from parenteral dosage forms can be prolonged by the preparation of suspension or oily solution dosage forms and by the use of devices such as infusion pumps.
   - Oily solutions are formulated in sterile vegetable oils which are not miscible with body fluids. This requires the drug to partition into body water and slows absorption.

Pharmaceutical example

Insulin products illustrate the use of solubility and dissolution rate to modify the onset and duration of drug activity (Figure 8.3, Table 8.5).

Release characteristics

The USP lists five labeling conventions for parenteral products (Table 8.6):

- injections that are solution dosage forms
- drugs for injection that are reconstituted into solution dosage forms with an appropriate vehicle
- injectable emulsions that are dispersions of drug in oil with aqueous continuous phase
- injectable suspensions
- drugs for injectable suspension that are reconstituted into suspension dosage forms with an appropriate vehicle.

The term large volume parenteral (LVP) applies to single dose injections intended for intravenous use and packaged in containers in volumes greater than 100 mL. Small volume parenterals (SVPs) are injections packaged in containers in volumes of 100 mL or less.

Table 8.3 Parenteral drug delivery devices for outpatient use

Device	Description	Examples
Prefilled syringes or pens	Device for injecting a fixed or adjustable dose of medicinal product from a prefilled syringe through a needle	Avonex prefilled syringe FlexPen (Novolog) Orencia SC (abatacept)
Auto injectors	Device that uses a spring loaded mechanism to inject drug liquid from a prefilled syringe to a predetermined depth below the skin surface using a needle. May be reusable or disposable.	EpiPen (epinephrine) Avonex Pen (interferon beta-1a) Rebismart (interferon beta-1a)
Jet (needleless) injectors	Gives an injection by means of a narrow high velocity fluid jet that penetrates the skin and delivers medicine or vaccine	Pharmajet, Cool Click 2, Serojet (somatropin in the Cool-Click technology), ZomaJet 2 Vision (somatropin), Bioject (nemifitide)
Infusion pumps – subcutaneous	Portable, programmable infusion pump that infuses drug subcutaneously via a catheter. May be integrated with blood glucose meter	Insulet OmniPod, Accu-Chek Spirit, OmniPod Solo micropump
Elastomeric pumps	Disposable, nonpowered pump with a balloon-like reservoir inside a protective shell and an administration set with flow restrictor system	HomePump, Intermate
Implantable venous access chamber	Sterile port inserted completely under the skin of the chest with attached catheter threaded through the subclavian vein to the superior vena cava. Lower risk of infection than with external catheter	Portacath

Immediate release dosage forms

The classification of parenteral dosage forms into immediate acting and sustained release (depot) dosage forms is useful for understanding how they may be delivered.

- Aqueous solutions are the most common parenteral formulation and can be used by any route. Because the drug is already in solution it is immediately available for movement across membranes to target tissues.
- Chemically unstable drugs prepared as powders for injection are re-dissolved immediately before use with a suitable vehicle to prepare a solution dosage form. As aqueous solutions these dosage forms can be used by any route.

Table 8.4 Devices for parenteral administration to inpatients

Device	Description
Vascular access devices	
Peripheral intravenous cannula	Sterile, plastic tubes with butterfly-like handles to aid insertion. Attached to connectors for infusion sets or injection ports. More comfortable than needles
Central catheter	Sterile, plastic tubing threaded from a smaller vein (usually subclavian) into the superior vena cava for the purpose of infusing prolonged or irritating drug therapy or nutrition solutions
Hickman catheter	Central catheter inserted surgically by tunneling subcutaneously for 5–10 cm before threading through the subclavian vein to the superior vena cava
Peripherally inserted central catheter (PICC line)	Central catheter inserted into a peripheral vein in the arm or leg and threaded into the superior vena cava
Parenteral containers	
Ampule	Sterile container made of and sealed with glass
Vial	Sterile plastic or glass drug container sealed with self-sealing septum which may be entered with a needle without compromising its sterility
Piggy backs	Intermediate volume parenterals that allow the administration of a second drug through a Y site in the same line
Infusion devices	
Gravity (drip) administration set	Connect a large volume parenteral to the patient and control the flow rate of the drug solution with a clamp and drip chamber
Y site	Y shaped port that allows the attachment of a second infusion to the primary infusion line running to the patient
Inline filters	Used to remove particles larger than the labeled pore size from LVP solutions before they reach the patient. Reduce phlebitis, sepsis, can bind drugs
Smart infusion pump	Programmable device that calculates the infusion rate then infuses drugs from a reservoir based on a drug library that specifies drug concentration, patient parameters and standard infusion protocols
Implantable infusion pump	Device surgically implanted into the patient entirely below the skin to provide continuous long-term drug administration via various routes

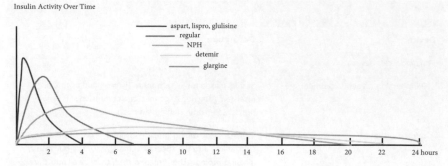

Insulin Activity Over Time

aspart, lispro, glulisine
regular
NPH
detemir
glargine

2   4   6   8   10   12   14   16   18   20   22   24 hours

Figure 8.3 Pharmacokinetics of insulins. (Reproduced with permission from JS Freeman. J Am Osteopath Assoc 2009; 109: 26–36.)

- Cosolvent solutions are formulated with cosolvent–water blends to make solution dosage forms from drugs with poor water solubility. These dosage forms can generally be used intravenously or intramuscularly but must be injected very slowly intravenously because the drug may precipitate as the solution is diluted by aqueous body fluids at the site of injection.
- Drugs in cosolvent solutions may also precipitate when used intramuscularly, and the undissolved drug precipitate may behave like a depot dosage form and be absorbed very slowly. Cosolvent solutions may be quite painful when used intramuscularly.
- Surfactant solubilized solutions contain nonionic surfactants to make solution dosage forms from poorly soluble drugs. These formulations can be used intravenously, subcutaneously or intramuscularly but most often are given intravenously where dilution by blood reduces their irritating properties.
- The surfactant ingredient will reduce the drop size of these solutions, which can result in underdosing if a drip administration set is used rather than a volumetric pump.
- Injectable emulsions are considered immediate acting dosage forms but because the drug is contained in the oil dispersed phase, it must partition into aqueous body fluids before absorption. How quickly a drug partitions out of the oil phase is dependent on the drug's chemistry but the rate can be considered intermediate between solution dosage forms and the sustained release products described below.
- Although few drugs are presently formulated as emulsions, there is strong interest in the formulation of water insoluble drugs in emulsion dosage forms for intravenous administration because they are less irritating to veins than cosolvent or surfactant solubilized solutions.
- Emulsions offer a less toxic alternative to Cremophor (surfactant) solubilized solutions, which can cause bronchospasm, hypotension, nephrotoxicity and anaphylactic reactions.

Table 8.5 Insulin formulation and effect on onset and duration of activity

Insulin; routes	Formulation	Onset (hours)	Duration of activity (hours)
Rapid acting insulins			
Insulin aspart (Novolog) amino acid substitution reduces aggregation; SC	Solution, pH 7–7.8	10–20 minutes	3–5
Insulin lispro (Humalog) amino acid substitution reduces aggregation; SC	Solution, pH 7–7.8	15–30 minutes	3–6.5
Insulin glulisine (Apidra) amino acid substitution reduces aggregation; SC, IV		25 minutes	4–5.3
Short acting insulins			
Regular (Humulin, Novolin); IV, SC	Solution of crystalline zinc human insulin, pH 7.0–7.8	½–1	6–8
Intermediate acting insulins			
Insulin aspart protamine (NPA); SC	Suspension of protamine insulin aspart, pH 7.0–7.8	1–4	12
Insulin lispro protamine (NPL); SC	Suspension of protamine insulin lispro, pH 7.0–7.8	1–4	10
Isophane insulin (NPH); SC	Suspension of protamine human insulin, pH 7.0–7.8	1–4	10–24
Insulin detemir (Levemir), increased aggregation and binding to albumin; SC	Solution of insulin detemir, pH 7.0–7.8	3–4	12–24
Long acting insulins			
Insulin glargine (Lantus); SC	Solution of insulin glargine, pH 4.0. Injection causes drug to form fine precipitate	5	24

SC = subcutaneous; IV = intravenous.

Adapted from *American Hospital Formulary Service Drug Information*, American Society of Health-system Pharmacists, accessed via MedicinesComplete.com at http://www.medicinescomplete.com/mc/ahfs/current and J Allen. Insulins. *Pharmacists Letter* 2003; 19: document 190806.

Table 8.6 Parenteral dosage forms and USP labeling conventions

Dosage form	USP label	Formulation	Release category
Aqueous solutions	Injection	Water for injection, buffers, antioxidants, antimicrobial preservatives, tonicity agents	Immediate
Cosolvent solutions	Injection	Water for injection, alcohol, propylene glycol, liquid polyethylene glycols, antioxidants	Immediate
Surfactant-solubilized solutions	Injection	Water for injection, nonionic surfactant, antimicrobial preservative, tonicity agents	Immediate
Powders for injection (reconstitution)	For injection	Protectants, buffers, tonicity agents	Immediate
Emulsions	Injectable emulsion	Water for injection, soybean or safflower oil, emulsifying agents, tonicity agents, antioxidants, antimicrobial preservatives	Immediate
Suspensions	Injectable suspension	Wetting agents, suspending agents, preservatives, buffers, antioxidants, water for injection *or* vegetable oil	Sustained
Powders for suspension (reconstitution)	For injectable suspension	Protectants, buffers, tonicity agents. Wetting agents, suspending agents, preservatives, buffers, antioxidants	Sustained
Oil-based solutions	Injection	Vegetable oil vehicle, local anesthetic (benzyl alcohol)	Sustained

- Emulsions can be used intravenously or intramuscularly.
- The dispersed phase droplets must be no larger than 0.5 μm because they are not miscible with blood and could otherwise produce small vessel occlusions.

Sustained release dosage forms

Sustained release parenteral dosage forms include suspensions and oily solutions and a number of novel delivery systems such as polymeric gels, liposomes, PEGylation and microspheres. Microspheres, polymeric gels, liposomes and PEGlyated polymers are used for the delivery of proteins, peptides and oligonucelotides and will be considered in Chapter 9.

- Suspensions are versatile dosage forms with greater chemical stability than solutions and the ability to release drug as it dissolves over time from a depot of injected drug particles.
- Suspensions are disperse systems with undissolved drug in a sterile aqueous or oily vehicle.
  - They must resuspend easily with gentle agitation.
  - The solid drug must be of a size that will draw through a needle.
- Suspensions can be used by intramuscular, subcutaneous, intra-articular, epidural and intrathecal routes.
- Drugs for injectable suspension are not sufficiently stable in water so they are prepared as powders to be suspended immediately before use in a suitable vehicle.
- Drugs that are formulated as solutions in oil have often been derivatized to decrease their water solubility.
- They are formulated in sterile vegetable oils, which are not miscible with body fluids. This requires the drug to partition into body water and slows absorption.
- Oily solution formulations are considered too irritating to use subcutaneously and are only used intramuscularly.

Pharmaceutical example

Testosterone cypionate, a lipid soluble derivative of testosterone is formulated in cottonseed oil and produces therapeutic levels for 2–4 weeks.

Preparation and handling of parenteral products

Key point

The sterility of parenteral products is ensured by a combination of properly trained personnel, use of appropriate equipment, supplies and procedures, in adequately maintained facilities.

Industrial processes

Parenteral products manufactured by the pharmaceutical industry are prepared with the highest quality ingredients using carefully developed and monitored protocols. Water for injection is distilled to remove pyrogens and particulate, and containers and closures are sterilized by moist or dry heat, irradiation or gas. Although the USP recognizes five sterilization methods, autoclaving and sterile filtration are used for the vast majority of drug formulations (Table 8.7). When stability allows, the drug formulation in its

Table 8.7 Common sterilization procedures

Method	Moist heat (autoclaving)	Sterile filtration
Description	121°C × 20 minutes (6.8 kg pressure)	Filtration through 0.22 µm filter (cellulose acetate, nylon, polyvinyl chloride, polycarbonate, Teflon)
Used for	heat stable drugs, aqueous solutions of heat stable drugs, aqueous vehicles, rubber closures, filters	Aqueous and nonaqueous solutions of nonheat stable drugs
Lethality	Spores, vegetative bacteria, fungi, viruses by heat denaturation of critical molecules	Removes particles exceeding pore size from solutions. Occasionally a 0.5 µm spore gets through. Viruses and endotoxins would get through
Validation	*Bacillus stearothermophillus* spores	*Brevundimonas (Pseudomonas) diminuta* at 107 (this practice currently under review), bubble point test

final container is terminally sterilized by moist heat using a carefully validated procedure that is specifically developed for the product. To validate a sterilization procedure the proposed formulation is inoculated with the microorganism most resistant to that method (for autoclaving, *Bacillus stearothermophilus* spores) and the product is sterilized until there are no viable organisms. Based on the time required to bring the product to zero viable organisms, a sterilization time is calculated for the formulation that would kill 10^6 more organisms than were present in the test vial. Often drugs are not sufficiently heat stable to be sterilized with autoclaving and must be filtered through sterilizing filters and metered into the sterile vials under aseptic conditions. Drugs that are not chemically stable in solution must be further processed by freeze drying (*lyophilization*), a process in which the water is removed as a gas while the aqueous formulation is frozen. Air filtration is used to remove airborne contaminants from areas where sterile products are prepared. Parenteral products that are not terminally sterilized must be aseptically processed in cleanrooms with ISO 5 air quality, an environment that contains no more than 100 particles per cubic foot of 0.5 µm and larger in size.

Preparation of sterile products in the pharmacy

Parenteral products manufactured in the industrial setting are in many cases ready to use. However, a significant number of medication orders for parenteral products require some compounding: dissolution of lyophilized powders, dilution of drug doses for infusion, mixing of dextrose, amino acids, vitamins and electrolytes for parenteral nutrition. Pharmacists have a professional responsibility to ensure that parenteral products compounded for individual patients retain their potency and sterility, and in order to provide for uniform practices in pharmacies nationwide, Chapter <797> of the *United States Pharmacopeia* provides detailed descriptions of the procedures, facilities and training required. The USP places aseptic compounding into risk categories for the purpose of defining the facilities or equipment, level of environmental monitoring and personnel training in each category. The risk categories are defined in Table 8.8.

- The compounding pharmacist is responsible for ensuring that sterile products are prepared by properly trained personnel, using appropriate equipment, supplies and procedures, in adequately maintained facilities.
- Contamination may be introduced from the environment, equipment and supplies, or personnel involved in the procedure.
- Sterile products must be prepared in an environment that provides ISO 5 air quality.
  - A *cleanroom* is a facility with filtered air and positive pressure to prevent introduction of contaminated air when a door is opened.
  - A *laminar flow hood* (Figure 8.4) is equipped with a high efficiency particulate air (HEPA) filter that provides ISO 5 air quality.
  - The air quality required for performing aseptic procedures is maintained by severely restricting the movement of personnel and supplies in the area that houses the hood and by following a prescribed cleaning and maintenance schedule.

Definition

Aseptic technique refers to carrying out a procedure under a controlled environment in a manner that minimizes the chance of contamination by microorganisms.

- Sterile, disposable syringes, needles and other supplies are used in all sterile procedures.
- Chapter <797> in the *United States Pharmacopeia* provides detailed information on these aspects of sterile product preparation.

Quality assurance

Quality assurance in a sterile products service includes assessment of air quality in the facility, the ability of staff to perform sterile transfers and the quality of the compounded sterile products prepared (Table 8.9).

Table 8.8 Compounding risk levels

Risk Level	Description	Facilities requirements
Low risk, Immediate use	Finished product is compounded with commercially available sterile components and involves measuring or mixing three or fewer components	Compounding occurs in a patient room, the emergency room or diagnostic area in response to an immediate patient need. Compounding does not exceed 1 hour
Low risk with buffer zone	Finished product is compounded with commercially available sterile components and involves measuring or mixing three or fewer components	Manipulations are entirely within an ISO 5 environment, with ISO 7 or better buffer area and ISO 8 or better ante area
Low risk, no buffer zone	Finished product is compounded with commercially available sterile components and involves measuring or mixing three or fewer components	Manipulations are entirely within an ISO 5 environment but there is no ISO 7 or better buffer area. There must be a segregated compounding area
Medium risk	Prepared from commercially available sterile products under one or more of the following conditions: (1) pooling of sterile drug products to prepare a compounded sterile product that will be administered to multiple patients or to one patient on multiple occasions; (2) complex, aseptic manipulations (other than single volume transfer) or lengthy compounding time	Manipulations are entirely within an ISO 5 environment, with ISO 7 or better buffer area and ISO 8 ante area
High risk	(1) Preparation of sterile products from nonsterile drug substances; (2) preparation of sterile products from ingredients that are exposed to air quality inferior to ISO 5 for more than 1 hour; (3) compounding personnel are improperly garbed or gloved; (4) non-sterile water-containing preparations are stored for more than 6 hours before sterilization	Manipulations are entirely within an ISO 5 environment, with ISO 7 or better buffer area and ISO 8 or better ante area. Buffer area should be a separate cleanroom

Data from USP Convention. United States Pharmacopeia, 34th edn, and National Formulary, 29th edn. Baltimore, MD: United Book Press, 2011.

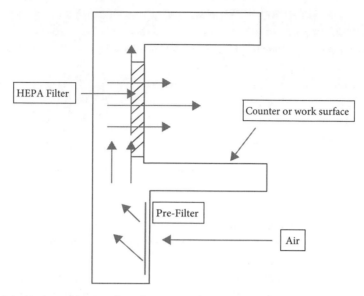

Figure 8.4 Horizontal laminar flow clean air workstation. (Reproduced with permission from L Felton (ed.), Remington: Essentials of Pharmaceutics. p. 534; published by Pharmaceutical Press, 2012.)

Table 8.9 Testing of finished compounded sterile products (CSPs)	
Test	**Description**
Visual inspection	Inspect solution against light and dark, for precipitation, cloudiness or leaks; check container or closure for defects
Ingredient verification	Inspect additive containers, syringes pulled back to volumes added for correct ingredients and amounts against original order
Volume checks	Weigh the CSP and determine accuracy of volume from density or specific gravity of ingredients
Sterility testing	Plate samples of high risk CSP on growth media, incubate and examine for growth
Bacterial endotoxin testing	Bacterial endotoxins in a sample of high risk CSP cannot exceed a threshold pyrogenic dose as measured by the limulus amoebocyte lysate test

Parenteral stability/compatibility

Beyond-use dates for parenteral products

To determine an appropriate beyond-use date for a sterile product, the pharmacist must consider both the microbial stability and the chemical

Table 8.10 Standard beyond-use dates for compounded sterile products			
USP temperature ranges	Low-risk CSP	Medium-risk CSP	High-risk CSP
Controlled room temperature	48 hours	30 hours	24 hours
Refrigerated	14 days	9 days	3 days
Frozen	45 days	45 days	45 days

Reproduced with permission from L Felton (ed.) Remington: Essentials of Pharmaceutics. Pharmaceutical Press, London, 2012.

stability of the product. The sterility of products is maintained by a combination of appropriately trained personnel, a properly maintained environment (hoods and clean rooms) and conservative policies in beyond-use dating of pharmacy prepared parenterals. The USP has developed standard beyond-use dates that may be used if the pharmacy prepares a compounded sterile product in a way that deviates from the manufacturer's labeling (Table 8.10). When a compounded sterile product is prepared according to the manufacturer's instructions, the beyond-use date provided in the labeling may be applied. In addition a published beyond-use date may be applied if the compounder has prepared the sterile product with the same ingredients and concentrations in the published report.

Multidose vials contain a preservative to inhibit growth in vials that may become contaminated during handling. Hospital policies require that vials be dated when opened and discarded after a set period of time. Single dose vials contain no preservative and must be discarded after the dose is withdrawn.

Pharmacists are often confronted by the need to provide multiple parenteral medications to the same patient, such that we are motivated to prepare drugs in solutions and combinations not addressed by the manufacturer's labeling. Pharmacists need to be aware of the kinds of incompatibilities that can occur when we combine two drugs in the same admixture bag, syringe or run them through the same IV tubing and how to prevent them (Table 8.11). Useful references on parenteral admixture compatibility include *King's Guide to Parenteral Admixtures* and *Trissel's Handbook on Injectable Drugs*.

Types of incompatibilities

Key Points

- Drugs that are weak acids in uncharged form will be more soluble at basic pH.
- Drugs that are weak bases in uncharged form will be more soluble at acidic pH.

Table 8.11 Compatibility problems and their management

Formulation	Description of problem	Management
Weakly acidic drug	Will precipitate as the solution becomes more acidic	Calculate pH of precipitation
Weakly basic drug	Will precipitate as the solution becomes more basic	Calculate pH of precipitation
Cosolvents	Will precipitate if diluted with water	Read labeling for recommendations on how to dilute. Advise nurses to inject slowly if used intravenously
Surfactants	Will precipitate if diluted with water	Read labeling for recommendations on how to dilute. Advise nurses to inject slowly if used intravenously
Powders for reconstitution	Unstable in solution	Do not mix with solutions with pHs significantly different than the pH of maximum stability or the pH of the formulation unless literature supports
Carbonate buffers	Generate carbon dioxide when mixed with acidic drugs	Avoid mixing with acidic drugs
Drugs solubilized by cosolvents or surfactants	Sorb to polyvinyl chloride containers and administration sets	Use polypropylene or polyethylene containers and administration sets
Protein drugs	Sorb to plastics and glass surfaces	Use human serum albumin to saturate binding sites on the plastic or glass surface

Incompatibilities can include physical changes in the appearance of a drug product, acceleration of the chemical degradation of a drug, leaching of packaging components into the drug solution and loss of drug onto the packaging material.

Physical changes

Key Points

- Always check pharmacy prepared products visually before they go up to the floor.
- Ask nurses to check again before administering.

- Precipitate or cloudiness in a drug solution indicates a change in the solubility of one of the components.
- Precipitates related to different pH requirements for drug solubility:

- Drugs that are weak acids in their uncharged form will be more soluble at basic pH, and drugs that are weak bases in their uncharged form are more soluble at acidic pH.
- When drugs with different pH solubility requirements are combined, the solubility of both drugs is reduced, at times to the point of precipitation.
- To prevent pH related incompatibilities, the pharmacist must check the pH of drug solutions proposed for admixture and review appropriate literature sources before mixing.
- If data are not available on the compatibility of the proposed mixture of a weak acid and weak base, the pharmacist can estimate the pH of precipitation using the drugs' concentrations, water solubilities and pK_a as illustrated in Chapter 4.

- Precipitates related to dilution of drugs solubilized by cosolvents:
  - When the concentration of cosolvent required to keep a drug in solution is reduced by dilution with an aqueous injection, the cosolvent solubilized drug will cloud the solution or form frank precipitates.
  - The pharmacist should not mix cosolvent solubilized drugs with drugs formulated in aqueous solutions. Often the labeling of drugs in cosolvents will specify that at some significantly more dilute concentration, you will reach the drug's water solubility and thus you may be able to administer some of these difficult medications in a large volume parenteral.
- Precipitates related to dilution of drugs solubilized by surfactants:
  - Dilution with a drug formulated in an aqueous vehicle may drop the level of surfactant below its effective concentration and the poorly soluble drug will precipitate.

Pharmaceutical example

Amiodarone formulated with 10% polysorbate 80 must be diluted with a compatible vehicle to 1–6 mg/mL before infusion. Dilutions to concentrations both higher and lower may result in precipitation.

- Thus far there are few reports of incompatibilities with drugs solubilized with cyclodextrins (aripiprazole and voriconazole injections).
- The combination of two electrolytes has the potential to result in precipitation due to the formation of insoluble salts. Carbonate and phosphate salts are much less soluble than chloride or acetate salts of the same cations.
- The most common example of this is the addition of phosphate injection to a solution containing soluble calcium salts such as calcium chloride. Calcium phosphate is formed and it is much less soluble than calcium chloride or potassium phosphate.

- It is important to note that any precipitate may be identified if the pharmacy staff examine the products they prepare so be sure to *look at all admixtures before you send them to the floor!*
- Evolution of gas:
  - Gas formation results from the reaction of acidic drugs with bicarbonate or carbonate releasing carbon dioxide.

Pharmaceutical example

Drugs affected include sodium bicarbonate injection, various cephalosporins that are buffered with carbonate, and various barbiturates that are buffered with carbonate. Mixing of these carbonate containing solutions with acidic drugs should be avoided to prevent the loss of the carbonate buffer.

Chemical incompatibilities

Key Points

- Drugs that are prepared as powders for reconstitution have stability problems in solution.
- As a general rule, an injection that is not a powder for reconstitution will be most stable at the pH at which it has been formulated.
- An entry in a reference that describes an admixture combination as physically compatible means that there are no observable changes in the solution. It does not rule out drug degradation.

- A chemical incompatibility is an admixture of drugs or drug and diluent that accelerates drug degradation. Any of the drug degradation pathways may be accelerated:
  - *Hydrolysis:* Ampicillin hydrolysis is accelerated by the addition of various drugs most of them in the acid range.
  - *Oxidation:* Changes in pH (occurs much more slowly below pH 4) and dilution of the antioxidants in gentamicin sulfate solutions contribute to its accelerated oxidation when admixed with a variety of drugs including the penicillins. Oxidation reactions are often accompanied by *color changes.* Epinephrine solutions turn pink and eventually brown as the phenolic groups on its aromatic ring oxidize. Nurses and patients should be advised not to use epinephrine solutions that have discolored.
  - *Photolysis:* Changes in pH will increase the rate of photolysis. In particular photo-oxidations occur more slowly below pH 4.
  - *Racemizations:* Proton exchange with the solvent occurs more readily at alkaline pH.

- Chemical degradations are difficult to detect except for oxidations accompanied by color change and often the only information in reference books relates to physical compatibility.
  - Powders for reconstitution indicate that the drug has stability problems. Do not mix these in the same admixture bag unless you have data that establish their *chemical* compatibility.
  - Check references for pH of maximum stability and if not available, avoid mixing oxidizable and racemizable drugs with basic solutions.
  - As a general rule, an injection that is not a powder for reconstitution will be most stable at the pH at which it has been formulated.

Packaging incompatibilities

Key Points

- Lipid soluble drugs that require cosolvents or surfactant solubilization often have packaging interactions with polyvinyl chloride containers and administration sets.
- Protein drugs can stick to glass and a variety of plastics.

- *Leaching:* Diethylhexylphthalate (DEHP) is a lipid soluble compound used as a plasticizer in polyvinyl chloride bags. Drug solutions formulated with surfactants, cosolvents and oily solvents leach DEHP out of the packaging and into the drug solution.
  - Solutions containing surfactants, cosolvents and oily solvents should be stored in glass or polyethylene containers, and administered with non-DEHP plasticized administration sets.
- *Sorption:* The drug may be lost from solution because of sticking on the surface of the container, administration set or inline filter.
  - Involves large protein drugs or very lipid soluble drugs that have an affinity for plastics or glass surfaces.
  - Plastics, particularly polyvinyl chloride, bind lipid soluble drugs.
  - Polypropylene and polyethylene that contain little or no plasticizer are less likely to adsorb drugs and are used to make specialized administration sets and containers for lipid soluble drugs.
  - Protein drugs may bind to plastics as well as glass surfaces and percent losses may be very high at dilute concentrations. Human serum albumin may be added to dilute protein drug solutions to saturate the sorption sites on packaging.

Parenteral compatibility checklist

A series of questions can be used to methodically review the potential problems that may be encountered when combining two drugs in the same syringe, Y site or IV container:

- What is the formulation (ingredients)? pH?
- Are there cosolvents or surfactants?
- Is the drug an acid or base in uncharged form?
- At what pH will the drug be most soluble?
- Is this a powder for reconstitution?
- What are the potential degradative pathways?
- At what pH is each drug most stable?
- What are the potential problems?

Pharmaceutical example

Use the checklist to determine if the two drugs that have been ordered must be drawn into two different syringes or can be placed in the same syringe.

Diazepam 10 mg, glycopyrrolate 0.3 mg IM

What is the formulation (ingredients)? pH?

Diazepam 5 mg/mL, propylene glycol 40%, alcohol 10%, sodium benzoate/benzoic acid 5%, benzyl alcohol 1.5%, in water for injection, pH 6.6 (Figure 8.5).

Glycopyrrolate 0.2 mg/mL, benzyl alcohol 0.9% in water for injection, pH 2.5 (Figure 8.6).

Are there cosolvents or surfactants?

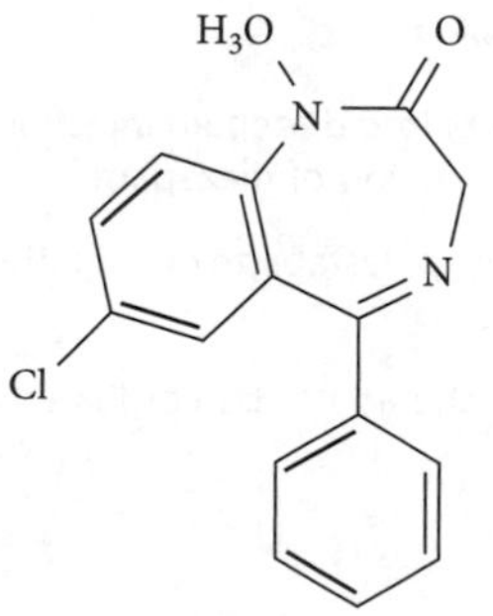

Figure 8.5   Diazepam

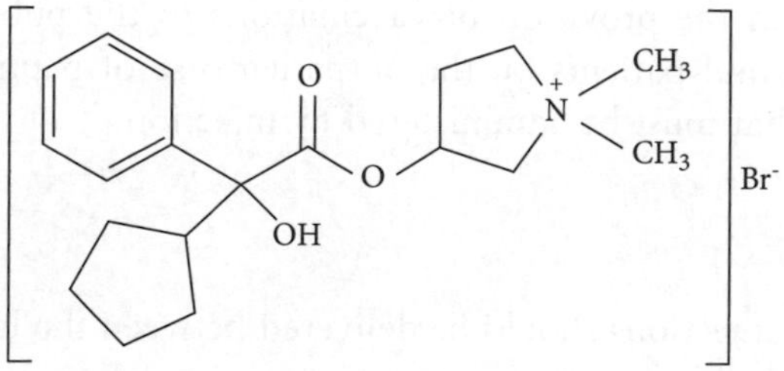

Figure 8.6   Glycopyrrolate

Diazepam injection contains alcohol and propylene glycol. There are no surfactants in either injection.

Are there carbonate buffers?

No carbonate buffers.

Is the drug an acid or base in uncharged form?

Diazepam is a weak base; glycopyrrolate is a quaternary ammonium compound with no basic electron pair available to pick up a proton. It has no acid–base properties but the pH of the injection is quite low.

At what pH will the drug be most soluble?

Diazepam will be more soluble at acidic pHs. Glycopyrrolate's solubility is not affected by pH.

Is this a powder for reconstitution?

Neither injection is a powder for reconstitution.

What are the potential degradative pathways?

Diazepam is hydrolyzable (cyclic amide) and oxidizable. Glycopyrrolate is potentially hydrolyzable (ester).

At what pH is each drug most stable?

This information can be found in *Trissel's Handbook of Injectable Drugs*. Diazepam is most stable between pH 4 and 8 and subject to acid catalyzed hydrolysis at pH less than 3.0. Glycopyrrolate is most stable at acidic pH and subject to hydrolysis at pH over 6.0.

What are the potential problems?

The dilution of cosolvent solubilized diazepam injection with glycopyrrolate, which is in water, can cause the precipitation of diazepam.

These two drugs are also likely to destabilize one another because of the difference in their pH of maximum stability.

Conclusion: These two drugs should not be combined in the same syringe.

Patient counseling

Pharmacists now receive formal training in the administration of intradermal, intramuscular and subcutaneous injections because of their increasing role in the provision of vaccinations to the public. In addition pharmacists counsel patients on the outpatient use of peptide and protein drug products that must be administered by injection.

Intradermal injections

- Intradermal injections should be delivered between the layers of skin.

- The volume injected intradermally must be very small, 0.1 mL, and if using a tuberculin syringe, the needle is inserted at a 10° angle to ensure that the injected fluid is placed in the skin.
- Injection of the 0.1 mL fluid will leave a raised wheal.

Pharmaceutical example

Intradermal influenza vaccines are placed in the skin over the deltoid. The needle used on the Fluzone intradermal system for influenza vaccine is about 1 mm long. This short needle is inserted perpendicularly into the arm to pierce the skin without compressing the tissue to ensure that the vaccine is injected into the skin and not subcutaneously.

Subcutaneous injections

- Subcutaneous injections are administered into the fatty tissue below the skin.
- The recommended site for vaccination of infants less than 12 months of age is the lateral thigh.
- For patients 12 months of age or older the upper outer triceps is used.
- Daily subcutaneous injections such as insulin therapy are given into a methodically rotated site in the abdomen (rapid acting insulins) or thigh (basal insulins) in order to maintain predictable absorption without repeating injection into the same site.
  - One easy to follow scheme with proven effectiveness divides the injection site into quadrants (or halves when using the thighs, buttocks or arms). One quadrant should be used per week moving always in the same direction, either clockwise or counter-clockwise, keeping the injections at least 1 cm apart to avoid repeat tissue trauma.

The standard administration procedure for subcutaneous injections:

- Organize supplies, wash hands and glove.
- Verify the drug and expiration date.
- Clean the septum of the drug vial with an alcohol swab.
- If the drug product is a suspension, shake or roll gently and then load dose into the syringe.
- Expose the injection site: posterolateral aspect of the upper arm, lateral thigh or abdomen.
- Wipe the site with an alcohol swab and allow to dry.
- Lift a fold of skin and insert the needle at a 45° angle in one quick, smooth motion.

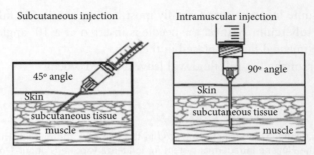

Figure 8.7 Needle placement for subcutaneous and intramuscular injections. (Reproduced with permission from selected reading 7.)

- Inject at a slow to moderate rate.
- Remove the needle at the same angle that it entered the skin.
- Activate the safety mechanism of the needle and discard the syringe and needle in the sharps container.
- Apply slight pressure to the injection site with a cotton ball to discourage bleeding. Apply bandage if needed.
- Provide patient reassurance and education. Document administration.

Note that the tissue should be pinched up and the needle inserted at a 45° angle to avoid injection into the muscle (Figure 8.7) unless the needle used is sufficiently short (4–6 mm) to reach only the subcutaneous area.

Intramuscular injections

- Intramuscular injections are administered into a large muscle mass below the subcutaneous tissue.
- The choice of needle size and site of injection should be made for each patient on the basis of the age, muscle size, thickness of adipose tissue and the volume of solution to be administered.
- The deltoid is used for vaccination of children older than 3 years of age and adults, and the vastus lateralis for children younger than 3 years old.
- The gluteals are used for intramuscular administration when the drug volume is greater than 2 mL.
  - To avoid the sciatic nerve, intramuscular injections into the gluteals must be placed superior and lateral to the posterior superior iliac spine and the greater trochanter.
  - The gluteals are never used to administer vaccines.
- Because there are no large blood vessels in the sites recommended for vaccination, aspiration before injection of vaccines is not necessary.
- For most adults the skin should be stretched taut before insertion of the needle to ensure that the injection is delivered deep into the muscle.

- In thin adult patients or children in whom there may be concern about hitting a bone, the muscle tissue may be bunched before insertion of the needle at a 90° angle.
- The needle chosen must be long enough to reach beyond the adipose tissue and into the muscle, and should be inserted all the way up to the hub. More detailed information on the choice of injection sites, needle sizes and technique can be found in Appendix D of the Center for Disease Control's *Pink Book*, *Epidemiology and Prevention of Vaccine-preventable Diseases*, at http://www.cdc.gov/vaccines/pubs/pinkbook/index.html#appendices.

The standard intramuscular injection protocol:

- Organize supplies, wash hands and glove.
- Verify the drug and expiration date.
- Clean the septum of the drug vial with an alcohol swab.
- If the drug product is a suspension, shake or roll gently and then load dose into the syringe.
- Expose the injection site: deltoid muscle of the upper arm or lateral thigh.
- Wipe the site with an alcohol swab and allow to dry.
- Spread tissue taut and insert the needle fully at a 90° angle in one quick, smooth motion.
- Inject at a slow to moderate rate.
- Remove the needle at the same angle that it entered the skin.
- Activate the safety mechanism of the needle and discard the syringe and needle in the sharps container.
- Apply slight pressure to the injection site with a cotton ball to discourage bleeding. Apply bandage if needed.
- Provide patient reassurance and education. Document administration.

Questions and cases

1. A physiological opportunity presented by the intravenous route is:
    (a) No drug metabolism
    (b) Rapid absorption
    (c) Predictable drug effects
    (d) Sustained drug effects

2. A patient on a total parenteral nutrition solution that is 1500 mosmol has developed inflammation of the median cubital vein. This hazard of intravenous therapy is called:
    (a) Hemolysis
    (b) Phlebitis
    (c) Extravasation
    (d) Air embolism

3. Phenytoin sodium is available as a cosolvent solubilized injection, which has a pH of 12. If the pH of muscle tissue is 7.0, what happens to phenytoin if the drug is injected intramuscularly instead of intravenously?

Phenytoin sodium

(a) Solubility increases, onset is faster and duration is shorter

(b) Solubility decreases, onset is faster and duration is longer

(c) Solubility increases, onset is slower and duration is shorter

(d) Solubility decreases, onset is slower and duration is longer

4. Isophane (NPH) insulin is considered an intermediate acting insulin because:

(a) It precipitates when injected

(b) It binds to albumin to slow its absorption

(c) Zinc and protamine are added to increase its tendency to aggregate

(d) Zinc and protamine are added to decrease its tendency to aggregate

5. You are preparing a sterile product for intrathecal use. Which of the following would make a good vehicle?

(a) Sterile water for injection

(b) Bacteriostatic water for injection

(c) Sterile saline

(d) Bacteriostatic saline

6. Use of alcohol and propylene glycol in the vehicle of an injection indicates that:

(a) The drug is not stable in water

(b) The drug is poorly soluble in water

(c) The drug will need to be shaken before use

(d) The drug is intended as a multidose vial

7. The use of trehalose in the formulation of an injectable drug indicates that:

(a) The drug is not stable in water

(b) The drug is poorly soluble in water

(c) The drug will need to be shaken before use

(d) The drug is intended as a multidose vial

8. High efficiency particulate air filtration:

(a) Produces a sterile air environment for sterile product preparation

(b) Reduces particle count per cubic foot to no more than 100

(c) Reduces particles but not pyrogens

(d) Reduces particles but not microbes

9. Which of the following is *not* a venous access device?

   (a) Winged infusion set

   (b) PICC line

   (c) Jet injector

   (d) PortaCath

10. You are advising a patient who will receive Nutropin (somatropin), a dry powder for reconstitution daily by subcutaneous injection. The patient should be advised to:

    (a) Reconstitute the drug with tap water

    (b) Rotate injection sites daily on the forearm

    (c) Stretch the skin taut before insertion of the needle

    (d) Insert the needle at a 45° angle

Example parenteral dosage formulations

Use Appendix Table A.2 and other references to answer the following questions about parenteral formulations.

11. What class of parenteral dosage form does each of the formulations below represent?

12. Is it is prompt or sustained release?

13. By which route(s) could it be administered?

14. What is the purpose of each ingredient in the formulation?

Diprivan injection

	Purpose of ingredient
Soybean oil 100 mg	
Glycerin 22.5 mg	
Lecithin 12 mg	
Disodium EDTA 0.05 mg	
Water for injection 1 mL	

Procaine penicillin injection

	Purpose of ingredient
Procaine penicillin 300 000 U	
Disodium citrate	
Trisodium citrate	
Lecithin 5 mg	
Carboxymethylcellulose 5 mg	
Povidone 5 mg	

	Purpose of ingredient
Propyl paraben 0.01 mg	
Methyl paraben 1 mg	
Water for injection qs 1 mL	

Haldol deconate injection

	Purpose of ingredient
Haldoperidol deconate 70.52 mg (50 mg haldol)	
Benzyl alcohol 12 mg	
Sesame oil 1 mL	

Herceptin vials

	Purpose of ingredient
Trastuzumab 440 mg	
L histidine HCl 9.9 mg	
L histidine 6.4 mg	
Trehalose 400 mg	
Polysorbate 20 1.8 mg	

Compatibility/stability cases

15. Can acyclovir sodium injection be combined with granisetron HCl injection?

Acyclovir sodium when reconstituted has a pH of 11.0 and contains 50 mg/mL. The water solubility of its uncharged form is 2.5 mg/mL and the pK_a of its enol is 9.25. Acyclovir also has a very weakly basic group with a pK_a 2.27. Because its pK_a is so low, it is not useful for the generation of B:H$^+$ to enhance water solubility, so the pK_a at 9.25 is the only one that is useful in a pH precipitation calculation. Granisetron HCl injection 1.12 mg/mL, which has a pH of 5.0 and a pK_a of 9.79 (estimated). The uncharged form of granisetron has a water solubility of 0.434 mg/mL.

For a weak acid, HA:

$$pH_P = pK_a + \frac{\log~[S_{A^-} - S_{HA}]}{[S_{HA}]}$$

Where pH$_P$ is the pH of precipitation, S_{A^-} is the solubility of the conjugate base and S_{HA} is the solubility of the free acid. If the conjugate base is not dissolved in the solution of interest at its solubility, S_{A^-} can be replaced by the desired concentration of the drug in the solution.

For a weak base, B:

$$pH_P = pK_a + \frac{\log[S_{B:}]}{[S_{BH+} - S_{B:}]}$$

Where pH_P is the pH of precipitation, $S_{B:}$ is the solubility of the free base and S_{BH+} is the solubility of the conjugate acid. If the conjugate acid is not dissolved in the solution of interest at its solubility, S_{BH+} can be replaced by the desired concentration of the drug in the solution.

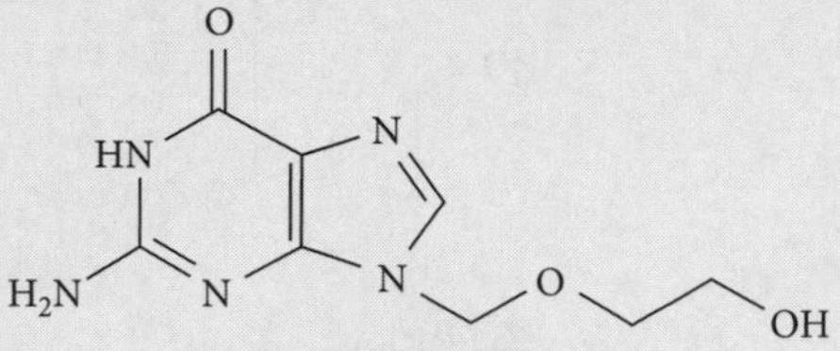

Acyclovir

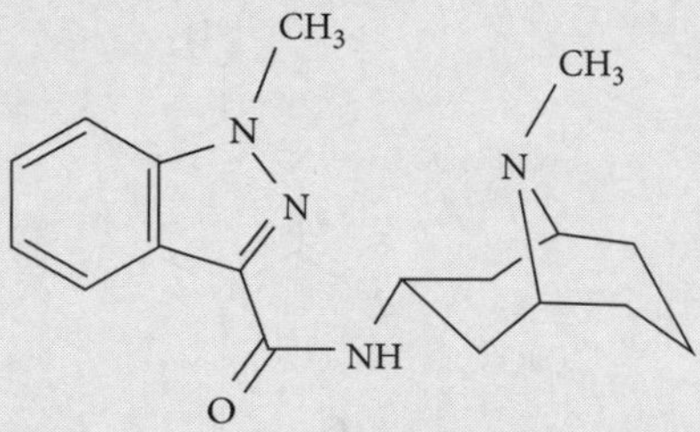

Granisetron

16. Use the parenteral compatibility checklist to consider the combination of dopamine hydrochloride injection and sodium bicarbonate Injection. What is the formulation (ingredients)? pH?

Dopamine:

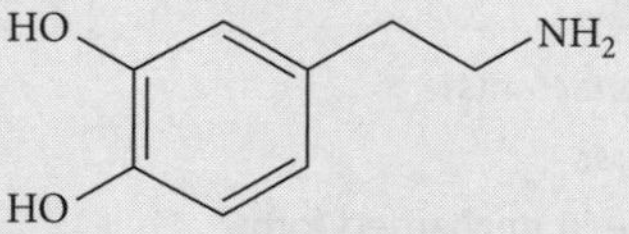

Dopamine

Dopamine formulation?:

Sodium bicarbonate:

NaHCO$_3$

Sodium bicarbonate formulation?:

Are there cosolvents or surfactants?

Are there carbonate buffers?

Is the drug an acid or base in uncharged form?

At what pH will the drug be most soluble?

Is this a powder for reconstitution?

What are the potential degradative pathways?

At what pH is the drug most stable?

What are the potential problems?

17. Ganciclovir sodium and gemcitabine HCl. Can these two drugs be combined in the same syringe, IV bag or Y site?

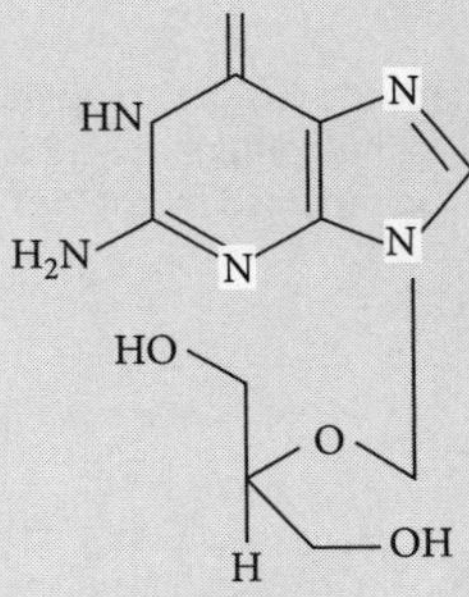

Ganciclovir

Ganciclovir formulation?:

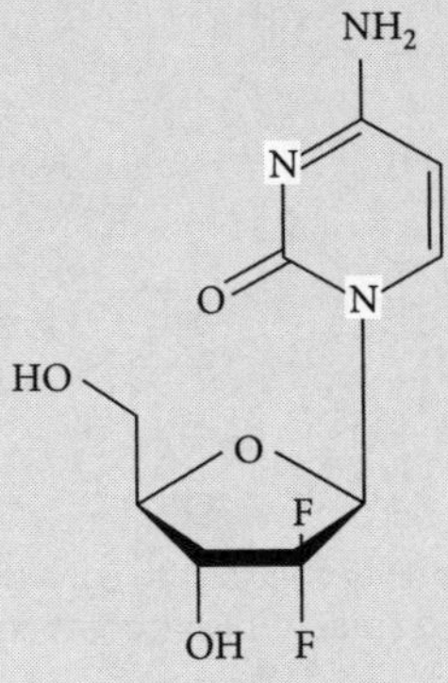

Gemcitabine

Gemcitabine formulation?:

Are there cosolvents or surfactants?

Are there carbonate buffers?

Is the drug an acid or base in uncharged form?

At what pH will it be most soluble?

Is this a powder for reconstitution?

What are the potential degradative pathways?

At what pH is the drug most stable?

What are the potential problems?

18. Amiodarone HCl and furosemide. Can these two drugs be combined in the same syringe, IV bag or Y site?

Amiodarone

Furosemide

Amiodarone formulation?:

Furosemide formulation?:

Are there cosolvents or surfactants?

Are there carbonate buffers?

Is the drug an acid or base in uncharged form?

At what pH will it be most soluble?

Is this a powder for reconstitution?

What are the potential degradative pathways?

At what pH is the drug most stable?

What are the potential problems?

Cases: Using parenteral pharmaceutics in practice

19. MB was recently diagnosed with multiple sclerosis and will be treated with Avonex (interferon beta-1a) 33 µg once weekly intramuscularly. It is available as a powder for reconstitution containing 33 µg interferon beta-1a and is prepared by diluting with 1.1 mL diluent. It is stable 6 hours after reconsitution and contains no preservative.

(a) Explain to the patient how she should prepare the medication

(b) Explain to the patient how she should administer the medication

(c) If therapy is initiated with 16.5 µg of Avonex twice weekly, can the unused portion of the vial provide the second dose?

20. HN is a patient who uses sumatriptan (Imitrex) as a subcutaneous solution for migraine headaches. How soon could he expect to feel the pain relieving effects of the drug? How long should he wait before he decides to use a second dose?

21. FR is a 60 year old diabetic patient who has been taking regular insulin mixed with NPH (70:30) before the morning and evening meals. His physician has switched him to regular insulin before meals and Lantus at bedtime. He wants to know if he can mix Lantus with regular insulin and administer the two together at the evening meal.

22. WT is a patient on cyanocobalmin injections for pernicious anemia. His dose is 100 µg intramuscularly every month. There are three different cyanocobalmin products available:

    (a) Cyanocobalmin 100 µg/mL in water for injection, 1 mL

    (b) Cyanocobalmin 100 µg/mL in water for injection with benzyl alcohol, 10 mL

    (c) Cyanocobalmin 100 µg/mL in water for injection with benzyl alcohol, 30 mL

Which of these products would be best to provide this patient with his outpatient therapy?

Selected reading

1. JE Thompson. A Practical Guide to Contemporary Pharmacy Practice, 3rd edn. Philadelphia PA: Lippincott Williams & Wilkins, 2009.
2. United States Pharmacopeia. Chapter <797> Pharmaceutical Compounding: Sterile Products, USP 34-NF-29, 2011.
3. TL Pearson. Practical aspects of insulin pen devices. J Diabetes Sci Tech 2010; 4: 522–531.
4. L Kroon. Overview of insulin delivery pen devices. J Am Pharm Assoc 2009; 49: 118–131.
5. LG Potti, ST Haines. Continuous subcutaneous insulin infusion therapy: A primer on insulin pumps, J Am Pharm Assoc 2009; 49: 1–13.
6. LF McElhiney. Misinterpretations of United States Pharmacopeia Chapter <797>. Int J Pharm Comp 2012; 16: 6–10.
7. Centers for Disease Control and Prevention. Epidemiology and prevention of vaccine-preventable diseases, Appendix D, 12th edn. Accessed at http://www.cdc.gov/vaccines/pubs/pinkbook/index.html#appendices on June 27, 2012.
8. LA Trissel. Handbook of Injectable Drugs, 17th edn. American Society of Health-system Pharmacists, Bethesda, MD 2013.

9

Delivery of biopharmaceuticals and the use of novel carrier systems

Are you ready for this unit?

● Review amino acids, peptides, proteins, nucleic acids, oligonucleotides, DNA, RNA, sustained release and controlled release.

Learning objectives

Upon completion of this chapter, you should be able to:

● Differentiate transgenes, antisense oligonucleotides, small interfering RNA, ribozymes, DNAzymes and aptamers.

● Explain how the physicochemical and biological properties of biopharmaceutical drugs influence the design of their delivery systems.

● Explain how the colloidal dimensions of biopharmaceutical macromolecules and their delivery systems influence their movement from site of administration to target.

● Explain the formulation approaches taken to protect macromolecules from enzymes, facilitate uptake and release, and direct these large molecules to their targets.

● Describe the design, ingredients, delivery and drug release characteristics of:
  — PEGylated macromolecules
  — microspheres and nanospheres
  — implants
  — dendrimers
  — liposomes and micelles
  — gels.

● Describe how viral vectors are used to deliver transgenes.

● Describe enhanced permeability and retention as a passive targeting approach and compare it with active targeting approaches.

● Define biosimilarity.

Introduction

The drugs discussed in this chapter arise from new technologies that permit the large scale synthesis or production of biologically active macromolecules.

Their large size and vulnerability to enzymatic degradation *in vivo* creates extra challenges for the development of systems for their delivery. Many of these new drug molecules are peptides or proteins, polymers of amino acids linked by amide bonds and weighing between 1 and 150 kilodaltons (kDa). The other groups of macromolecules considered in this chapter are the oligonucleotides and transgenes, polymers of nucleotides that can weigh as much as 13 000 kDa (20 kilobase pairs). Interest in delivery of these macromolecules has driven most of the innovative work in parenteral formulation in the past 25 years. These delivery systems use polymers and phospholipids to protect macromolecules from enzymes, facilitate uptake, release and, in some cases, direct these large molecules to their targets. These two basic categories of materials are assembled into liposomes and lipoplexes, micelles, nanoparticles and nanocapsules, dendrimers, nanogels, and nanoemulsions for the delivery of peptides, proteins and nucleic acid-based therapeutics.

Physicochemical and biological properties

Peptide and protein drugs (Table 9.1)

The properties of peptide and protein drugs are:

- Molecular weight
  - *Peptide* drugs include molecules between 1 and 5 kDa
  - *Protein* drugs include two broad categories: the recombinant protein products and the monoclonal antibodies, which can range between 6 and 150 kDa.
- Drug targets
  - Most peptide drugs bind to cell surface receptors.
  - Cyclosporine, a hydrophobic, immunosuppressant peptide is the sole example of a successfully marketed peptide with an intracellular receptor.
  - Several of the peptide hormone analogs have receptors in the brain as do various peptide drugs under development.
  - Although most of the targets of currently available protein products are cell surface receptors, these receptors may be located in inaccessible tissues such as the brain or retina.
  - Therapeutic proteins are generally highly specific for the receptor that mediates their biological function and require only small amounts to produce an effect.
  - Monoclonal antibodies are immunoglobulin proteins produced by cell lines derived from a single B lymphocyte selected to react to a specific antigenic site with high affinity.
  - One goal of macromolecular drug delivery research has been to enhance the ability of these large drugs to cross biological barriers.

Table 9.1 Selected protein and peptide biopharmaceutical drugs

Drug	Formula weight (kDa)	Indication	Target location	Formulation
Calcitonin	3.431	Osteoporosis	Calcitonin receptors on the surface of osteoclasts	Nasal spray, solution for injection
Corticotropin	4.541	Diagnosis of adrenocortical deficiency	Cell surface receptors on adrenal cells	Injectable gel
Cyclosporine	1.2	Organ transplant prophylaxis	Bind cyclophilin, an intracellular protein regulator in T cells	Surfactant solubilized solution for injection, capsules, ophthalmic emulsion
Desmopressin	1.069	Diabetes insipidus	Vasopressin v1 and v2 receptors on renal epithelial cells and vascular smooth muscle	Oral tablet, intranasal, solution for injection
Alteplase, tPA (Activase)	59.04	Pulmonary embolism, acute myocardial infarction, ischemic stroke	Activates plasminogen in a vessel clot	Powder for injection
Interferon alfa-2a (Roferon)	19.24	Chronic myelogenous leukemia, hairy cell leukemia	Type 1 interferon receptor on infected cells, tumor cells, immune cells	Prefilled syringes of solution for injection
Etanercept (Enbrel)	150	Rheumatoid arthritis	Binds tumor necrosis factor in synovial fluid, plasma	Powder for injection, solution for injection
Pegaspargase	31.73	Acute lymphocytic leukemia	Depletes L-asparagine around lymphoid tumor cells lacking aspargine synthetase	Pegylated protein solution for injection
Adalimumab (Humira)	144.190	Ankylosing spondylitis, Crohn's disease, psoriasis, rheumatoid arthritis	Binds tumor necrosis factor α in synovial fluid, plasma and other inflamed tissues	Solution for injection, prefilled syringes, pens

(continues)

Table 9.1 (*continued*)

Drug	Formula weight (kDa)	Indication	Target location	Formulation
Bevacizumab (Avastin)	149	Cancers, macular degeneration, diabetic retinopathy	Binds to vascular endothelial growth factor on tumor or retinal endothelial cells (and other vascular endothelial cells)	Solution for injection
Certolizumab Pegol	91	Crohn's disease, rheumatoid arthritis	Binds tumor necrosis factor α in synovial fluid, plasma and other inflamed tissues	Pegylated powder for injection
Trastuzumab (Herceptin)	145.531	Breast cancer	HER2 protein on the surface of breast, colon and ovarian cancer cells	Powder for injection

- Stability *in vitro*:
  - Large size increases their tendency to aggregate.
  - Susceptible to hydrolysis, oxidation, photolysis and racemization due to the variety of functional groups.
- Stability *in vivo*
  - rapid degradation by peptidases.

Oligonucleotide and transgene drugs (Figure 9.1, Tables 9.2 and 9.3)

The properties of oligonucleotide and transgene drugs are:

- Molecular weight
  - Transgenes are double stranded rings of DNA, which can be as large as 20 kilobase pairs and weigh 13 000 kDa.
  - Other oligonucelotide drugs weigh between 6 and 150 kDa.
- Drug targets
  - The mRNA target of antisense oligonucleotides can be found in the cell's cytoplasm.
  - Small interfering RNAs, ribozymes and DNAzymes work in the cell cytoplasm rather than the nucleus.
  - An aptamer's target protein could be intracellular or extracellular.
- Stability *in vitro*
  - Large size increases their tendency to aggregate.
  - Susceptible to hydrolysis, oxidation, photolysis and racemization due to variety of functional groups.

Table 9.2 Types of nucleic acid-based therapeutic agents

Type	Description
Transgenes (gene therapy)	Segment of DNA that codes for the protein that will produce the therapeutic effect when transferred to the patient's cell for translation. The gene sequence of interest is spliced into an independently replicating plasmid along with promoter, and occasionally tissue specific translational factors and delivered to the nucleus of the body cell that will translate the genetic material into the therapeutic protein
Antisense oligonucleotides	Short, single-stranded sequences of DNA that bind to complementary segments of messenger RNA and prevent their translation
Small interfering RNA (siRNAs)	Short, double-stranded segments of RNA that are complementary to the mRNA of the protein to be turned off. When siRNA interacts with its complementary mRNA, the resulting complex recruits a ribonuclease that destroys the mRNA, preventing its translation into protein
Ribozymes	RNA molecules that have their own nuclease activity, binding to a targeted mRNA sequence and destroying it
DNAzymes	Sequences of deoxyribonucleic acids that have ribonuclease activity. They bind to cytoplasmic mRNA sequences and destroy them so they cannot be translated
Aptamers	Single-stranded or double-stranded oligonucleotide that binds a target other than RNA or DNA such as a protein and interferes with the target protein's action. Can be prepared to bind quite selectively like antibodies but are nonimmunogenic

Table 9.3 DNA/RNA based products

Drug	Formula eight (kDa)	Indication	Target location	Dosage form
Fomivirsen (Vitravene)	6.682	Cytomegalovirus retinitis (currently off market)	Binds to mRNA sequence in the cytomegalovirus particle at site of infection	Intravitreal solution for injection
Pegaptanib (Macugen)	50	Age-related macular degeneration	Binds extracellular vascular endothelial growth factor 165 at retinal endothelial cells	Pegylated oligonucleotide solution for injection, prefilled syringes

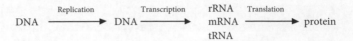

Figure 9.1 Flow of genetic information. DNA, deoxyribonucleic acid; RNA, ribonucleic acid; rRNA, ribosomal RNA; mRNA, messenger RNA; tRNA, transfer RNA

- Stability *in vivo*
  - rapid degradation by nucleases.

Movement from administration site to target

The movement of these macromolecular drugs across biological barriers is hampered not only by their large size, but also frequently by the necessity to deliver the large molecule to a target inside the cell. Most of these large molecules must be injected to produce a large enough concentration at their site of action.

The particulate delivery systems presented in this chapter as well as large protein drugs of colloidal dimensions will adsorb opsonins (immuno-globulins, complement) resulting in rapid removal from circulation due to engulfment by liver and spleen macrophages. Escape from the mononuclear phagocytic system depends on particle size, surface charge and surface hydrophobicity. Particles 100 nm or smaller with hydrophilic surface are opsonized less than larger more hydrophobic or cationic particles. Particles that adsorb dysopsonins (albumin, apolipoprotein) experience little or no uptake by macrophages. This phenomenon is important not only to the clearance of these macromolecules but also to their distribution, as the engulfed particles are found disproportionately in the liver and spleen.

The movement of macromolecules through endothelial membranes may occur via the paracellular route in the liver where the fenestrae or pores between cells are as large as 180 nm but, in other tissues, pore size prevents the paracellular movement of these drugs. Likewise most of these macromolecules do not move through the transcellular pathway of the lipid bilayer to a significant extent. There are saturable carriers for a few neuropeptides at the blood–brain barrier that facilitate the movement of these molecules across that epithelial barrier. However, the main mechanism for the transport of large proteins and oligonucleotides across cell or epithelial membranes is via endocytosis. A number of endogenous proteins: insulin, epidermal growth factor, luteinizing hormone, thyroid hormone and immunoglobulins are transported by receptor mediated endocytosis, a multistep process that is initiated by binding the macromolecule to a specific receptor (Figure 9.2). The receptor and its ligand are engulfed into a vesicle that carries the contents into the cell. The ligand may dissociate from its receptor in the resulting endosome, and the free receptor is recycled to the

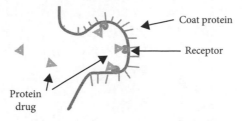

Figure 9.2 Receptor mediated endocytosis. A coated pit containing receptors and protein drug forms a vesicle and internalizes its contents

cell surface. Alternatively the contents of the endosome may be sent to an intracellular organelle for degradation. It is important for macromolecular drugs transferred by endocytosis to be released from the endosome rather than degraded. Components of the delivery system for macromolecules may be chosen to assist with binding to a particular receptor and to cause release of the macromolecule ligand from the endosome.

The enzymes that deactivate proteins and oligonucelotide drugs, the proteases and nucleases are ubiquitous, unlike most enzymes that metabolize small drug molecules that are concentrated in the liver and kidneys. At the kidney any molecules smaller than 30 to 50 kD (10 to 15 nm) depending on their charge and shape, will be filtered into the urine. Thus the peptides and some of the smaller proteins and oligonucletides may be removed from action by the kidneys.

Key points

- Peptides, proteins and oligonucleotides are large, biologically active molecules that are vulnerable to rapid enzymatic degradation and often have intracellular receptors.
- Most of these large molecules must be injected to produce a large enough concentration at their site of action.
- The size of these drug molecules when combined with their delivery systems can result in rapid removal from circulation because of engulfment by liver and spleen macrophages.
- Many of these drugs are too large to diffuse transcellularly or paracellularly through biological membranes, moving instead though endocytosis.

Design of delivery systems

Most of the macromolecular products available today are formulated as aqueous solutions or reconstitutable powders for injection (see Tables 9.1, 9.3 and 9.4) However, the design of delivery systems that protect macromolecules from enzymes, facilitate uptake and endosomal release and, in some cases, direct these large molecules to their targets is the subject of intense research.

Table 9.4 Sustained release formulations of proteins and peptides

Drug dosage form	Comfort	Stability	Sustained release mechanism	Vehicle
Pegademase (Adagen)	Sodium chloride	Sodium phosphate, mono- and dibasic	Pegylation	SWI
Peginterferon alfa-2a (Pegasys)	Sodium chloride	Benzyl alcohol, polysorbate 80, acetic acid, sodium acetate	Pegylation	SWI
Pegfilgrastim (Neulasta)	Sorbitol	Polysorbate 20, sodium acetate	Pegylation	SWI
Pegvisomant (Somavert) powder for reconstitution	Mannitol	Glycine, sodium phosphate, dibasic and monobasic	Pegylation	Reconstitute with SWI
Exenatide (Bydureon), powder for suspension for injection	Diluent: sodium chloride	Sucrose – diluent: carboxymethyl cellulose, polysorbate 20, sodium phosphate, mono- and dibasic	PLGA microspheres	Prefilled syringe containing diluents: SWI with excipients
Octreotide (Sandostatin LAR), powder for suspension for injection	Diluent: mannitol	Diluent: carboxymethyl cellulose	PLGA microspheres, mannitol	Diluent: SWI with excipients
Leuprolide (Eligard) suspension for reconstitution two prefilled syringes (subcutaneous)			PLGA in N-methyl-2-pyrrolidone forms a gel on subcutaneous injection	Atrigel: PLGA in N-methyl-2-pyrrolidone
Goserelin (Zoladex) biodegradable Implant	Not applicable	Not applicable	1 mm implantable, cylindrical matrix made of PLGA	Not applicable

PLGA = poly(lactic-co-glycolic) acid; SWI = sterile water for injection.

Adapted from the product labelling accessed from the FDA Approved Drug Products website at http://www.accessdata.fda.gov/scripts/cder/drugsatfda/index.cfm?fuseaction=Search.Search_Drug_Name

- A number of polymer based delivery systems have been developed to enmesh macromolecular drugs and protect them from enzymatic degradation and provide sustained release.
- A second approach has been to enclose the macromolecular cargo in phospholipid vesicles called *liposomes* or *lipoplexes*.

Polymer based systems

Key Points

- Drug delivery polymers function to sterically hinder the approach of drug degrading enzymes and prevent filtration of the drug–polymer complex through the kidneys with their large mass.
- Polyethylene glycol polymers may be linked to macromolecules or their drug delivery system to reduce the adsorption of opsonin proteins and subsequent uptake by the mononuclear phagocytic system.

- Polymers and lipid complexes are used to prepare drug carrier systems of colloidal dimensions, generally between 10 and 100 nm in diameter, which protect proteins and nucleic acid based macromolecules from enzymatic degradation.
- The polymer based systems include PEGylation, microspheres and nanoparticles, implants, micelles, dendrimers and gels.
- The lipid based delivery systems include the liposomes for peptide and protein drugs and lipoplexes for nucleic acid macromolecules.
- These systems provide enzymatic protection for macromolecules as well as prolonged release.
- Use of appropriately selected polymer or lipid components can also facilitate cell uptake, promote endosomal release and, in some cases, direct these large molecules to their targets (Table 9.5).
- Viral vectors are under development as delivery systems for transgenes.
- Although these delivery systems have been developed for delivery of protein and nucleic acid macromolecules, many of the first products to come to the market have been used to reduce the toxicity of small molecule, anticancer drugs.

PEGylation

- Drug macromolecules are covalently linked to the polyethylene glycol (PEG) polymer:
  - using stable links with spacers that allow the drug molecule to interact with its target though generally with lower affinity than the free drug
  - or by a hydrolysable bond that can be cleaved in the body to allow drug binding to target.

Table 9.5 Selected ingredients used in biopharmaceuticals

Ingredient class	Description	examples
Lipids	Form liposomes that protect drug from enzymes, glomerular filtration. Provide passive targeting through EPR	Cholesterol, soy phosphatidyl choline, distearoyl-phosphatidyl glycerol, Distearoyl phosphatidylethanolamine (DSPE), Dioleoyl phosphatidylcholine (DOPC), dipalmitoyl phosphatidylglycerol (DPPG), tricaprylin, triolein
Polymers	Attach or encase drug to protect from enzymes, glomerular filtration. Provide passive targeting through EPR	Polyethylene glycol (PEG), poly D,L-lactic-coglycolic acid (PLGA), Polycaprolactone (PCL), Chitosan, Polifeprosan 20, Albumin, Alginic acid, Polyethylenimine (PEI), Polyamidoamine (PAMAM)

EPR = enhanced permeability and retention.

Table 9.6 Pharmacokinetics of peginterferon alfa-2a

Interferon	T_{max} (hours)	$T_{1/2}$ (hours)	Frequency of dosing
Peginterferon alfa-2a	72–96	80	Weekly
Alfa-2a interferon	7.3 hours	5.1	Daily

T_{max} = time to peak concentration; $T_{1/2}$ = drug half-life.

- Because the polymers are highly hydrated, the addition of PEG to a drug macromolecule increases its diameter by 5–10-fold more than would be predicted by the increase in molecular weight.
- The increase in size reduces the drug's glomerular filtration, protects it from proteases, increases drug half-life and maintains the drug in the vascular compartment.
- PEGylation reduces the adsorption of opsonin proteins to a macromolecular drug or particulate delivery system via steric repulsion. This prolongs circulation time by reducing uptake of the drug or delivery system by the mononuclear phagocytic system.

Pharmaceutical example

Table 9.6 illustrates the effect of PEGylation on the pharmacokinetics of interferon alfa-2a.

Microspheres and nanoparticles

Key Point

- Microspheres and nanoparticles are spherical matrices of polymer-embedded drug that prevent enzymatic degradation or glomerular filtration of the macromolecular drug.

- The main distinction between microspheres and nanoparticles is their size; both are tiny spherical matrices of polymer embedded drug that prevent enzymatic degradation or glomerular filtration of the macro-molecular drug. In some cases the drug may be attached to the surface of the particle to be readily available at the initiation of therapy.
  - Microspheres range from 1 to 250 μm.
  - Nanospheres and nanocapsules, collectively known as nanoparticles, measure between 0.01 and 1 μm.
- The most commonly used polymers for peptide and protein drugs are esters of polylactic acid (PLA) and its copolymer with glycolic acid (PLGA).
- Nanoparticles for nucleic acid based macromolecules are generally prepared with cationic polymers. They may be PEGylated or targeted to reduce the toxicity associated with cationic polymers.
- Microspheres and nanoparticles have also been used to prepare sustained release products for small drug molecules.
  - Release rate is also influenced by the size of the particle with larger particles releasing drug more slowly than smaller particles.
  - Most particles are designed in the smaller range, 0.01–0.1 μm in diameter, since this is considered the ideal size for gaining entry into the leaky vasculature of tumor cells.
  - The release rate of smaller particles may be fined tuned by crosslinking, copolymerizing or blending with other polymers.
- Microspheres are given by subcutaneous or intramuscular routes and are generally presented as powders for suspension that must be mixed with an accompanying diluent containing suspending agent, buffers and tonicity agents.
- Nanoparticles may be given intravenously.

Parenteral gel forming products

Key Point

- Parenteral gels are liquids that can be injected subcutaneously and form semisolid structures that release drug as they slowly degrade.

Parenteral gels are liquids that can be injected subcutaneously and form semisolid structures that release drug as they slowly degrade. They can be used to 'target' a particular site by limiting the dose that is injected to a particular site.

Pharmaceutical example

Atrigel is a gel forming delivery system composed of poly D,L-lactic-coglycolic acid (PLGA) in the organic solvent, N-methyl-2-pyrrolidone, and forms a solid gel upon contact with body water. The gel releases drug as it degrades over 1–6 months. Atrigel techonology has been used to formulate a line of leuprolide acetate products for prostate cancer, and a sustained release doxycycline gel that is placed into the periodontal pocket through a cannula and forms a solid implant for local drug release over a week.

Implants

Implants for the purpose of drug delivery require surgical implantation or the use of trocars, sharp-pointed rods that pierce the skin and wall of a body cavity to place the implant.

Pharmaceutical example

The Zoladex (goserelin) biodegradable implant is placed subcutaneously by trocar in the anterior abdominal wall. The peptide drug is dispersed uniformly in the implant matrix composed of D,L-lactic-coglycolic acid (PLGA) and is released as the device degrades over 28 days.

Pharmaceutical example

Another implant design is the Gliadel Wafer, a polymer device containing carmustine (small molecule, anticancer drug), and used in the treatment of malignant glioma. Carmustine is homogeneously distributed in a matrix of polifeprosan 20. The wafers are placed in the affected area after the tumor is surgically removed. They degrade very slowly by surface erosion producing a constant release of carmustine over 3 to 4 weeks.

Dendrimers

Key Point

- Dendrimers are highly branched spherical polymers that are bound to or complex with biopharmaceuticals to prevent enzymatic degradation or glomerular filtration.

- Dendrimers are highly branched spherical polymers built from a central core molecule with each added layer of branching referred to as a generation (Figure 9.3). To serve as drug carriers, dendrimers need

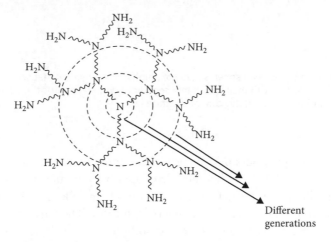

Figure 9.3 Dendrimer, e.g. Superfect

between four and six generations of branching, which gives them a formula weight of 10–50 kDa, and a diameter of 4–7 nm.
— The more generations added to the dendrimer structure, the slower the release.
● Drugs are linked to the peripheral groups either covalently or by electrostatic complexation or encapsulated in the internal cavity. The covalent links can be cleaved *in vivo*.
— Hydrophobic drugs are attracted to the hydrophobic central cavity of dendrimers by hydrophobic interactions and charged drug molecules may be attracted by electrostatic interactions to oppositely charged groups on the dendrimer periphery.
— Drugs that are covalently linked demonstrate first order release while drugs that are noncovalently attracted to dendrimers are released in 'bursts'.
● Dendrimer technology has not yet come to market but there is a product in clinical trials.

Pharmaceutical example

Vivagel developed by Starpharma, Australia, is a formulation of polylysine G4 (four generations) dendrimers with an anionic surface of naphthalene disulfonate in a carbomer (crosslinked polyacrylate) gel. The product exhibits antiviral activity against HIV and HSV for the prevention of sexually transmitted diseases and is given by intravaginal administration.

Micelles

Key Point

- Micelles are single layer spheres made of amphiphilic polymers that self-assemble in aqueous solution with their hydrophilic segment oriented towards the water and the hydrophobic section in the core of the micelle.

- Amphiphilic polymers may be produced from copolymerizing a hydrophilic polymer with a more hydrophobic polymer in blocks. These amphiphilic polymers may be used to prepare micelles for the delivery of protein and nucleic acid based macromolecules. They are safer than smaller surface active agents such as sodium lauryl sulfate or cetyltrimethylammonium bromide.
- Amphiphilic polymers will self-assemble in aqueous solution with their hydrophilic segment oriented towards the water and the hydrophobic section in the core of the micelle.
  - The hydrophilic surface minimizes opsonization and allows the micelles to evade capture by the mononuclear phagocytic system.
- Polymeric micelles are between 0.02 and 0.1 μm, large enough to prevent glomerular filtration by kidneys and the approach of proteases or nucleases thus prolonging circulation time of entrapped drug molecules.
- Drug molecules are attached by attraction or covalent link to the polymer core. Drugs that are linked covalently to micelles can provide sustained release, or if acid sensitive bonds are used, the drug can be selectively released in the acidic environment of tumor cells or endosomes.

Pharmaceutical example

Genexol is a polymeric micelle formulation of paclitaxel under development that has a PEG surface and a polylactide core. Paclitaxel is an anticancer injectable drug with very low water solubility. The original formulation of this drug is made with a surfactant that causes severe hypersensitivity reactions thus the micellar delivery system has the potential to significantly improve drug therapy with paclitaxel.

Liposomes and lipoplexes

Key Points

- Liposomes are spherical bilayers of phospholipid loaded with drug either in their aqueous core or in the bilayer that protect proteins from *in vitro* aggregation and from proteases or glomerular filtration *in vivo*.
- Lipoplexes are self-assembling complexes formed between negatively charged nucleic acids and cationic lipids that protect oligonucelotides from enzymatic degradation.

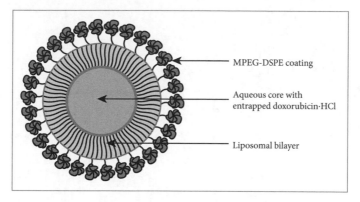

Figure 9.4 Stealth liposome. (Reproduced with permission from AT Florence, An Introduction to Clinical Pharmaceutics, p.78; published by Pharmaceutical Press, 2010)

- Liposomes are spherical bilayers of phospholipid loaded with drug either in their aqueous core or in the bilayer itself.
- A stealth liposome has PEG attached to the surface of the liposome to prevent its detection by liver and spleen macrophages and lengthen its circulation time relative to uncoated liposomes (Figure 9.4).
- Liposomes protect proteins from *in vitro* instability (in particular, aggregation) as well as from proteases *in vivo* resulting in longer half-lives. They also reduce immunogenicity of protein molecules.
- Because there is no covalent link to the protein, liposomes do not reduce the activity of the drug macromolecule as covalent binding with PEG does.
- Liposomes are believed to interact with cells via three different mechanisms:
  - endocytosis by phagocytic cells of the mononuclear phagocytic system
  - adsorption to the cell surface by nonspecific hydrophobic or electrostatic or specific interactions with cell surface components
  - fusion with the plasma cell membrane by insertion of the liposome's lipid bilayer.
- The small drug molecules that have been formulated using liposome technology are presented in Table 9.7. Small drug molecules may be released from liposomes by diffusion across the liposomal membrane.
- Drug macromolecules are released as the liposome is degraded *in vivo* or when it fuses with a cell membrane.
- The release rate of drug from the liposome *in vivo* can be varied by changing the membrane fluidity of the liposome.
- The smaller liposomes (<0.1 µm) are less likely to be phagocytosed and should have longer circulation times than larger ones.

Table 9.7 Novel sustained release formulations of small drug molecules

Dosage form; route	Comfort	Stability	Release/targeting ingredients	Vehicle
Carmustine (Gliadel Wafers)			Polifeprosan 20 implantable wafers	
Risperidone (Risperdal Consta) powder for suspension; IM	Sodium chloride	Diluent: polysorbate 20, carboxymethyl cellulose, sodium phosphate dibasic, citric acid	PLGA microspheres	Diluent: SWI plus excipients
Ferumoxytol (Feraheme) injection; IV	Mannitol		Iron oxide nanoparticle coated with polyglucose sorbitol carboxy methylether, 17–34 nm	SWI
Paclitaxel (Abraxane) suspension for reconstitution; IV	Diluent: sodium chloride		Albumin bound nanoparticle (130 nm) sodium caprylate, sodium acetyltryptophanate	Reconstitute with normal saline
Amphotericin B (AmBisome) lyophilized powder; IV		Alpha-tocopherol, sucrose, disodium succinate	Liposomes (cholesterol, soy phosphatidyl choline, distearoyl-phosphatidyl glycerol) unilamellar bilayer, diameter <100 nm	Reconstitute with SWI
Doxorubicin (Doxil); IV dispersion	Sucrose	Histidine	Stealth liposomes: N-carbonylmethoxyPEG, DSPE, hydrogenated soy phosphatidyl choline, cholesterol, 100 nm	SWI
Morphine (DepoDur); epidural suspension	Sodium chloride		Liposomes: DOPC, DPPG, cholesterol, tricaprylin, triolein, 17–23 μm	0.9% sodium chloride
Cytarabine (DepoCyt); intrathecal suspension	Sodium chloride		Liposomes: DOPC, DPPG, cholesterol, triolein	0.9% sodium chloride

IM = intramuscular; IV = intravenous; SWI = sterile water for injection.

Adapted from product labeling at FDA Approved Drug Products website.

Table 9.8 Viral vectors for gene delivery

Vector (family)	Genes	Host cell preference (tropism)	Duration of gene transfer	Immunogenicity	Biosafety level
Adenovirus (adenovirus)	dsDNA	Broad (epithelial cells, lymphoid cells)	Transient	High	Level 2
Murine leukemia virus (oncoretrovirus)	RNA	Broad (macrophages, T cells, dendritic cells, microglia, brain cells) dividing cells only. May be altered	Stable	Low	Level 2
Lentivirus (retrovirus)	RNA	Broad (central nervous system, muscle, liver) dividing and nondividing cells. May be altered	Stable	Low	Level 2/3
Adeno-associated virus (parvovirus)	ssDNA	Broad	Mostly transient	Low, not associated with human disease	Level 1
Herpes simplex 1 (alpha herpesvirus)	dsDNA	Neurons, epithelial cells	Stable in neurons	High	Level 2

- Cationic lipoplexes may be useful as a non-viral delivery system for gene therapy.
  - Negatively charged DNA forms self-assembling complexes with cationic lipids that protect, transfer and release DNA for cellular expression.
  - So far lipoplexes do not achieve a sufficient level of transfection efficiency (percent of cells transformed in a targeted population) for potential therapeutic use.

Viral vectors (Table 9.8)

- Viruses in many ways are the ideal vector for the transfer of genes into living cells.
  - They have evolved to bind specific receptors on a preferred host cell and penetrate the cell membrane with their intact viral particle or their genetic information.
  - The virus then takes over the cell's reproductive mechanisms to reproduce its genetic material.

— Retroviruses carry their own enzymes (reverse transcriptase and integrase), which allow their RNA to be converted to double stranded DNA that can be replicated and translated by the host cell. Integrase transports the DNA copy to the cell nucleus where the viral DNA is spliced into the host cell DNA.

— Because the viral DNA is inserted into the host cell genetic information, the host cell passes the viral DNA down to its progeny resulting in enduring expression of viral genes.

— The host cell machinery reproduces the retroviral RNA, protein coat and enzymes.

— Adenoviruses contain double stranded DNA that requires no reverse transcription. They enter the nucleus through a nuclear pore and direct their replication using host cell enyzmes.

- The disadvantages of viral vectors are likewise considerable.

— They are disease-causing agents that if not properly deactivated could produce serious illness.

— The most frequently used viral vector, adenovirus, is a common human pathogen to which most people have been previously exposed. Thus use of the adenovirus as a delivery system for transgenes can evoke a systemic immune response.

— Adenovirus gene expression is transient because it does not insert its genes into the host genome.

— Retroviruses and lentiviruses do integrate their genes into the host genome, producing enduring expression. However, these viruses may insert genes in a site that produces uncontrolled cell division.

— There remains much to be learned about the molecular biology and safety of viral vectors.

Pharmaceutical example

In 1999 a young patient in a clinical trial died as a result of a severe immune response to the adenovirus vector used in his gene therapy. The next year, nine children were successfully treated with stem cells modified by a retrovirus but subsequently three of the nine children treated have developed a leukemia-like illness.

- Despite these disadvantages considerable progress has been made in developing viral vectors to transfer nucleic acids to somatic cells to produce a therapeutic effect.

— One approach has been to remove the genes responsible for replication from the viral genome.

— The gene of interest is spliced into the viral genome along with promoters and the virus is grown in human embryonic kidney cells and purified with chromatographic methods.

Pharmaceutical example

About 50% of tumor cells are deficient in the tumor suppressor protein, p53. Restoration of p53 function by replacement of the gene expressing this protein induces cell death or cell cycle arrest in various p53 deficient cancers. The first gene therapy virus approved for clinical use was Gendicine, a replication incompetent adenovirus loaded with the p53 transgene. The product was approved in China in 2003 for the treatment of head and neck, squamous cell carcinoma. A similar product, Advexin (Introgen; Austin, TX) is under development for use in the US and Europe. Another approach is to make the adenovirus 'conditionally replicative' such that it is only capable of replicating in target cells. Oncorine is the first commercially available conditionally replicative adenovirus. It also carries the p53 gene and was approved in China in 2006 for head and neck cancers. It is an adenovirus vector that only replicates in p53 deficient cells. The product selectively suppresses the growth of transfected tumor cells by transferring the gene for a functional p53 protein.

Passive targeting (Figure 9.5)

Key Point

- Normal tissues restrict the diffusion of large molecules or particles through their capillary systems while solid tumors have poorly differentiated and disorganized capillary systems that allow drug carriers between 0.01 and 0.1 μm into the tumor.

Nano-sized drug carriers may be targeted or directed to their drug targets using enhanced permeability and retention in tumor cells with leaky vasculature.

Normal tissues restrict the diffusion of large molecules or particles through their capillary systems while solid tumors have poorly differentiated and disorganized capillary systems that are 'leaky'.

Drug carriers between 0.01 and 0.1 μm are able to diffuse into the tumor and are retained within the interstitium because of inadequate lymph drainage in the tumor. Other pathological states that may demonstrate enhanced permeability and retention of nano-sized drug carriers include infarcted tissue, chronic inflammation and infections.

Active targeting (Table 9.9)

Key Point

- Active targeting of drug delivery systems is accomplished either by ligand mediated interaction with target cells or by release of the drug from the carrier in response to a condition found primarily in the target tissue.

- One approach to actively targeting drug delivery systems is ligand mediated interaction with target cells.

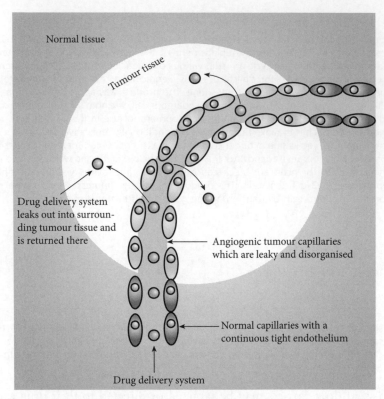

Figure 9.5 Enhanced permeability and retention targeting in tumor tissue. (Reproduced with permission from Y Perrie, T Rades, FASTtrack, Pharmaceutics – Drug Delivery and Targeting, 2 edn, p. 156, published by Pharmaceutical Press, 2012.)

Table 9.9 Targeted formulations

Delivery system	Drug	Challenges	Location of target	Targeting mechanism
Nanoparticles – albumin bound; powder for suspension	Paclitaxel (Abraxane); used intravenously	Toxicity, solubility	Albumin binding protein, cell surface	Ligand targeted to albumin binding protein overexpressed in breast cancer cells. Transcytosis
Monoclonal antibody	Trastuzumab (Herceptin)	Mononuclear phagocyte system uptake	HER2 protein	Receptor overexpressed in HER2 positive breast cancer cells
Pegylated, folate targeted	Doxorubicin (Ebewe)	Toxicity	Folate receptor	Increased expression in tumor cells

— A 'ligand', such as a monoclonal antibody, binds preferentially to a specific site on the targeted cell.

Pharmaceutical example

Trastuzumab binding selectively to the HER2 protein on tumor cells is one example of ligand mediated active targeting.

— Another approach using monoclonal antibodies has been to link an antineoplastic drug to a monoclonal antibody directed at a tumor associated antigen. These drug carriers are called *immunoconjugates* or antibody–drug conjugates. However, addition of an antibody to a drug delivery system may create a particle size that is recognized and engulfed by macrophages before it escapes the circulation to bind its target cell.

— Many tumor associated antigens and receptors have been identified that are expressed more frequently on the surface of tumor cells than on normal cells. These ligands may be attached to liposomes, micelles, nanoparticles and dendrimers to provide for delivery to the cell type expressing the receptor.

● Active targeting also may be accomplished by developing a carrier that releases the drug in response to a particular condition found primarily in the target tissue.

— Drugs may be covalently attached to the carrier via a hydrazone bond that is stable at the pH of the serum but is cleaved in the mildly acidic endosomes (pH 5.0–6.5) or lysosomes (pH 4.5–5.0).

— Peptide bonds can be quite stable in the serum as proteases are less active at physiological pH than in the lysosome.

— Disulfide bonds can be used to link drugs to their carrier and will be released intracellularly by glutathione.

— Micelles and liposomes have been prepared that disassociate at lower pH or higher temperature.

● Cell penetrating peptides may be attached that trigger cell uptake of macromolecules by endocytosis and release from the endosome and penetration of the nuclear membrane.

Biosimilars

As part of the Affordable Care Act of 2010, Congress directed that an abbreviated pathway be developed for the licensure of biopharmaceutical drugs shown to be similar to products previously approved in order to encourage price competition and affordable care. The FDA has developed three guidances to provide information to companies and the public about how biosimilar products will be evaluated.

> *Definition*
>
> *Biosimilarity* is defined as a biological product that is highly similar to the reference product, in particular that there are no clinically meaningful differences between the biosimilar and the reference product in terms of safety, purity and potency.

Reference products are given a 12 year exclusivity period from the date of first approval. A biosimilar product may have different inactive ingredients but would be expected to have the same route of administration, indications for use and performance such that it could be safely used interchangeably with the reference product. A company desiring to license a biosimilar product will submit pharmacokinetic and pharmacodynamic studies directly comparing the performance of the reference product with the biosimilar product. As you enter pharmacy practice prepare yourself to evaluate the available literature in order to appropriately select interchangeable biosimilar drug products.

Questions and cases

1. Delivery systems for biopharmaceutical macromolecules create steric hindrance around the macromolecular drug because:

   (a) This enhances their absorption

   (b) This reduces their elimination

   (c) This improves receptor binding

   (d) This facilitates their uptake by mononuclear phagocytes

2. The design characteristic that facilitates uptake by the receptor dependent pathway is:

   (a) Attachment of polyethylene glycol

   (b) Attachment of monoclonal antibodies

   (c) Attachment of cationic molecules

   (d) Use of lipids with high phase transition temperatures

3. The design characteristic that reduces recognition by mononuclear phagocytes is:

   (a) Attachment of polyethylene glycol

   (b) Attachment of monoclonal antibodies

   (c) Attachment of cationic molecules

   (d) Use of lipids with high phase transition temperatures

4. Which of the following polymers is not biodegradable?

   (a) Albumin

   (b) Poly D,L-lactic-coglycolic acid

   (c) Polyethylene glycol

   (d) Polylysine

5. Attaching polyethylene glycol to a protein drug will:

    (a) Decrease its half life

    (b) Increase its absorption rate

    (c) Reduce the frequency of dosing

    (d) Facilitate its uptake by receptor-mediated endocytosis

6. The release of drug from nanoparticles:

    (a) Is faster from small particles than large particles

    (b) Is increased by PEGylation

    (c) Must occur before gaining entry to the tumor cell

    (d) Occurs more slowly than from microspheres

7. You have a protein drug that has been formulated as a stealth liposome. You would predict that this liposome would:

    (a) Not be engulfed by liver and spleen macrophages

    (b) Reduce protein aggregation

    (c) Release its drug cargo by fusion with the plasma membrane

    (d) All of the above

8. The best viral vector for replacement of a defective gene in a hereditary disorder would be:

    (a) Adenovirus because it has broad cell preferences

    (b) Lentivirus because its duration of gene transfer is stable

    (c) Adeno-associated virus because it is not associated with human diseases

    (d) Herpes simplex virus because its cell preference is narrow

Cases
Students may need to refer to outside references to answer these questions.

9. You have a patient who is starting drug therapy with paclitaxel albumin bound nanoparticles. Given how colloidal delivery systems distribute, what organ is most likely to be the target of toxicity for a colloidal anticancer drug? What lab tests should be used to monitor for potential general toxic insult to this organ?

10. Compare the conventional dosage forms of doxorubicin and morphine with the liposomal form in terms of the following pharmacokinetic parameters. What can you conclude about the effect of the colloidal product on the absorption, distribution, and elimination of the drugs they enclose?

Product	Doxil 1.5 mg/kg	Doxorubicin injection 1.2 mg/kg
Parameter	Mean	Mean
Clearance (mL/hour)	23	25 300
Volume (L)	3	365
AUC (mg hour/L)	4082	3.5
$T_{1/2}$ (hour)	84	0.06/10.4

*Biphasic.

From the DepoDur labeling on the FDA Drugs website.

Product	DepoDUR 5 mg	Morphine sulfate injection
Parameter	Mean (SD)	Mean (SD)
C_{max} (ng/mL)	7.1 (3.4)	23.3 (12.8)
T_{max} (hour)	1 (0.3–4)	0.3 (0.3–2)
AUC (ng hour/mL)	38.8 (10.4)	42.8 (8.4)
$T_{1/2}$ (hour)	3.8 (1.0)	2.2 (0.5)

11. What effect would these changes have on the toxicity of the drugs?

Selected reading

1. R Airley. Cancer Chemotherapy: Basic Science to the Clinic, Chichester, UK: John Wiley & Sons, 2009.
2. DS Pisal, MP Kosolski, SV Balu-lyer. Delivery of therapeutic proteins. J Pharm Sci 2010; 99: 2257–2575.
3. J Goodchild. Therapeutic oligonucleotides. Methods Mol Biol 2011; 764: 1–15.
4. JK Raty, JT Pikkarainen, T Wirth et al. Gene therapy: the first approved gene-based medicines, molecular mechanisms and clinical indications. Curr Molecul Pharmacol 2008; 1: 13–23.
5. K Singha, R Namgumg, WJ Kim. Polymers in small-interfering RNA Delivery. Nucl Acid Ther 2011; 21: 133–147.
6. BD Ulery, LS Nair, CT Laurencin. Biomedical applications of biodegradable polymers. J Polym Sci B Polym Phys 2011; 49: 832–864.
7. A Bolhassani. Potential efficacy of cell-penetrating peptides for nucleic acid and drug delivery in cancer. Biochem Biophys Acta 2011; 1816: 232–246.
8. SL Ellison, DL Hunt. Perceived versus real risks of handling gene transfer agents in the pharmacy environment. Am J Health-Syst Pharm 2010; 67: 838–848.
9. K Maruyama. Intracellular targeting delivery of liposomal drugs to solid tumors based on EPR effects. Adv Drug Del Rev 2011; 63: 161–169.
10. MV Pasquetto, L Vecchia, D Covini et al. Targeted drug delivery using immunoconjugates: Principles and application. J Immunother 2011; 34: 611–628.

10

Drug delivery to the eye

Learning objectives

Upon completion of this chapter, you should be able to:

- Predict the effect of each of the following drug properties on the rate and extent of drug travel to its target in the anterior or posterior segment of the eye: water solubility, log *P*, molecular weight, stability in solution, enzymatic degradation and location of drug receptors.

- Identify the challenges and opportunities presented by each of the following route related characteristics to the rate and extent of drug travel to its target from the anterior or posterior segment of the eye: accessibility, surface area, permeability and type of absorbing membrane, nature of the body fluids that bathe the absorbing membrane, retention of the dosage form at the site of absorption, location and number of drug metabolizing enzymes or extremes of pH, and blood flow primacy of the eye.

- List the categories of ingredients that are added to ophthalmic dosage forms, and identify how the ingredient contributes to the manageable size of the dosage form, its palatability or comfort, its stability (chemical, microbial and physical), the convenience of its use and the release of drug from the dosage form.

- Describe the dosage forms/drug delivery systems used by this route and specify how well they are retained.

- Explain how the qualities of the dosage forms are evaluated *in vitro* and/or *in vivo*.

- Explain to patients and/or other healthcare providers how to use, store and/or prepare dosage forms for administration via the ophthalmic route.

Introduction

The delivery of drugs to the eye is a rapidly changing area of dosage form design. Medications applied to the front of the eye to treat superficial infections or conditions of the structures underlying the cornea are substantially lost because of tearing, drainage and spillage from the eye. Improvements in dosage forms to treat conditions in the anterior portion of the eye have reduced the need for frequent administration and the occurrence of systemic side effects. However, there remains a strong need to educate patients about proper use of eye drops to optimize their therapeutic effects. Much effort is currently directed towards development of dosage forms for the delivery of drugs, particularly products of biotechnology, to the posterior segment of the eye. At present the posterior segment is inaccessible to treatments applied to the front of the eye and requires intraocular injection. With the amount of research activity in this area, the pharmacist may anticipate significant new delivery systems to treat sight threatening disease in the posterior segment.

Characteristics of the ophthalmic route

See Appendix Table A.1, Table 10.1, Figures 10.1 and 10.2.

Drug travel from dosage form to target

Drug delivery scientists divide the eye into the anterior segment and the posterior segment since reaching the targets in these two areas of the eye presents different challenges.

- The challenge of drug delivery to the anterior segment has been to maintain a high concentration of drug in the tear film for long enough to provide therapeutic drug levels at the target site.
- It is estimated that 90–95% of drug applied to the eye is lost before it reaches the target.
  - A large proportion of drug is diluted by the tears and spills or drains out of the eye. Between 50% and 80% of the dose drains out of the eye into the nose where it can be absorbed into the systemic circulation with no first pass.
  - Because this systemic absorption can cause side effects, the pharmacist should recommend pressure on the lacrimal ducts to reduce losses.
- The cornea is moistened by tears produced by glands and cells in the upper and lower eyelids. (pH 7.0–7.4)
  - The tear film on the cornea consists of water, mucus and an oily layer that prevents evaporation.
  - The eye normally maintains a tear volume about 7–10 µL and at most can hold 30 µL without overflow.
- Two possible routes of absorption from the front of the eye exist: one through the cornea and the other through the conjunctiva and sclera.
  - The cornea is the desired route for permeation of many drugs applied to the eye because their receptors are located beneath it.
  - Corneal surface is relatively small, 130 mm^2, and is less permeable than most other epithelial membranes.
  - The conjunctiva and underlying sclera are more permeable to drug molecules including hydrophilic and high molecular weight drugs although, because of their greater blood flow, the drug may be carried away from the target area.
  - Drug is absorbed via both routes primarily by passive diffusion.

Table 10.1 Structures of the eye and their drug targets

Structure	Function	Drug targets
Anterior segment		
Cornea	Provides initial refraction or bending of light as it enters the eye	Corneal injuries: anti-infective drugs
Conjunctiva	Outermost surface membrane that covers the white of the eye and lines the eyelids	Allergic conjunctivitis: antihistamines, steroids and mast cell inhibitors. Bacterial conjunctivitis: anti-infective drugs
Sclera	Protective outer support extending from the edge of the cornea to encase the posterior 5/6 of the eyeball. The 'white' of the eye	
Lower conjunctival sac	Portion of the conjunctiva that covers the lower eye and eyelid	As above for conjunctiva
Lacrimal gland	Produces tears that moisten the cornea and flush foreign objects and irritating substances from the eye	Keratoconjunctivitis sicca: cyclosporin
Pupil	Admits light entering through the cornea to the inner eye	
Uveal tract	Includes the iris, ciliary body and choroid	Uveitis: antivirals, steroids
Iris	Colored part of the eye that controls the size of the pupil	Mydriatics stimulate alpha (α) adrenergic receptors to dilate pupil. Miotics stimulate muscarinic (M) cholinergic receptors to constrict pupil
Ciliary body	Ciliary muscle adjusts the lens for near and far vision. Secretory processes produce aqueous humor	Aqueous humor production. Reduced by beta-receptor blockers (timolol)
Aqueous humor	Fluid between the cornea and the lens that maintains the shape of the anterior portion of the eye	
Trabecular meshwork	Drains the aqueous humor from the anterior chamber into the canal of Schlemm and thence to the episcleral venous plexus	
Uveoscleral route	Outflow route for aqueous humor through the ciliary muscles into suprachoroidal space	Aqueous humor outflow. Increased by prostaglandin F receptor agonists (latanoprost). Increased by alpha-receptor agonists (epinephrine)
Lens	Focuses light on the retina	

(continues)

Table 10.1 (*continued*)

Structure	Function	Drug targets
Posterior segment		
Vitreous	Colorless gel-like substance that fills eye behind the lens and maintains the shape of the posterior eyeball	Vitreal replacements
Retina	Inner layer of the posterior segment containing the photoreceptors (rods and cones) responsible for vision	Retinitis: anti-infectives. Diabetic retinopathy
Retinal pigmented epithelium	Layer of pigmented cells that prevent light reflection in the eye	Macular degeneration: steroids, pegaptanib, ranibizumab, bevacizumab
Choroid	Layer of blood vessels between the retina and the sclera providing nutrients and removing wastes. Continues with the sclera into the anterior segment	Chorioretinitis: anti-infectives. Macular degeneration: steroids, pegaptanib, ranibizumab, bevacizumab
Sclera	Protective outer support extending from the edge of the cornea to encase the posterior 5/6 of the eyeball. The 'white' of the eye	
Optic nerve	Transmits light signals as electrical impulses to the brain	

- Drug metabolizing enzymes include esterases and peptidases. In addition, cytochrome P450 activity has been estimated to be about 4% of liver activity.
- The structures of the eye are not readily accessible to drug from the systemic circulation. There are two major blood supplies to the eye: one to the uvea and another to the retina. Both are covered with tight junctions and prevent substantial penetration from systemic sources.
- The back of the eye is isolated from absorbed materials originating in the front of the eye or from the systemic circulation making this area inaccessible to drugs unless injected directly into the eye.
  - The retinal pigmented epithelium that separates the posterior segment from its blood supply is a single epithelial layer with poor permeability.
  - Currently drug therapy is delivered to drug targets in the posterior segment using intravitreous or periocular routes of injection.
- The vitreous humor, with its considerable volume and viscosity, represents a significant barrier to drug diffusion from the front of the eye.
- A few drug transporters have been identified in human ocular structures but they are not well characterized (Table 10.2).

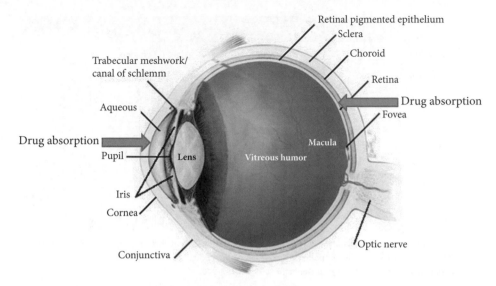

Figure 10.1 Structures of the anterior and posterior segments and their drug delivery routes. (Modified from National Eye Institute, National Institutes of Health, NEI catalog number NEA04.)

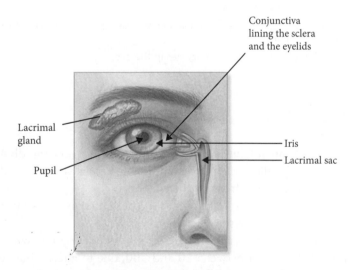

Figure 10.2 The front of the eye. (Reproduced with permission from National Library of Medicine Medical Encyclopedia at http://www.nlm.nih.gov/medlineplus/ on July 3, 2012.)

Table 10.2 Transporters in the eye

Transporter	Location	Function
p-glycoprotein	Conjunctiva, RPE	Efflux
MRP 1, 5	Cornea (1,5) conjunctiva (1)	Efflux
BCRP	Cornea	Efflux
LAT1	Cornea, RPE	Influx
LAT2	RPE	Influx
PEPT1	Cornea	Influx
PEPT2	Cornea	Influx

MRP = multi-resistant polypeptide; BCRP = breast cancer resistance protein; LAT = large neutral amino acid transporter; PEPT = peptide transporter; RPE = retinal pigmented epithelium.
Information from Eur J Pharmaceut Biopharm 2005; 60: 227–240.

Key Points

- The front of the eye is accessible but liquid dosage forms drain quickly to the nose.
- The cornea is multiple layers of stratified squamous epithelium with moderate permeability.
- Drug is absorbed via passive diffusion.
- The posterior segment where the retina and macula are located is difficult to access with drug treatments.
- There are few drug metabolizing enzymes in the eye.

Drug properties and the eye

- If a drug is not already in solution, it must dissolve very rapidly to be absorbed through the cornea due to rapid and extensive losses of the drug after application to the front of the eye.
- Drugs with poor water solubility may be presented as oily solutions, ointments, emulsions or suspensions.
- Drugs absorbed through the cornea are modestly lipophilic with a partition coefficient of between 10 and 100, due to alternating layers of 'fat-water-fat' in the corneal barrier.
- Small, hydrophilic drugs (molecular weight ≤500) may enter via paracelllular tight junctions.
- Drugs traversing the pigmented retinal epithelium in the posterior eye must be small (molecular weight ≤400) and relatively lipid soluble.
- Drugs that are not stable in solution may be prepared as powders for reconstitution or solid inserts.

Table 10.3 Formulation of commercially available ophthalmic dosage forms

Dosage form	Comfort	Stability	Drug retention	Vehicle
Anterior segment				
Latanoprost 0.005% ophthalmic solution	Sodium chloride	Benzalkonium chloride, sodium phosphate, monobasic and dibasic		SWI
Timolol 0.25% gel-forming solution	Mannitol	Benzododecinium bromide, boric acid, polysorbate 80	Gellan gum, tromethamine (pH adjustment)	SWI
GenTeal severe dry eye ophthalmic gel	Sorbitol	Phosphoric acid, sodium perborate	Hypromellose, carbomers, sodium hydroxide	Distilled water
Dexamethasone 0.01% ophthalmic suspension	Sodium chloride	Benzalkonium chloride, citric acid, sodium phosphate, dibasic, EDTA, polysorbate 80	Hypromellose (hydroxyl-propylmethylcellulose)	Purified water
Erythromycin 0.5% ophthalmic ointment			White petrolatum	Mineral oil
Bioerodible insert (Lacrisert)			Sterile rod of hydroxypropylcellulose (1.27 × 3.5 mm long)	
Posterior segment				
Pegaptanib (macugen sodium) solution for intravitreous injection	Sodium chloride	sodium phosphate, monobasic and dibasic		SWI
Ganciclovir intravitreal implant (Vitrasert)		Magnesium stearate	Polyvinyl alcohol ethylene vinyl acetate polymers	Diffusion
Dexamethasone intravitreal implant (Ozurdex)			poly (D,L-lactide-co-glycolide) PLGA	Bioerodible

SWI = sterile water for injection.

- Drugs applied to the eye are subject to relatively low enzymatic activity. Drugs may be administered as prodrugs and be activated by esterases (dipivefrin and latanoprost).
- The drug targets in the anterior segment are found on the surface of the eye and beneath the cornea and are accessible to drugs applied to the front of the eye. Drug receptors in the posterior segment are inaccessible from the front of the eye and must be injected (see Table 10.1).

Dosage forms for the ophthalmic route (Table 10.3)

Design of dosage forms for the anterior segment

1. One dose in a manageable size unit
   - The drug dose must be contained in a 30 μL drop or smaller.
   - Most drugs can be prepared as ophthalmic solutions but these have poor contact time and require frequent administration.
   - Drugs with poor water solubility can be prepared as ointments, emulsions, suspensions or inserts.
2. Comfort
   - Ophthalmic liquids are preferably isotonic with a pH of 4.5–8.
   - The particle size of ophthalmic suspensions must be less than 10 μm to prevent tearing and irritation to the eye.
3. Stable
   - All products for the eye must be sterile.
   - Multidose products must be preserved to be resistant to growth.
   - Drugs with poor chemical stability may be prepared as powders for reconstitution or solid inserts.
4. Convenient/easy to use
   - Because of the significant losses of drug, topical application results in short peaks and long valleys of drug concentration and may require frequent dosing unless the dosage form can be designed to provide adequate contact time.
   - An elegant approach has been to prepare ophthalmic products that are solutions in their container but form gels in the eye. These products are called gel forming solutions and contain gelling agents that increase viscosity via a variety of mechanisms when they contact the tears (Table 10.4).

Pharmaceutical example

The gel forming agent in Timoptic (timolol) increases its viscosity in the presence of the higher ionic strength of the tears.

Table 10.4 Viscosity enhancers for ophthalmic products

Viscosity enhancer	Description	Gelling properties
Methylcellulose, hydroxypropyl methylcellulose	Will come out of solution when autoclaved and redisperse on cooling	Mucoadhesive, liquid at room temperature but gel at higher temperature. Conc. 0.2–2.5%
Polaxamer (Pluronic)	Copolymer of polyethylene oxide and polypropylene oxide	Liquid at room temperature but gels at body temperature, conc. 0.2–5%
Alginate	Mixture of polyuronic acids derived from algae	Gels in the presence of polyvalent cations
Gellan gum	Linear anionic heteropolysaccharide	Gels in the presence of cations in the tears, mucoadhesive
Carbomer	Crosslinked poly acrylic acid	Liquid at low pH but gels at pH >7.0, conc. 0.05–2.5%
Polyvinyl alcohol	Water soluble polymer of varying molecular weight	Concentration 0.1–5%, viscosity is concentration dependent, gels in the presence of borax
Polyvinylpyrrolidine (povidone)	Linear polymer of 1-vinyl-2-pyrrolidinone with solubilizing properties	Viscosity is concentration dependent, conc. 0.1–2%
Xanthan gum	Pseudoplastic, high molecular weight polysaccharide	Mucoadhesive, gel-forming

- Ophthalmic ointments provide good contact time and the added benefit of emolliency for patients with dry eye.
- Most ophthalmic ointments are formulated with white petrolatum and mineral oil (liquid petrolatum), the water free product providing good chemical and microbial stability.

Pharmaceutical example

Cyclosporine is formulated as an emulsion using carbomer, castor oil, glycerin, polysorbate 80 and distilled water.

- Suspensions are unlikely to provide longer residence time than solutions since their drainage rate may exceed the dissolution rate of their particles.
- Ocular inserts intended for treatment of conditions of the anterior segment are small, solid disks or rods placed in the lower conjunctival sac that permit less frequent drug administration.

Pharmaceutical example

Ocusert is an example of a nonerodible insert that contains the drug enveloped in a thin, ethylene vinyl acetate membrane. The insert must be removed before placement of the next dose; at present Ocusert is not commercially available.

- Erodible ocular inserts are prepared from soluble matrices composed of hydroxypropylcellulose, hyaluronic acid, carbomer or poly lactic-coglycolic acid (PLGA). The bioerodible insert is placed in the lower conjunctival sac and draws water from the tears to form a gel-like mass that releases the drug over time.
- Inserts or gels made with bioadhesive materials are promising new products to enhance drug retention in the anterior segment.

5. Release of drug
   - The most common ophthalmic dosage form is the solution.
   - Oily solutions, ointments and emulsions require diffusion of the drug from the vehicle to the cornea.
   - Ocular inserts can maintain release of the drug at a constant level over hours or days by diffusion (nonerodible) or by slow dissolution (erodible inserts).

Design of dosage forms for the posterior segment

1. One dose in a manageable size unit
   - The drug doses for ophthalmic injection are contained in 0.05–0.09 mL aqueous solution and must have water solubility that supports this small volume.
2. Comfort
   - Intraocular irrigating and injection solutions are preferably isotonic with a pH of 7.4.
3. Stable
   - Ophthalmic injections must be sterile, preservative-free and nonpyrogenic.
   - Drugs with poor chemical stability may be prepared as powders for reconstitution or solid inserts.
4. Convenient/easy to use
   - Injection into the posterior segment requires skilled practitioner administration.
     - Frequent injections into the posterior segment increase the risk of retinal detachment, endophthalmitis and vitreous hemorrhage.
   - Erodible and nonerodible inserts permit less frequent administration of drug.
     - Inserts must be placed surgically or with microtrochars.
     - Nonerodible inserts must be removed.

Table 10.5 Ingredients in compounded ophthalmic solutions, suspensions and injections

Ophthalmic solutions	Ophthalmic suspensions	Ophthalmic injections
Tonicity agents	Tonicity agents	Tonicity agents
Antioxidants	Antioxidants	Buffers
Antimicrobial preservative	Buffers	Solvent
Buffers	Antimicrobial preservative	
Viscosity agents	Wetting agent	
Solvent	Suspending agent	
	Solvent	

5. Release of drug
    *   Most of the currently available products for intravitreous injection are solutions or powders for solution that produce immediate release of drug.
    *   Erodible and nonerodible inserts produce release of drug over months or years.

Many of the drugs developed for the treatment of posterior segment disease (age related macular degeneration, diabetic retinopathy and retinal venous occlusive disease) are high molecular weight biopolymers. Developments that the pharmacist may expect in this area include sustained release formulations such as inserts, liposomes, microparticles and nanoparticles to reduce the frequency of administration, and products that enhance the accessibility of the posterior segment from the front of the eye.

Pharmaceutical example

Pegaptanib sodium (Macugen), is an oligonucleotide that binds vascular endothelial growth factor preventing its stimulation of vascular growth and inflammation. It is used for the treatment of wet type, age related macular degeneration. PEGylation of the macromolecule prolongs its activity such that it can be injected intravitreally every 6 weeks.

Pharmaceutical example

Retisert is a nonerodible polymeric disc that releases fluocinolone acetonide for up to 3 years for the treatment of chronic noninfectious uveitis in the posterior segment. It must be removed because it does not degrade *in vivo*.

Pharmaceutical example

Ozurdex is a bioerodible implant that releases dexamethasone for 6 months and is used to treat macular edema associated with retinal vein occlusion. It is a rod-shaped matrix composed of PLGA that degrades through non-enzymatic ester hydrolysis. The insert is placed in the vitreous using a microtrochar. There are several other insert devices under development for delivery of drugs to the posterior segment.

Ingredients

The general formulas for ophthalmic dosage forms are presented in Table 10.5 and common ingredients in Appendix Table A.2. Eye drops are buffered at the pH of maximum stability and for the comfort of the patient with a compromise in favor of drug stability often being necessary. The pH of the instilled solution will have a significant influence on the absorption of the drug because it will determine the pH of the local environment. For drugs that are bases in uncharged form, adjustment of the pH to match the tears will generally enhance absorption. The buffer capacity should be kept low so that the buffer system in the eye is allowed to reinstate the normal pH of the eye after the drug is administered.

Benzalkonium chloride is the most commonly used antimicrobial preservative for ophthalmic products being highly effective, well tolerated, stable and rapid acting. However, it disrupts the oily layer of tears that prevents their evaporation, binds soft contact lenses and can cause corneal irritation. Preservatives chosen for ophthalmic formulations vary in their ability to eliminate a broad range of bacteria and are fungistatic or at best are not rapidly fungicidal. Thus it is essential that patients be taught to handle the ophthalmic product to minimize the possibility of contamination.

Viscosity agents give ophthalmic formulations longer contact time with the eye providing more time for drug absorption. These agents may be added to ophthalmic solutions and suspensions or used to make ophthalmic gels. Some of the available gel products are liquids that gel when dropped into the eye. The viscosity of these gel forming ingredients may be controlled by temperature, pH or ionic strength. Some of the viscosity agents are mucoadhesive, meaning that they bind to the epithelial surface of the eye providing enhanced contact time. Ophthalmic solutions will generally have a viscosity range of 10–25 centipoises (mPa · s) and gels, 25–50 centipoises (mPa · s).

Preparation

Ophthalmic products must be sterile and thus must be prepared with the same attention to quality of ingredients, training of personnel and environmental control as parenteral products. Suspensions and ointments can be challenging to sterilize if the drug is not heat stable. Please see selected reading 1 for a detailed discussion of the compounding of ophthalmic products.

Quality evaluation and assurance

In addition to drug potency requirements and limitations on impurities, the USP specifies that commercially prepared ophthalmic products have the following specific qualities:

- Samples of all ophthalmic products must be tested and found to be sterile. (USP <71>).
- The USP requires that manufacturers test the efficacy of their ophthalmic preservative system against representative Gram positive, Gram negative and fungal organisms for 28 days. Manufacturers use the test as a minimum standard and attempt to formulate their products with a safety margin for preservation.
- Ophthalmic solutions may not have more than 50 particles >10 μm in diameter per mL, 5 particles >25 μm in diameter per mL, or 2 particles >50 μm in diameter per mL.
- Ophthalmic ointments may not have more than 5 metal particles >50 μm in diameter per tube for a lot of 10 tubes (USP <755>).
- Ophthalmic injections must meet the same particulate standards as parenteral injections.
- Individual drug monographs specify the pH range permitted for the individual drug solutions.

Counseling patients

Administration to the front of the eye

A number of studies have investigated the proper technique for the administration of eye drops to the front of the eye. Proper technique has been defined as placement of the drop on the corneal/conjunctival surface or in the lower conjunctival sac (Figure 10.3). The patient should then close the treated eye for about 5 minutes to allow drug absorption. There is some controversy about this point in the literature because the open eye has a larger capacity for liquid; however, closing the eye spreads the drug solution over a larger surface and prevents the patient from blinking, which can force the drug solution out of the eye through the lacrimal duct. In addition the patient should occlude the lacrimal duct by gently placing the index finger at the medial corner of the eye. This practice prevents loss of the medication into the nasopharynx and the possibility of systemic side effects. Doses to the same eye should be separated by at least 5 minutes since two drops will exceed the volume capacity of the eye. To prevent microbial contamination of the product, the patient should wash his/her hands before use and the tip of

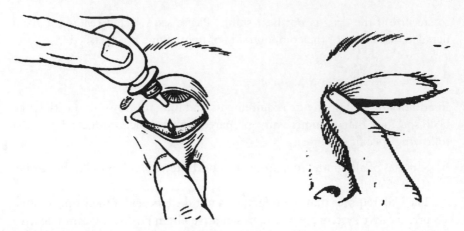

Figure 10.3 Administration of eye drops. Instilling eye drops and then punctal occlusion: good technique maximizes efficacy and can reduce systemic absorption by up to 50%. Reproduced with permission from The American Academy of Ophthalmology.

the dropper should not make contact with the patient's fingers or eyes. A study of the effect of technique on absorption into the anterior chamber found that 5 minutes of eyelid closure or nasolacrimal occlusion increased the peak fluorescein concentration in the anterior chamber by 69% and 46% respectively. This same group found that timolol *plasma* concentrations were significantly reduced by 5 minutes of nasolacrimal occlusion and eyelid closure. In a second study, 3 minutes of nasolacrimal occlusion significantly increased the effect of timolol on intraocular pressure in healthy volunteers. A recent study found no effect of 1–3 minutes of eyelid closure on the intraocular pressure lowering effect of topically applied prostaglandins. Common errors in patient administration technique include failure to wash hands, improper placement of drops and touching the dropper to the eye surface or lid.

Solutions, suspensions and gel forming solutions

The patient should be given the following instructions.

- Wash your hands.
- Tilt your head back and make a pouch by pulling out the lower lid.
- Place the eye drop into the pouch.
- You can lie down if you have difficulty making a pouch.
- To prevent drainage of the drug into the nose, hold your index finger next to the inner corner of your eye for about 5 minutes.
- Close your eye to spread the medicine across the surface and prevent drainage due to blinking.
- Allow about 5 minutes between two drops to the same eye to prevent overwhelming the volume capacity of the eye.

- Do not touch the dropper with your fingers or to your eye or any other surface.
- Eye drops should be used 30 minutes before putting in contact lenses or remove lenses before administration.

The patient may be given the following advice.

- If using antibiotic drops, do not wear contact lenses.
- The onset of most drug effects is within 30–60 minutes of application to the eye.
- Some eye drops require refrigeration (pilocarpine, latanoprost unopened).
- Shake suspensions well.
- Do not reuse unit dose eye drops once opened.
- Discard preserved eye drops after 60 days or manufacturer's recommendation.

Ophthalmic ointment and gels

The patient should be given the following instructions.

- Ointments blur vision because they are not miscible with tears, so administration before sleep is best to avoid disruption of your normal activities.
- Wash your hands.
- Apply the prescribed ribbon of ointment or gel to the pouch in the lower lid.
- Close your eye to spread the drug.
- Administer eye drops before ointments or gels.
- Wait 10 minutes after drops before administering ointments or gels.

Questions and cases

1. The major challenge in the administration of drugs to the front of the eye is:
    (a) The eye does not absorb water soluble drugs
    (b) The eye is extremely sensitive to locally applied drugs
    (c) A large proportion of drug applied to the eye is lost before it reaches the receptors
    (d) The absorbing membrane is not accessible
    (e) The eye has high drug metabolizing capability
2. To move into the subcorneal structures of the eye, a drug requires:
    (a) Adequate lipid and water solubility
    (b) High molecular weight

(c) Penetration enhancers

(d) Occlusive ingredients

(e) All of the above

3. A patient should be advised to press the index finger against the inner corner of the eye after administering eyedrops. The reason for this counseling advice is:

   (a) It reduces tearing and subsequent flushing of the drug from the eye

   (b) It will reduce losses of the drug through the tear ducts into the gastrointestinal tract

   (c) It will reduce systemic side effects through drainage of the drug into the gastrointestinal tract

   (d) All of the above

   (e) Answers (b) and (c) only

4. To not exceed the capacity of the eye, the patient should be advised to:

   (a) Press the index finger against the inner corner of the eye

   (b) Avoid touching the dropper to the eye

   (c) Place the eyedrop in the lower conjuctival sac

   (d) Wait 5 minutes before applying a second drop to the same eye

5. Research the chemical and biological properties of the following drug. DrugBank is a good source for pharmaceutical properties of drugs with the exception of *in vitro* stability and side effects (http://www.drugbank.ca/). *Handbook of Injectable Drugs* is a good source for information on drug stability in solution. What challenges do these drugs present to the formulator in terms of dissolution in body fluids, movement through membranes, stability in dosage forms, loss of drug due to metabolism, access to receptors or selectivity of the target?

Timolol maleate for administration to the anterior portion of the eye

Design issue	Drug property	Analysis
Drug must be dissolved before it can cross membranes		
Drug must have sufficient lipophilicity to cross membranes		
Large molecular weight drugs have difficulty moving across membranes		
Drugs may hydrolyze, oxidize, photolyze, or otherwise degrade in solution		
Is the enzymatic degradation of this drug substantial enough to prevent its use by this route?		
Intracellular drug receptors or receptors in the brain or posterior eye are challenging to target		
Some drugs cause serious toxicity as a result of poor selectivity of drug action		

6. Research the inactive ingredients in the following dosage form. This information can be found in the labeling which can sometimes be found on the FDA website, National Institute of Health's *DailyMed* or from the *Physicians' Desk Reference* which is a collection of product labeling published by Thomson Healthcare. Determine the function of the excipients using Appendix Table A.2. Categorize them as contributing to the manageable size of the dosage form, its palatability or comfort, its stability (chemical, microbial, or physical), the convenience of its use, or the release of drug from the dosage form. Note that ingredients may be included in many dosage forms for more than one purpose.

Timolol maleate ophthalmic gel (Timpoptic XE)

Ingredient	Category	Purpose of ingredient

7. Use this table to summarize the characteristics of drug delivery to the front of the eye and its travel to the drug receptor in the anterior segment. Think about what you know about the anatomy and physiology of the eye as it applies to each of the route-related characteristics for dosage form design and whether the characteristic represents a challenge or an opportunity for drug delivery.

Ophthalmic route – anterior segment

Questions	Answer	Challenge or opportunity?
What is the membrane that will absorb the drug from this route and how accessible is it?		
Is the absorbing membrane single or multiple layers and how permeable is it?		
How large is its surface?		
What is the nature of the body fluids that bathe the absorbing membrane?		
Will the dosage form remain at the absorbing membrane long enough to release its drug?		
Will the drug encounter enzymes or extremes of pH that can alter its chemical structure before it is distributed to its target?		
What is the blood flow primacy to the absorbing membrane and the distribution time from it?		

8. What would you expect to happen to a pseudoplastic liquid when exposed to the shear forces of blinking?

9. A patient taking Tears Naturale II lubricant ophthalmic drops (which contain Dextran 0.01%, hydroxypropylmethylcellulose 0.3%, polyquaternium 1, potassium chloride, sodium borate, sodium chloride, and distilled water) is on a camping vacation. She has had her 'eye drops' in a small bag in the car. She reports that she could not get the drug to drop out of the dropper last night although she can see medication in the bottle. It seemed to be in a 'blob' instead of liquid like usual. Explain to the patient how her drops work and how they should be stored while on vacation.

10. A middle aged man comes into the pharmacy with a visibly reddened right eye. He tells the pharmacist that he poked his eye with a weed stalk while gardening and would like a recommendation for a product to prevent infection and soothe the irritation. The pharmacist's initial reaction is to refer him to a physician, however, the man explains that he doesn't have health insurance and he isn't going to pay a lot of money to have a physician 'fool with something little like this'. The man selects a tube of triple antibiotic ointment. Explain to the patient the difference between products for the skin and products made for the eye. Are there free clinics in your community or patient assistance programs through the pharmaceutical manufacturers that could help this patient? Discuss your own position on 'frontier medicine' in which patients express a desire to take care of themselves with minimal assistance from the healthcare system.

Selected reading

1. JE Thompson, LW Davidow. A Practical Guide to Contemporary Pharmacy Practice, 3rd edn. Lippincott, Williams & Wilkins, Philadephia, 2009.
2. MR Prausnitz, JS Noonan. Permeability of cornea, sclera and conjunctiva: A literature analysis for drug delivery to the eye. J Pharm Sci 1998; 87: 1479–1488.
3. PM Hughes, O Olejnik, JE Chang-Lin et al. Topical and systemic drug delivery to the posterior segments, Adv Drug Del Rev 2005; 57: 2010–2032.
4. R Gaudana, HK Ananthula, A Parenky et al. Ocular Drug Delivery. AAPS J 2010; 12: 348–360.
5. E Eljarrat-Binstock, J Pe'er, AJ Domb. New techniques for drug delivery to the posterior eye segment. Pharm Res 2010; 27: 530–543.
6. R Gupta, B Patil, B Shah et al. Evaluating eye drop instillation technique in glaucoma patients, J Glaucoma 2012; 21: 189–192.
7. P Furrer, JM Mayer, R Gurny. Ocular tolerance of preservatives and alternatives. Eur J Pharmaceut Biopharm 2002; 53: 263–280.

11

Drug delivery to and from the oral cavity

Learning objectives

Upon completion of this chapter, you should be able to:

- Identify the challenges and opportunities presented by each of the following route related characteristics to the rate and extent of drug travel to its target: accessibility, surface area, permeability and type of absorbing membrane, nature of the body fluids that bathe the absorbing membrane, retention of the dosage form at the site of absorption, location and number of drug metabolizing enzymes or extremes of pH and blood flow primacy of the oral cavity.

- Predict the effect of each of the following drug properties on the rate and extent of drug travel to its target from the oral cavity: water solubility, log P, molecular weight, stability in solution, enzymatic degradation and location of drug receptors.

- List the categories of ingredients that are added to dosage forms for the oral cavity, and identify how the ingredient contributes to the manageable size of the dosage form, its palatability or comfort, its stability (chemical, microbial and physical), the convenience of its use and the release of drug from the dosage form.

- Describe the dosage forms/drug delivery systems used by this route and whether immediate or extended release.

- Outline the movement of a drug from its site of administration to its target including:
  - release
  - absorption
  - distribution to target.

- Explain how the qualities of the dosage form are evaluated *in vitro* and/or *in vivo*.

- Explain to patients and/or other healthcare providers how to use, store and/or prepare dosage forms for administration via the oral cavity.

Introduction

There is considerable interest in developing dosage forms for the delivery of drugs from the oral cavity to drug targets in the systemic circulation. The absorbing membrane is easily accessible and there are significantly fewer drug metabolizing enzymes in the mouth than are encountered in the stomach and small intestine. The considerable blood flow to the area drains to the systemic circulation circumventing the first pass losses to the liver that occur when drugs are swallowed. The constant flow of saliva and the substantial mobility of the mouth while talking, eating or drinking create a challenge to the retention of dosage forms and the drug released in the oral cavity. Thus dosage forms for application to the buccal mucosa should be designed with sustained release and mucoadherent properties whether the drug is intended to provide systemic concentrations or to treat a local condition in the mouth. In general the membranes of the oral cavity are less permeable than the mucosa of the small intestine, necessitating the use of penetration enhancers for drugs with targets outside the oral cavity.

Characteristics of the oral cavity route

- Drugs are applied to the entire oral mucosa for treatment of local conditions of the mouth or to the sublingual area and the buccal mucosa for treatment of conditions outside the mouth (Figure 11.1).

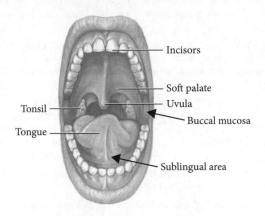

Figure 11.1 Buccal and sublingual areas of the oral cavity

Key Point

- The oral cavity has a readily accessible membrane with modest permeability.

- The sublingual membrane is relatively thin and the blood flow is high, creating favorable conditions for rapid absorption of drugs.
- The gums and the hard palate are subjected to hot and abrasive foods and are keratinized epithelia. These cells are packed with the structural protein, keratin, which produces a dense, low permeability barrier that protects underlying tissues.
- The buccal mucosa (Figure 11.2) is a nonkeratinized region composed of 40–50 cell layers of stratified squamous epithelium. Tight junctions are rare in the oral mucosa. Instead the lipophilic membranes are surrounded by an intracellular matrix composed of relatively polar lipids.

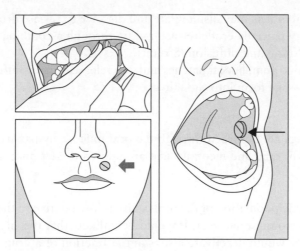

Figure 11.2 Fentanyl buccal tablet placement. (Reproduced with permission from CA Langley, D Belcher, Applied Pharmaceutical Practice, p. 248; published by Pharmaceutical Press, 2009.)

The mucous layer covering oral epithelial cells consists of water and mucins, negatively charged, linear glycoproteins.

- The total surface area of the oral cavity is estimated to be about 200 cm^2 while the unkeratinized portion is about 100 cm^2.
- Total salivary secretion is between 0.5 and 2 L per day resulting in about 1 mL of saliva in the mouth at any one time for drug dissolution. While the volume is small, water soluble drugs may have a large concentration gradient to drive their diffusion. The pH of saliva is between 5.8 and 7.4.
- The oral mucosa is highly vascularized, and as a result the rate of drug absorption is not limited by blood flow but by permeability.

Drug travel from dosage form to target

Liberation

- A variety of mouthwashes containing anti-infectives or local anesthetics are available as solutions or suspensions, the latter requiring dissolution before drug is available for action.
- Sublingual:
  - Tablets release drug by dissolution in saliva.
  - Aerosol product is presented as a solution in alcohol.
- Buccal:
  - Lozenges and troches are designed to dissolve slowly to release medication.

— Buccal tablets may disintegrate and dissolve to release drug (Fentora) or may partially erode releasing drug through a combination of dissolution and diffusion (Striant).

— Nicotine gum, buccal patches and gels release drug by diffusion out of the dosage form into the saliva.

Absorption

- Most drugs are absorbed through the oral mucosa by passive diffusion.
  - Salicylic acid and nicotinic acid are known to be transported.

Metabolism

- The enzyme population of the oral cavity is low relative to the remainder of the gastrointestinal tract. It would be unlikely that the enzymes in the oral cavity would deactivate a substantial fraction of a drug dose.
  - Peptidases have been identified in the oral cavity that could deactivate peptide or protein drugs
  - Esterases are present that can hydrolyze the ester bond which is so common in drug molecules.

Distribution

- Drug absorbed is drained into the facial vein and thence to the internal jugular vein where it is rapidly distributed back to the heart and out to the systemic circulation.
  - Drugs administered to the oral mucosa can produce relatively rapid peak concentration and onset of effects within 15 minutes.

Key Point

- Drugs applied to the oral cavity are rapidly absorbed bypassing hepatic first pass.

Drug properties and the oral cavity

See Appendix Table A.1. The following properties affect the design of dosage forms for the oral cavity.

- Drugs in products that will be placed sublingually must have good water solubility because of the relatively small volume of saliva there.
- Drugs applied to the buccal mucosa may be dissolved in the matrix of a tablet or film and diffuse into the lipid matrix between the buccal epithelial cells.
- Favorable log P for the oral cavity is between 2.0 and 4.0.
- Preferred formula weight for moving through the multiple layers of epithelia is less than 500 Da.
- Drugs with stability problems may be formulated as solid dosage forms.

- Drugs with high first pass loss when swallowed and absorbed through the small intestine will have improved bioavailability when applied to the oral cavity.
  - There is considerable interest in the delivery of peptide drugs through the nonkeratinized portion of the oral mucosa because there are few peptidase enzymes there.
- Drugs with local or systemic receptor targets may be used by this route.
  - Local targets: an infection of the mouth, inflammatory conditions of the teeth, irritation in the throat or muscarinic receptors in the salivary gland.
  - Systemic: cardiovascular system, brain and hormone responsive tissues.

Oral cavity dosage forms

Design

Key Point

- Drugs applied to the oral cavity must be rendered tasteless or masked by sweeteners and flavors.

The liquid dosage forms applied to the oral cavity include solutions, suspensions and sprays, Solid dosage forms may be presented as tablets, lozenges, troches or gum. Patches and films are thin, polymeric devices containing drug for sustained release. Dosage forms designed for the oral cavity must overcome limited saliva for drug dissolution, difficulty with retaining the dosage form in the mouth, and the modest permeability of the buccal mucosa.

1. One dose in a manageable size unit
   - The dimensions of solid dosage forms are limited to 1–3 cm^2 and the dose contained generally needs to be 25 mg or less to maintain patient comfort with the size of the unit.
2. Palatable
   - Because of its proximity to taste buds the drug must be rendered tasteless or masked with sweeteners and flavors.
3. Stable
   - Multidose liquid products must be preserved to prevent bacterial growth.
   - Drugs with poor chemical stability may be prepared as solid dosage forms.
   - Buffering agents may be added to adjust pH to a suitable value for the chemical stability of the drugs.

- Solutions may require cosolvents to prevent precipitation.
- Suspensions may require wetting and suspending agents and aerosols may require surfactants for uniformity.
- Tablets may require binders and flow enhancers for an intact, uniform unit.

4. Convenient/easy to use
   - Use of dosage forms for the oral cavity is relatively easy for patients; however, it is helpful if the pharmacist provides some short instructions when first dispensing a product.
   - Solid dosage forms are portable.
   - Liquids may require frequent administration because of poor retention. Viscosity enhancers or mucoadhesive agents may facilitate retention in the mouth.

5. Release of drug
   - Immediate release is provided by solutions, suspensions, sprays, gels and sublingual tablets.
   - Buccal tablets, gum and films or patches can provide rapid, sustained release of drug if longer duration of action is desired.
   - Penetration enhancers may be used to improve the modest permeability of the buccal mucosa.

Ingredients

Ingredients used to enhance palatability include sweeteners and flavors. Generally non-cariogenic sweeteners such as saccharin, sucralose, aspartame and acesulfame are preferred to sucrose for dosage forms that will have prolonged contact with the oral cavity. Flavors are available as oils, such as peppermint and orange oil, liquid concentrates or powders. Oils require some alcohol or other cosolvent to solubilize them in aqueous solutions. Cocoa, vanillin and a variety of fruit flavors are available as powders. The selection of a flavor is based on the taste of the drug that must be masked (Table 11.1). Drugs may also be rendered tasteless by derivatization with

Table 11.1 Choosing flavors	
Taste	**Flavor to mask**
Salty	Cinnamon, orange, cherry, cocoa, wild cherry, raspberry, licorice
Sour	Raspberry, lemon, lime, cherry, berry
Bitter	Cocoa, raspberry, cherry, cinnamon, licorice, orange, mint
Oily	Wintergreen, peppermint, lemon, orange, anise

Information from selected reading 5.

functional groups that reduce their solubility, use of complexing agents or coating.

A number of polymers have been developed that increase the retention of liquids through increasing viscosity and solid dosage forms through mucoadhesion (Table 11.2). Mucoadhesion is a result of attractive forces between the mucous layer of the oral cavity and the polymer agent, generally through hydrogen bonding with hydroxyl, carboxylic acid or amine groups of the mucoadhesive agent. If mucoadhesive agents are used to retain a dosage form in the oral cavity, residence time of the dosage form is limited by the mucin turnover time, which in humans is between 12 and 24 hours. The buccal epithelium itself turns over in 3–8 days.

Penetration enhancers may be used in delivery systems developed for the oral cavity especially for drugs with high molecular weights that must be absorbed into the systemic circulation (Table 11.3). The mechanisms by which these ingredients enhance membrane penetration may include extraction of lipids from the bilayer or intercellular space or increasing the

Table 11.2 Polymers with mucoadhesive properties

Polymer category	Mucoadhesive mechanism	Examples
Cellulose derivatives	Hydrogen bonding	Methylcellulose (MC), hydroxypropyl methylcellulose (HPMC, hypromellose), carboxy-methylcellulose (CMC), hydroxyethyl cellulose (HEC)
Carbomers (polyacrylic acid crosslinked with allyl sucrose or pentaerythritol)	Hydrogen bonding	Carbomer 934, carbomer 940, carbomer 941, carbomer 971P
Polycarbophil (polyacrylic acid crosslinked with divinyl glycol)	Hydrogen bonding	Molecular weight 700–3 000 000 kDa (estimated)
Chitin-based	Electrostatic	Chitosan, trimethyl chitosan
Polyvinyl alcohol	Hydrogen bonding	MW 20–200 kDa
Natural polymers	Hydrogen bonding	Gelatin, hyaluronic acid, xanthan gum, pectin, sodium alginate
Povidone (polyvinylpyrrolidone)	Interpenetration of mucin and polymer side chains	MW 1.5–3000 kDa
Thiolated polymers	Disulfide links with mucin	Thiolated polyacryate, thiolated chitosan, thiolated polycarbophil
Lectin-mediated bioadhesive polymers	Binding of lectin protein to Glu-NAc	Tomato lectin

Glu-NAc = N-acetylglucosamine.
Information from VV Khutoryanskiy. Advances in mucoadhesion and mucoadhesive polymers. Macromol Biosci 2011; 11: 748–764.

Table 11.3 Penetration enhancers

Category	Examples
Fatty acids	Oleic acid, eicosapetaenoic acid, docosahexaenoic acid, sodium caprate, sodium laurate
Bile acids	Sodium deoxycholate, sodium glycocholate, sodium taurocholate, sodium glycodeoxycholate
Surfactants	Sodium lauryl sulfate, isopropyl myristate, isopropyl palmitate, polyoxyethylene-9-lauryl ether, polyoxyethylene-20-cetyl ether, benzalkonium chloride, polyglycol mono- and diesters of 12-hydroxystearate
Polymers	Chitosan, trimethyl chitosan
Complexing agents	Sodium N-[8-(2-hydroxybenzoyl)amino]caprylate (SNAC), 8-(N-2-hydroxy-5-chloro-benzyl)-aminocaprylic acid (CNAC), cyclodextrins
Lipid mimics with bulky polar groups	Cyclopentadecalactone, 3-methylcyclopentadecanone, 9-cycloheptadecen-1-one, cyclohexadecanone, cyclopentadecanone and oxacyclohexadecan-2-one. Azone (laurocapram,1-dodecylazacycloheptan-2-one) and derivatives

Adapted from selected reading 1, 3, and 4.

fluidity of the membrane. Another approach to enhancing the penetration of charged molecules is complex formation. A complex is formed between the drug and an oppositely charged organic ion that shields the charge of the two ions and allows them to cross the membrane transcellularly.

Dosage forms for local treatment of the mouth (Table 11.4)

- Solution and suspension dosage forms are easily applied to the entire surface of the oral cavity using a swish and expectorate or swish and swallow procedure.
  - The aqueous vehicle compensates for the relatively small volume of saliva available for drug dissolution.
  - Liquids must be used frequently because drug does not persist long after administration.

Key Point

- Mucoadhesive agents are added to dosage forms for the oral cavity to enhance retention of drug for absorption through the membranes of the mouth.

Table 11.4 Commercially available dosage forms for local targets

Dosage form	Palatability	Stability	Retention	Release	Penetration
Lidocaine viscous oral solution	Saccharin	Methylparaben, propylparaben	Carboxy methylcellulose	Distilled water	
Nystatin suspension	Cherry mint, glycerin, sucrose	Alcohol, citric acid, sodium citrate, methylparaben, propylparaben	Carboxy methylcellulose sodium	Distilled water	
Benzocaine bioerodible strips (20/20)	Glycerin, peppermint oil flavor, sucralose, sodium copper chlorophyllin (color)		Carboxy methylcellulose sodium, gum arabic	Hydroxy propyl methylcellulose, Polysorbate 80	Polysorbate 80
Triamcinolone in Orabase (Kenalog paste)			Gelatin, pectin, carboxy methylcellulose sodium	Polyethylene and mineral oil base (Plastibase)	
Clotrimazole lozenge (Troche)	Dextrates	Magnesium stearate, microcrystalline cellulose, povidone	Carboxy methylcellulose	Carboxy methylcellulose	
Menthol cherry cough drops (CVS)	Cherry, eucalyptus oil, FD&C Red No. 40, glucose, sucrose	Soybean oil, sucrose, water	Soybean oil, sucrose, water	Soybean oil, sucrose, water	
Lidocaine patch for dental anesthesia (Dentipatch)	aspartame, spearmint flavor glycerin		Karaya gum	Polyester film laminate and polyester–rayon fabric	Dipropylene glycol, lecithin, propylene glycol
Atridox doxycycline gel (injected into periodontal pockets)			poly-(DL-lactide) (PLA) gels on contact with oral cavity fluids	poly-(DL-lactide) (PLA) dissolved in N-methyl-2-pyrrolidone (NMP)	N-methyl-2-pyrrolidone (NMP)

The pastes and gels are semisolid dosage forms with mucoadhesive, viscosity enhancing agents to prolong their retention in the mouth. Despite the addition of mucoadhesive ingredients, gels and pastes must be reapplied up to four times daily.

- Troches or lozenges are solid dosage forms designed to dissolve slowly in the mouth to release drug.
  - Despite the small amount of fluid available in the mouth for dissolution and the dosage form's consequent slow diminution, these dosage forms generally must be administered frequently (4–8 times daily).
  - Cough drops (lozenges) are made by heating sucrose and water to high temperature (150°C) such that the mixture is converted to a glass-like solid. These dosage forms do not disintegrate but dissolve slowly from their surface.
  - Troches are molded dosage forms prepared using lower temperatures (50–70°C) or compression. When prepared by the compounding pharmacist, troches are formulated with either a gelatin or polyethylene glycol base. Please see selected reading 5 for details on the preparation of troches.
- Local anesthetic drugs have been formulated as patches or films that release drug from a polymer matrix.
  - Patches are laminated dosage forms with an impermeable backing and a mucoadhesive matrix containing drug. Currently available oral patches must be removed after use.
  - A film is a thin, flexible layer of mucoadhesive polymer containing drug that erodes slowly in the saliva.

Pharmaceutical example

The benzocaine film provides rapid anesthesia when applied to the affected area but may require use up to four times daily.

- In the US, there are several anti-infective drugs commercially available in biodegradable controlled delivery systems for administration into periodontal pockets for the treatment of adult periodontitis. These delivery systems are designed using polymers that release drug slowly over time.

Pharmaceutical example

Chlorhexidine gluconate is available as a rectangular chip made from a biodegradable matrix of hydrolyzed gelatin cross-linked with glutaraldehyde (PerioChip). When applied to the periodontal pocket in the dentist's office, the chip releases a burst of chlorhexidine followed by a sustained level over 7–10 days.

Table 11.5 Formulation of dosage forms for systemic targets

Dosage form	Palatability	Stability	Retention	Release	Penetration enhancer
Nitrostat sublingual tablets		Calcium stearate, silicon dioxide		Prompt: pregelatinized starch	Glyceryl monostearate
Nitrolingual sublingual spray	Peppermint oil	Medium-chain triglycerides, sodium lactate, lactic acid		Dehydrated alcohol	Medium-chain triglycerides
Striant buccal tablets		Magnesium stearate, silicon dioxide	Carbomer 934, polycarbophil	Sustained: hydroxypropyl methylcellulose, starch	
Actiq buccal lozenges	Confectioner's sugar, berry flavor, citric acid	Magnesium stearate, dibasic sodium phosphate	Handle	Hydrated dextrates, modified food starch	
Nicorette chewing gum	Acesulfame, sucralose, xylitol, menthol, peppermint oil	Titanium dioxide, magnesium oxide	Acacia, carnauba wax, polysorbate 80	Hypromellose, gum base	Sodium bicarbonate, sodium carbonate
Suboxone film	Lime flavor, maltitol, acesulfame potassium, FD&C yellow #6	Citric acid, sodium citrate	Polyethylene oxide, hydroxypropyl methylcellulose	Polyethylene oxide, hydroxypropyl methylcellulose	

Dosage forms for systemic targets (Table 11.5)

Sublingual

- Sublingual tablets are made with water soluble ingredients to ensure their rapid disintegration, dissolution and release of drug.
- The sublingual aerosol is an innovative method to provide a rapid and metered dose in liquid form. The drug is formulated as a solution or suspension that is expelled from the dosing chamber of a metered dose device when it is activated. The patient should be instructed on the priming of the metered dose device when the drug is first dispensed.
- Occasionally tablets or solutions designed to be swallowed are used sublingually in hospice patients who have difficulty swallowing or in emergency situations. The pharmacist should be prepared for oral tablets to disintegrate more slowly than sublingual tablets and to adjust the dose downward if there is normally a substantial first pass effect.

- Suboxone film is a delivery system containing buprenorphine and naloxone that is used sublingually to assist in the treatment of opioid dependence. It is used once daily initially and then tapered.

Buccal

- Buccal dosage forms release drug relatively rapidly, and they are intended to continue drug release in a sustained fashion in order to produce less frequent administration times than sublingual products.
- Buccal tablets are small, compressed dosage forms that are either designed to be removed and replaced after a set time, or to erode completely and be replaced at a set interval.
- Buccal dosage forms are generally placed between the gum and the buccal mucosa.
- They may release drug both towards the gum and cheek (multidirectional release). Multidirectional designs release drug from all exposed surfaces. The main drawback to this design is a significant amount of drug is lost into the saliva and swallowed.
- They may release only towards the cheek which is more permeable. Unidirectional devices have an impermeable material that prevents release from one or multiple sides of the device except the one facing the buccal mucosa.
- Newer buccal delivery system designs include mucoadhesive ingredients to aid in their retention while talking, drinking or eating. Because they can cause irritation to the mucosa, the site of application should be rotated.

Quality evaluation and assurance

- *In vitro* dissolution and *in vivo* bioequivalence testing is recommended but not required for dosage forms applied to the oral cavity.
- Disintegration testing and determination of the Uniformity of Dosage Units is required by the United States Pharmacopeia for official products. Tablets are stirred in a medium specified in the drug monograph at 37°C and must disintegrate within a specified time limit.

Pharmaceutical example

Nitroglycerin sublingual tablets must disintegrate within 2 minutes when stirred in water at 37°C in an official disintegration apparatus.

- The uniformity of drug content in the individual dosage units must be determined using either the weight variation test or the content uniformity test, which are presented in more detail in Chapter 12.

Counseling patients

Key Point

- Patients will require brief instruction on the use of dosage forms for the oral cavity so they understand that they are not intended to be swallowed.

While the administration of drugs to the oral cavity is relatively easy, the pharmacist may want to communicate a few simple instructions that clarify how these dosage forms differ from those that are swallowed. The dose of pastes and gels may be described in relation to the space they occupy on a fingertip. A ribbon of gel that stretches from the tip of the index finger to the first palmer crease of the distal interphalangeal joint is referred to as a fingertip unit and weighs approximately 0.5 g. The dose of gel or paste required to treat an aphthous ulcer in the mouth would be about a quarter of a fingertip unit and would weigh about 0.125 g. Depending on whether the tablet is designed to release in one or all directions, a buccal tablet may need to be positioned with one particular side against the more permeable buccal membrane and with the mucoadhesive side against the least mobile membrane of the gum. For the placement of films or patches for systemic absorption, note that the permeability of the mucosa of the mouth is greatest in the sublingual membrane, followed by the buccal, and the least permeable are the gums or hard palate.

Solutions and suspensions

The patient may be given the following advice.

- Confirm the volume of the dose and the frequency of use.
- To reduce losses to the gastrointestinal tract, use after meals and at bedtime.
- Whether to swallow or expectorate the mouthwash.

Pastes and gels

The patient may be given the following advice.

- Point out the size of the dose in relation to a fingertip and specify the frequency of use.
- To reduce losses to the gastrointestinal tract, use after meals and at bedtime.

Patches and films

The patient may be given the following advice.

- Confirm the number of units to apply and the frequency of use.

- To reduce losses to the gastrointestinal tract, use after meals and/or at bedtime.
- Confirm whether the unit should be removed or will erode.
- The application site for treating local conditions is not rotated.
- When using a patch or film for systemic drug targets, the site of application should be rotated to reduce irritation.

Sublingual aerosols

The patient may be given the following advice.

- The metered dose pump needs to be primed (pumped) several times before use.
- A metered dose pump should be held vertically with the dose chamber on top and the spray opening as close to the mouth as possible.
- Apply the spray under the tongue. Do not inhale or swallow the spray.
- Avoid drinking or eating for several minutes until the drug has been absorbed.

Sublingual tablets

The patient may be given the following advice.

- Place the tablet under the tongue. Do not chew or swallow the tablet.
- Avoid drinking or eating for several minutes until the drug has been absorbed.
- Confirm frequency or timing of use.

Buccal tablets

The patient may be given the following advice.

- Place the tablet between the cheek and the gum. Do not chew or swallow the tablet.
- Confirm whether the unit should be removed or will erode.
- The site of application should be rotated to reduce irritation.
- Confirm frequency or timing of use.

Gum

The patient should be given the following instructions.

- This dosage form is intended to be chewed but not swallowed.
- Chew the gum slowly until it tingles. Then park it between your cheek and gum. When the tingle is gone, begin chewing again, until the tingle returns.

- Repeat this process until most of the tingle is gone (about 30 minutes); then discard.
- Do not eat or drink for 15 minutes before chewing the nicotine gum, or while chewing a piece.

Questions and cases

1. Research these properties for palonosetron

Design issue	Drug property	Analysis
Drug must be dissolved before it can cross membranes		
Drug must have sufficient lipophilicity to cross membranes		
Large molecular weight drugs have difficulty moving across membranes		
Drugs may hydrolyze, oxidize, photolyze or otherwise degrade in solution		
Some drugs cannot be used by mouth because of substantial losses to enzyme degradation		
Intracellular drug receptors or receptors in the brain or posterior eye are challenging to target		
Some drugs cause serious toxicity as a result of poor selectivity of drug action		

2. Saphris (Asenapine) sublingual tablets. Identify the category of each of the inactive ingredients in Saphris sublingual tablets and the quality each contributes to the dosage form

Ingredient	Category	Quality
Gelatin		
Mannitol		
Sucralose		
Black cherry flavor		

3. Discuss the oral cavity route of administration with a group of classmates. From what you know about anatomy and physiology, determine the answer to the questions below about the route and whether the characteristic represents a challenge or an opportunity for drug delivery. How does this table compare with the one you constructed in the first question?

Oral cavity route

Questions	Answer	Challenge or opportunity?
What is the membrane that will absorb the drug from this route and how accessible is it?		
Is the absorbing membrane single or multiple layers and how permeable is it?		
How large is its surface?		
What is the nature of the body fluids that bathe the absorbing membrane?		
Will the dosage form remain at the absorbing membrane long enough to release its drug?		
Will the drug encounter enzymes or extremes of pH that can alter its chemical structure before it is distributed to its target?		
What is the blood flow primacy to the absorbing membrane and the distribution time from it?		

4. Oral solutions (intended to be swallowed) for lorazepam, haloperidol and diazepam are used sublingually in hospice patients to reduce terminal restlessness and anxiety. Do these drugs experience large first pass losses? What can the pharmacist advise regarding the dosing of these medications sublingually using dosage forms designed to be swallowed? What monitoring should be done to determine if the patient has received too much medication?

Drug, strength	Oral bioavailability (%)	Plan for using sublingually
Lorazepam 2 mg/mL		
Haloperidol 2 mg/mL		
Diazepam 5 mg/mL		

5. You have a patient with dry mouth. What effect will this have on the absorption of her ergotamine sublingual tablet for her cluster headaches? What can you suggest to ameliorate this problem?

Parameter	Intravenously	Sublingual pH 4	Sublingual pH 9
C_{max} (ng/mL)	185	64.88	95.24

Parameter	Intravenously	Sublingual pH 4	Sublingual pH 9
T_{max} (hours)	0	0.33	0.33
$T_{1/2}$ (hours)	1.13	1.00	1.11
AUC (ng · minute/mL)	12 791	5807	8965.3
AUC_{SL}/AUC_{IV}	1.00	0.454	0.701

Data from AAPS PharmSciTech 2006; 7: 1.

6. Which oxycodone formulation had the greater area under the curve (extent of absorption) and why?

   (a) Absorption was greater from the pH 4 formulation because there was more membrane soluble form at this pH

   (b) Absorption was greater from the pH 4 formulation because there was more water soluble form at this pH

   (c) Absorption was greater from the pH 9 formulation because there was more membrane soluble form at this pH

   (d) Absorption was greater from the pH 9 formulation because there was more water soluble form at this pH

7. Which of the following two formulations would be preferred for the treatment of painful aphthous ulcers? Orajel Severe Pain Formula gel (contains 20% benzocaine in cellulose gum, gelatin, menthol, methyl salicylate, pectin, polyethylene glycol, saccharin) or oral pain relief topical gel (contains 20% benzocaine in PEG-75 lanolin, polyethylene glycol, polyoxyethylene stearates, saccharin, sorbic acid)

   (a) Oral pain relief topical gel because it contains a sweetener and Orajel Severe Pain Formula does not

   (b) Orajel Severe Pain Formula because it contains a sweetener and oral pain relief topical gel does not

   (c) Oral pain relief topical gel because it contains a mucoadhesive agent and Orajel Severe Pain Formula does not

   (d) Orajel Severe Pain Formula because it contains a mucoadhesive agent and oral pain relief topical gel does not

8. Which of the following describes the release of drug from a buccal tablet?

   (a) Rapid, short term release

   (b) Rapid, sustained release

   (c) Slow, short term release

   (d) Slow, sustained release

9. Buccal tablets may be designed to release drug:

   (a) Into the epithelium of the gums

   (b) Into the buccal mucosal

   (c) Towards the gums and buccal mucosa simultaneously

   (d) Answers (b) and (c)

10. A desirable property for viscosity agents used in the oral cavity is:

    (a) Thixotropy

    (b) Lipophilicity

(c) Low molecular weight

(d) Mucoadhesion

11. Patients should be advised to apply buccal tablets to the gums because:

(a) The epithelium of the gums is highly permeable

(b) There are too many tight junctions in the buccal mucosal

(c) The gums are the least mobile surface

(d) The gums are unlikely to react to the adhesive

12. Find a formulation for compounded troches in selected reading 5. Determine the purpose of each of the ingredients in the formulation and which quality of the dosage they contribute to.

Lorazepam 0.5 mg sublingual troches

Ingredient	Category	Quality
Gelatin base		
Silica gel		
Aspartame		
Acacia powder		
Citric acid monohydrate		
Orange oil		

13. You need to make a sublingual troche containing 0.25 mg palonosetron. Determine the ingredients that you will use to make the dosage form and their purpose in the formulation.

Palonosetron 0.25 sublingual troches

Ingredient	Category	Quality

Selected reading

1. VF Patel, F Liu, MB Brown. Advances in oral transmucosal drug delivery. J Contr Rel 2011; 153: 106–116.
2. V V Khutoryanskiy. Advances in mucoadhesion and mucoadhesive polymers. Macromol Biosci 2011; 11: 748–764.
3. NVS Madhav, AK Shakya, P Shakya et al. Orotransmucosal drug delivery systems: A review. J Control Rel 2009; 140: 2–11.
4. BJ Aungst. Absorption enhancers: applications and advances. AAPS J 2012; 14: 10–18.
5. LV Allen. The Art, Science and Technology of Pharmaceutical Compounding, 3rd edn, Washington DC: American Pharmacists Association, 2008.

12

Oral delivery of immediate release dosage forms

Learning objectives

Upon completion of this chapter, you should be able to:

- Identify the challenges and opportunities presented by each of the following route related characteristics to the rate and extent of drug travel to its target: accessibility, surface area, permeability and type of absorbing membrane, nature of the body fluids that bathe the absorbing membrane, retention of the dosage form at the site of absorption, location and number of drug metabolizing enzymes or extremes of pH and blood flow primacy of the small intestine.

- Predict the effect of each of the following drug properties on the rate and extent of drug travel to its target from the small intestine: water solubility, log P, molecular weight, stability in solution, enzymatic degradation and location of drug receptors.

- List the categories of ingredients that are added to dosage forms intended to be swallowed, and identify how the ingredient contributes to the manageable size of the dosage form, its palatability or comfort, its stability (chemical, microbial, physical), the convenience of its use and the release of drug from the dosage form.

- Describe the dosage forms/drug delivery systems used by this route.

- Outline the movement of a drug from its site of administration to its target including:
  - release
  - absorption
  - distribution to target.

- Explain how the qualities of the dosage form are evaluated *in vitro* and/or *in vivo*.

- Explain to patients and/or other healthcare providers how to use, store and/or prepare dosage forms for administration via the oral route.

Introduction

About 90% of marketed medicines are administered orally; thus it is important to understand how drugs in their dosage forms are handled via this route. Oral dosage forms require little patient education for appropriate use although frequent administration can cause problems with patient

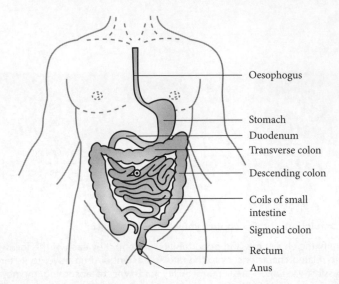

Figure 12.1 The alimentary canal. (Reproduced with permission from RJ Greene, ND Harris, Pathology and Therapeutics for Pharmacists, 3rd edn, p. 69; published by Pharmaceutical Press, 2008.)

compliance. Extended release formulations that reduce the frequency of administration to once or twice a day have addressed this issue satisfactorily and will be considered in Chapter 13. One of the major challenges for the oral route is the variability in the rate and amount of drug that emerges from the gastrointestinal tract into the systemic circulation. Some of this variability arises from the differences in drug transporters and metabolic enzymes from patient to patient. As the availability of pharmacogenetic information about individual patients expands, the pharmacist will be able to predict individual responses more accurately. Another source of variability that may be impacted with advice from the pharmacist is the use of dosage forms with or without food. Thus a good understanding of the gastrointestinal tract and the dosage forms used by this route is essential to the practicing pharmacist.

Characteristics of the gastrointestinal route

See Figure 12.1 and Appendix Tables A.1 and A.4.

Drug travel from dosage form to target

- Drugs that are swallowed are absorbed from the epithelial membrane of the small intestine.
- Drug dosage forms intended to be swallowed begin in the mouth where there is little fluid, the pH is acid to neutral and there are few enzymes.

- From the mouth, the dosage form must travel through the stomach where the pH is extremely low and transit may be delayed by food or non-disintegration of the dosage form.
- With the exception of oral disintegrating tablets (ODTs), solid dosage forms are taken with water, which aids in moving them through the esophagus and into the stomach.
- Swallowing becomes difficult in several chronic disease states such as Parkinson's disease, stroke and myasthenia gravis, necessitating the use of ODTs or liquids for these patients.
- A variety of drugs such as nonsteroidal anti-inflammatory drugs, potassium chloride, tetracycline and quinidine can become lodged in the esophagus and damage the esophageal mucosa.

Dosage forms in the stomach

- When a dose of liquid dosage form arrives in the stomach, it is retained only briefly, as liquids and small particles less than 1 mm are emptied more rapidly than solids.
- An oral solid dosage form will begin to disintegrate in stomach fluids.
  - Immediate release capsules disintegrate within 6 minutes.
  - Immediate release tablets disintegrate within 15 minutes.
- The stomach is not an area of the gastrointestinal tract where food or drugs are significantly absorbed. Factors that speed or delay stomach emptying are presented in Table 12.1.
- The presence of food in the stomach will delay its emptying into the small intestine so that the contents can be processed to the appropriate consistency for further digestion and absorption in the small intestine.
- Food slows the movement of dosage forms into the small intestine and in the case of drugs such as the bisphosphonates that are irritating to the

Table 12.1 Factors that affect rate of stomach emptying

Increase rate	Decrease rate
Empty stomach	Food in the stomach
Low viscosity	Higher viscosity
Higher pH	Lower pH
Isoosmotic	Hyper- or hypotonic
Fewer calories	More calories
Low fat	High fat
Larger volume	Small volume

esophagus, increases the likelihood that the stomach contents will splash up into the esophagus where the tissue is less well protected.

- Undigested solids including tablets that do not disintegrate will remain in the stomach until a migrating myoelectric complex pushes them into the small intestine.
- The migrating myoelectric complex passes through moving contents from the stomach and small intestine towards the colon about every 2 hours in the unfed state. The cycle is interrupted when food is eaten, and the stomach becomes occupied with the physical breakdown of the meal.
- The gastric pH increases initially as food mixed with stomach fluids brings the pH of stomach contents higher. The pH returns to around 2.0 in approximately an hour because of increased acid secretion in response to the meal.
- When the stomach contents have attained the appropriate temperature, viscosity, pH and osmotic pressure, they are emptied into the duodenum.

Dosage forms in the small intestine

- When the drug enters the small intestine it encounters a very large surface area created by a combination of mucosal folding, villi and the fingerlike projections on apical epithelial cell membranes called microvilli (Figure 12.2).
  - Each villus has an arteriole, venule and lymph vessel positioned between its ascending and descending epithelia to collect nutrients and drugs absorbed from its surface and to transport them to the liver via the portal vasculature.
- The absorbing membrane of the small intestine is a single layer of very permeable columnar epithelia.

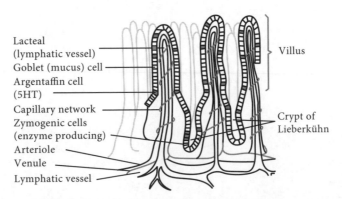

Figure 12.2 The intestinal villus. (Reproduced with permission from RJ Greene, ND Harris, Pathology and Therapeutics for Pharmacists, 3rd edn, p. 71; published by Pharmaceutical Press, 2008.)

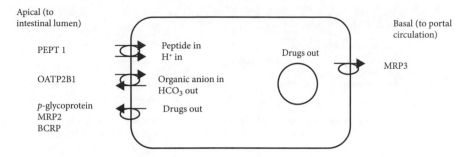

Figure 12.3 Selected transporters in the enterocyte (small intestine)

- The liver has the highest concentration of drug metabolizing enzymes in the body and can eliminate a substantial percent of a drug dose as it makes its 'first pass' through the liver to the systemic circulation.
- The pH of the fluids in the duodenum increases to between 5.0 and 7.0 and will rise slightly as the intestinal contents move through the jejunum and ileum.
- The liver secretes bile into the duodenum when a meal is ingested.
  - Bile functions to emulsify fats and facilitate their absorption from the small intestine.
- The three segments of the small intestine differ in their populations of enzymes and transporters.
  - The influx transporters that facilitate the absorption of vitamins and drugs resembling nutrients, such as levodopa, aminopenicillins and aminocephalosporins, are found in the duodenum and jejunum, creating an absorption window for these few substances that can limit their absorption to that specific region (Figure 12.3).
  - Several drugs are metabolized by Cyp3A4, a cytochrome P450 enzyme found in enterocytes in the duodenum and jejunum. Many of these same drugs are effluxed by *p*-glycoprotein back into the lumen of the small intestine. Both metabolism in the enterocyte and efflux by *p*-glycoprotein reduce the amount of these drugs absorbed.
- Transit through the small intestine is also slowed by food from 3 to 4 hours in the unfed state to 8 to 12 hours in the fed state.
- Delayed movement to and past the absorbing membrane of the small intestine has several different effects on the extent and rate of drug absorption.
  - These effects are most noticeable with ingestion of a high fat meal because the high calorie load slows transit most dramatically (Table 12.2).
- Many drugs can affect gastrointestinal transit or absorption of other drugs (Table 12.3).

Table 12.2 Effects of food induced slow gastrointestinal transit on drug absorption

Drug description	Food mechanism	Food effect
High solubility, high permeability drugs	Delay gastric emptying	Slows rate but affects the extent of absorption little or not at all
Low solubility, high permeability drugs	Delay gastric emptying, stimulate bile flow	Increases the extent of absorption by providing more time and bile emulsification for dissolution
Acid-labile drugs	Delay gastric emptying	Decreases extent of absorption due to greater exposure to stomach acid
High solubility, low permeability	Decrease transporter activity Delay gastric emptying	Slows rate, decreases the extent of absorption (binding to food)

Information from selected reading 6.

Effect of disease states on drug absorption

- Dysphagic patients have difficulty moving the drug to the absorbing membrane but since the difficulty can be resolved by selecting an alternative dosage form, the condition is generally not associated with drug absorption problems.
- Several disease states have been identified that reduce or increase gastrointestinal drug absorption and can increase the unpredictability of drug effects by this route (Table 12.4).
- The pharmacist should remain alert for signs of therapeutic failure or side effects when monitoring oral drug therapy in patients with these conditions.

Drug properties and the gastrointestinal route

- The physicochemical properties that distinguish orally active (absorbed) drugs have been summarized using the following 'Rule of Five'.
  - Compounds with a log P of 5.0 or less should have adequate water solubility to dissolve in the fluids of the gastrointestinal tract. Compounds with a log P of more than 5 have good membrane solubility but poor water solubility and therefore poor absorption. Highly permeable drugs have a log P of 1.9 or more.
  - Compounds with five or fewer hydrogen bond donors (counted as OH and NH) will have adequate membrane permeability for good oral absorption. Compounds with more than five hydrogen bond donors have good water solubility but inadequate membrane permeability to have good oral absorption.

Table 12.3 Food and drug interactions in the gastrointestinal tract

Food or drug	Effect on drug absorption	Management
Anticholinergic drugs (scopolamine, hyocyamine, tricyclic antidepressants, antipsychotic drugs, muscle relaxants)	Slow gastrointestinal transit resulting in reduced rate of absorption	Avoid use. Does not impact daily medications significantly
Opiate pain relievers	Slow gastrointestinal transit resulting in reduced rate of absorption	Monitor. Does not impact daily medications significantly
Extenatide	Slows gastrointestinal transit resulting in reduced rate of absorption	Monitor. Does not impact daily medications significantly
Proton pump inhibitors and histamine 2 blockers	Raise gastric pH: reduced absorption of ketoconazole, iron salts, digoxin. Premature release of enteric coated drugs	Take affected drugs with orange juice or other acidic beverage
Metoclopramide	Increases stomach emptying rate and increases intestinal motility	Monitor for toxicity and therapeutic failure of affected drugs
Cholestyramine, colestipol	Bind acidic (warfarin, NSAIDs, furosemide, thiazides, penicillin, tetracyclines) and a few basic (or hydrophobic) drugs	Separate administration by 1 hour before or 6 hours after cholestyramine
Antacids containing calcium, magnesium or aluminum	Bind tetracyclines, quinolones (ciprofloxacin, levofloxacin)	Separate administration by 2 hours before or 2 to 6 hours after the antacid
Iron salts	Bind tetracyclines, quinolones (ciprofloxacin, levofloxacin)	Separate administration by 2–3 hours before or 2–6 hours after the iron salt
Rifampin, rifabutin, rifapentine	Eliminate colonic bacteria that deconjugate hormonal drugs for enterohepatic recycling	Use alternative methods during antitubercular therapy
Grapefruit juice	Inhibits p-glycoprotein transport and Cyp3A4 metabolism of alprazolam, atorvastatin, buspirone, carbamazepine, clonazepam, cyclosporine, felodipine, indinavir, lovastatin, midazolam, nifedipine, propafenone, quinidine, simvastatin, triazolam, verapamil, zolpidem	Avoid concurrent use
Food containing calcium or magnesium	Binds tetracyclines, quinolones (ciprofloxacin, levofloxacin)	Separate administration by 2 hours before or 2 hours after the food

— Drugs with 10 or fewer hydrogen bond acceptor groups should have adequate membrane permeability, while those with more than 10 hydrogen bond acceptors will have good water solubility but poor membrane permeability. The sum of nitrogen and oxygen atoms is used to count hydrogen bond acceptor groups.
— Drugs should have a molecular weight of 500 Da or less to have adequate membrane permeability based on their size. Unless a drug molecule is very small (<200 Da), paracellular transport will have a minor role in its absorption.
- Orally active drugs that do not fit the Rule of Five have characteristics that allow them to be substrates for transporters.
- Most drugs are absorbed transcellularly (through the lipid bilayer) by passive diffusion.
- Most drugs administered by the oral route have targets that require they be absorbed into the systemic circulation and distributed to sites in other tissues.
— Sufficient drug will need to emerge from the first pass through the liver intact.
— The bioavailability of drugs used orally for systemic purposes can range from 0.6% for ibandronate to 100%.
- There are a few drugs that act within the gastrointestinal tract:
— the antacids and proton pump inhibitors in the stomach
— the bile acid binding resins, cholestyramine and colestipol, which act in the small intestine
— drugs for inflammatory bowel conditions that act in the colon.

Immediate release dosage forms: oral liquids

The dosage forms designed for immediate release of drug in the gastrointestinal tract include solutions (syrups and elixirs), suspensions, orally disintegrating tablets, immediate release tablets, soft gel and immediate release, hard gelatin capsules. These products can be expected to provide onset of drug effects in 30–60 minutes and if the drug is rapidly eliminated from the body, will require frequent administration (Table 12.5).

Table 12.4 Disease states that affect drug absorption from the gastrointestinal tract

Disease state	Effect on gastrointestinal physiology
Achlorhydria (gastric pH ≥6.5)	Decreased solubility of weakly basic drugs
Diabetes	Delayed gastric emptying
Hyperthyroidism	Rapid small intestinal transit resulting in decreased absorption of drugs with small absorption window (riboflavin)
Hypothyroidism	Rapid small intestinal transit resulting in increased absorption of drugs with small absorption window (riboflavin)
Bariatric surgery	Reduced surface area for absorption, in particular absorption window for vitamins. Higher pH may decrease solubility of weakly basic drugs. Reduced bile may reduce dissolution of hydrophobic drugs
Congestive heart failure	Reduces rate of drug removal from small intestine and rate of absorption
Diarrhea	Rapid transit of contents through the gastrointestinal tract
Celiac disease	Increased rate of stomach emptying, inflammation increases permeability, destruction of enterocytes reduces absorption
Dysphagia (myasthenia gravis, the elderly)	Difficulty delivering drug to site of absorption
Achalasia (inability to empty esophagus)	Difficulty delivering drug to site of absorption
Liver disease	Slow gastric emptying, slow oral drug absorption
Vomiting	Loss of medication from the route of administration

Design

1. One dose in a manageable size unit
   - The dose of an oral liquid is generally contained in 1–15 mL. Those drugs that do not have sufficient solubility to provide the dose in this volume of solution can be formulated with cosolvents or as suspensions.
2. Palatable
   - Oral solutions require sweeteners and flavors.
   - Viscosity enhancers may reduce the contact of drug with the taste buds.
3. Stable
   - Multidose liquid products must be preserved to be resistant to growth.

Table 12.5 Example oral immediate release formulations

Dosage form	Size	Palatability	Stability	Convenience	Release
Diphenhydramine (Benadryl) syrup	Glycerin, purified water	Anhydrous citric acid, D&C red #33, FD&C red #40, flavors, monoammonium glycyrrhizinate, poloxamer 407, sucrose, sodium chloride	Sodium benzoate, sodium citrate	Glycerin, purified water	Glycerin, purified water
Dexamethasone 0.5 mg/mL elixir	Alcohol, propylene glycol	Saccharin, sorbitol, raspberry flavor, citric acid, FD&C Red 40	Benzoic acid, citric acid, EDTA	Alcohol, propylene glycol, distilled water	Distilled water
Amoxicillin powder for oral suspension	Distilled water is added on dispensing	Sucrose, FD&C Red No. 3, bubble-gum flavor	Sodium citrate, sodium benzoate, edetate disodium, xanthan gum, colloidal silicon dioxide	Distilled water is added on dispensing	Distilled water is added on dispensing
Montelukast sodium (Singulair) ChewTab	Microcrystalline cellulose, hydroxypropyl cellulose	Mannitol, red ferric oxide, cherry flavor, aspartame	Microcrystalline cellulose, hydroxypropyl cellulose, magnesium stearate		Croscarmellose sodium
Tapentadol (Nucynta) tablets	Microcrystalline cellulose, lactose, povidone	Film-coating (polyvinyl alcohol, titanium dioxide, polyethylene glycol, talc, aluminum lake coloring)	Microcrystalline cellulose, lactose, povidone, magnesium stearate		Croscarmellose sodium

- To provide sufficient chemical stability for an oral liquid may require that the drug be prepared as a powder for reconstitution.
  - The powder contains flavors, sweeteners, buffers and suspending agent. The liquid product is prepared by the pharmacist just before dispensing.
  - Other approaches to manage chemical stability problems include addition of buffers and antioxidants, storage of liquids in the refrigerator and light-occluding packaging.
- Oral solutions formulated with cosolvents to prevent precipitation are known as elixirs. They:
  - have good microbial stability
  - are less palatable than aqueous solutions because of their alcohol content.
- Wetting agents, flocculating agents and suspending agents are added to enhance the uniformity of suspensions, which can be prone to aggregation and caking.

4. Convenient/easy to use
   - The oral liquids are easily swallowed and provide exceptional flexibility in dosing for children who may require a fraction of an adult dose.
   - Oral liquids are bulky to carry and can require refrigeration.
   - Liquids may require frequent administration because of rapid drug elimination.

5. Release of drug
   - Immediate release is provided by solutions and most suspensions.
   - Release of drug from suspensions is easily achieved through dissolution.

Ingredients

See Appendix Table A.2.

Palatability is an important quality for oral liquids. A variety of noncariogenic sweeteners are available. However, because they are significantly sweeter than sucrose, none is used in sufficient concentration to produce viscosity in the solution as well as sweetness. High viscosity reduces the contact of the drug in solution with the taste buds, enhancing the palatability as well as the mouth feel of oral liquids. Sucrose is widely used in syrup dosage forms because it has a pleasing sweetness and produces sufficient body to aid palatability. Aspartame is insufficiently stable to be used in liquid dosage forms. Mannitol and sorbitol produce a cooling sensation, which can enhance palatability, as well as sweetness. When cosolvents are used to solubilize a drug in an oral solution, sucrose may not be sufficiently soluble in the resulting vehicle to provide the necessary sweetness or viscosity to the solution. Elixirs may be sweetened with saccharin or a saccharin sucrose blend and flavored with essential oils such as orange, lemon or peppermint oil.

Natural flavors such as essential oils or other extractive from a spice, fruit, plant or animal source, are generally incorporated into water based formulations with surfactant or cosolvent. Artificial flavors are synthetic compounds with flavors that mimic those found in nature. The taste of a drug is most effectively masked by a flavor that would be expected to include the drug taste as an undertone. For example, a drug with a sour taste can be blended with fruit flavors or a drug with a bitter taste can be blended with sour flavors that naturally have slight bitterness (cherry). Some flavors will overshadow the taste of the drug with a very intense flavor or sensation such as mint or cinnamon. Generally a compatible dye is chosen that provides a color concordant with the flavor. Please see selected reading 3 for a detailed discussion of flavoring and sweetening.

Preparation

The compounding pharmacist may dissolve the drug in a small amount of purified water before the addition of a sweetened, flavored vehicle because the high viscosity of the vehicle will slow the dissolution rate of the drug. The chemical stability of the drug should be researched in the available literature to ensure that the drug molecule will remain intact during the course of treatment.

Compounded suspensions are frequently requested for patients who have difficulty swallowing solid dosage forms. The pharmacist may prepare an extemporaneous suspension by reducing the drug particle size in a mortar, wetting the particles and transferring the mixture into a graduated cylinder using small portions of suspending vehicle. Alternatively suspensions can be made by placing an appropriate number of immediate release tablets in a bottle with a small amount of purified water. The tablets are allowed to disintegrate and then brought to volume with suspending vehicle. See references 3 or 4 for a detailed discussion of the preparation of compounded solutions and suspensions.

Beyond-use dates

The USP devotes Chapter <795> to the description of good compounding practices. If you are interested in practicing compounding pharmacy, you should become familiar with the guidance on facilities, equipment, ingredient selection and quality control. USP <795> also provides general guidance on the beyond-use dates or the date after which the preparation should not be used. The USP recommendations for beyond-use dates for compounded products are presented in Table 12.6.

Table 12.6 Beyond-use dates for nonsterile compounded preparations	
Formulation	**Beyond-use date**
Nonaqueous liquids and solid formulations	Not later than the time remaining until the earliest expiration date of any active pharmaceutical ingredient or 6 months, whichever is less
Water containing oral formulations	Not later than 14 days when stored at cold temperatures
Water containing topical/dermal and mucosal liquid and semisolid formulations	30 days

Reproduced with permission from USP Convention. United States Pharmacopeia, 34th edn, and National Formulary, 29th edn. Baltimore, MD: United Book Press, 2011.

Quality assurance

- Check the volume in a graduated vessel against the theoretical volume.
- Check the pH with paper indicators or a meter.
- Specific gravity may be measured with a hydrometer in the appropriate range.
- Quantify active drug concentration.
- Check clarity of solutions.
- Check height and density of suspension sediment; observe the suspension to see if it cakes. Most commercial suspensions have relatively large, loose sediments.
- Check redispersibility of suspensions.
- USP standards for preservatives for oral liquids are:
  - ≤100 viable aerobic organisms per mL after a 48–72 hour incubation at 30–35°C
  - freedom from *Staphylococcus aureus*, *Pseudomonas aeruginosa*, *Salmonella* spp. and *Escherichia coli*.

Immediate release dosage forms: oral solids

Key Point

- Immediate release tablets and capsules are designed to disintegrate rapidly to increase the surface exposed to the body fluids that will dissolve them.

Design

1. One dose in a manageable size unit
   - Tablets and capsules that exceed 500 mg finished weight can be difficult to swallow, however, this drawback can be overcome by the development of orally disintegrating tablets.

- The pharmaceutical industry has largely mastered the production of tablets and capsules in reproducible doses although this remains challenging for the compounding pharmacist who prepares custom capsules.

2. Palatable
   - Drug in solid dosage forms can only be tasted if it begins to dissolve in the mouth.
   - Capsules keep the drug away from taste buds with a gelatin shell. Tablets can be coated to prevent dissolution until the drug reaches the stomach.
   - Chew tabs and orally disintegrating tablets have contact with taste buds and must include sweeteners and flavors.
     - Some drugs in orally disintegrating tablets cannot be masked and must be prepared as coated pellets to render them tasteless.

3. Stable
   - Solid dosage forms provide exceptional chemical and microbial stability.
     - Coatings have been developed that reduce drug exposure to air, humidity and light.
   - Gelatin capsules contain 13–16% moisture and:
     - can support microbial growth when they exceed 16% moisture
     - become tacky if exposed to excessive moisture and brittle if exposed to dry air too long.
   - The physical stability of tablets is enhanced by:
     - binding agents that ensure they remain intact after compression
     - lubricants and glidants are added to the powder to improve flow properties and to prevent caking, or the particle size of powders is increased by granulation, before they are pushed into tablet dies or capsule shells. This ensures the uniformity of these dosage forms from unit to unit.

4. Convenient/easy to use
   - Oral solids dosage forms require minimal patient education to use properly.
   - They can be difficult to swallow and do not offer good flexibility in dosing.
     - Tablets can be cut in half although the equality of halves can be suboptimal.
     - Capsules cannot be split although a few are made to be opened and sprinkled on food so they can be more easily swallowed.
   - Solid dosage forms can be easily pocketed for use at work or leisure.

5. Release of drug
   - Hydrophilic ingredients facilitate rapid and reproducible release of drug from immediate release tablets and capsules by disintegration

and subsequent dissolution upon contact with fluids in the gastrointestinal tract.
- Orally disintegrating tablets disintegrate in the mouth, and those that are not coated to prevent taste may dissolve and start the absorption process.
  - ODTs prepared as coated pellets will dissolve and release drug in the stomach.

The size and shape of oral solid dosage forms has been the subject of numerous studies. Tablets and capsules are typically between 50 and 750 mg finished weight. Very large drug doses may require the use of two units (tablets or capsules) and very small doses must be mixed with diluents so that the dosage form can be easily handled. Patient preference studies indicate that patients find capsules easier to swallow than tablets and oval tablets easier than oblong tablets. Round tablets are the most difficult to ingest although coatings can be used to enhance the swallowability of any tablet design.

Immediate release oral solids *disintegrate* quickly to increase the surface of drug exposed to the body fluids that will *dissolve* it. For drugs with low water solubility, dissolution rate may be the rate limiting step in absorption and therefore in producing the onset of the drug. An immediate release drug product is considered *rapidly dissolving* when no less than 85% of the labeled amount of the drug in the dosage form dissolves within 30 minutes. Disintegrants are added to immediate release tablets to enhance dissolution rate. They are water soluble tablet ingredients that wick water rapidly into the tablet so that it falls apart into many particles and surface is increased. Orally dissolving tablets are lyophilized to prepare a porous, highly soluble mixture that dissolves in the mouth without additional water. This property also makes them adsorb water readily requiring individual unit packaging. Gelatin capsules dissolve easily in the contents of the stomach at body temperature. Milling to reduce particle size and surfactants to facilitate wetting of the drug solid will also increase surface exposed to dissolution media and consequently dissolution rate.

Tablet designs and ingredients

Key Points

- Designs that facilitate immediate release include lyophilization of tablet solids, use of water soluble disintegrants, release of gas upon contact with stomach acid and dissolution of the drug in a liquid filled capsule.
- Orally disintegrating tablets and chewable tablets can be used for patients who have difficulty swallowing.

See Appendix Table A.3.

- Compressed tablets are a mixture of drug, diluents, binders, disintegrants and lubricants that is fed into the cavity (die) of a tableting machine and compressed to remove the air.
  - Tablets may be coated with sugar, polymers or gelatin to provide a barrier to taste and facilitate swallowing.
  - Sugar coatings for tablets are quite water soluble but bulky compared with film coats.
  - Film coatings for tablets are thin elastic polymers that may rupture rather than dissolve.
  - Gelcaps are gelatin-coated, capsule shaped compressed tablets that are smaller than a capsule filled with the equivalent dose.
- Multiply compressed tablets are subjected to more than one compression step creating a layered dosage form. They may be used to separate drugs that are incompatible or to reduce drug dust (antineoplastics).
- Effervescent tablets are designed to be dropped into a glass of water where a carbonate or bicarbonate salt and citric or tartaric acid generate carbon dioxide.
- Chewable tablets are compressed solid units sweetened with mannitol and flavored to be palatable while chewed.
- Orally disintegrating tablets are lyophilized or soft, direct compressed units designed to break apart into small particles in the mouth, which gives them the potential to dissolve and to be absorbed to some extent in the mouth.
  - ODTs provide a faster rate and potentially greater extent of absorption than their conventional immediate release counterparts.
  - Some ODTs are designed with coated granules that disintegrate but do not dissolve until they are past the taste buds (Advantab, Flashtab). These ODT designs do not differ significantly in their absorption characteristics from conventional immediate release tablets but are easier to swallow.
  - Orally disintegrating tablets are brittle and must be removed from their individual dose packaging gently and protected from moisture.

Capsule designs

- Most capsules dosage forms consist of a hard gelatin shell enclosing solid drug and excipients including diluents, disintegrants, lubricants and glidants.
  - Diluents, lubricants and glidants provide good flow properties for the drug powder mixture that must be metered into the body of a hard gelatin capsule.

- Disintegrants provide rapid swelling of contents, exposing more surface area to gastrointestinal fluids.
- Capsule contents may be granulated to ensure better flow through encapsulating equipment.
- Because of the superior flow properties of liquids, the soft capsule design is particularly useful for small doses of drug that are difficult to distribute uniformly in solid form or drugs that only dissolve slowly in gastrointestinal fluids.
  - The shell of soft gelatin capsules is made with gelatin and glycerin or sorbitol to soften it.
  - The drug is dissolved or suspended in a solvent that does not soften the gelatin capsule further.

Preparation of tablets and capsules

Pharmacist compounded capsules are occasionally necessary in order to provide a customized dose, to wean a patient from a hypnotic drug or to assist a patient who needs to avoid a particular excipient in the commercially available dosage form. These dosage forms require the reduction of particle size, careful blending of excipients with drug powders and attention to consistent filling of the capsule bodies. The extent of mixing of capsule powders may be monitored by addition of a compatible dye to the drug powder as it is mixed with diluents. Please see selected readings 3 or 4 for a thorough discussion of the preparation of compounded capsules.

Tablet splitting

Many patients split their tablets to facilitate swallowing, to adjust the dose or to reduce cost. The practice was reviewed by a pharmacopeial task force in 2009 in order to stimulate discussion about standards for accuracy of subdivision of scored tablets. They found that significant variation occurs in the weight of tablet halves regardless of the splitting method or person doing the splitting. Many of the tablets studied did not split into two equal size parts although use of a tablet splitter improved accuracy. In some cases the tablets lost mass owing to crumbling, particularly if the tablets were split with a razor blade or knife. The task force proposed that weights of the halves resulting from splitting scored tablets not differ by more than ±15% and the loss of mass be limited to less than 3% of intact tablet weight.

Quality evaluation and assurance

The quality and performance of oral immediate release tablets and capsules are evaluated using a number of tests that are presented in more detail in Table 12.7.

Table 12.7 Quality control tests for oral solid dosage forms

Test	Description	Standard
Weight variation	Weigh 10 tablets individually, calculate average weight and relative standard deviation of active substance	Relative standard deviation should not exceed 2%
Content uniformity	Assay 30 individual units for drug content	Content within the limit specified in the monograph (i.e. ±10% of the label claim).
Disintegration testing	Time to reach soft mass when dipped in simulated gastric fluid at 37°C	Specified in USP monograph
Dissolution testing	Release of drug from dosage form into dissolution media over time	Rapidly dissolving defined as no less than 85% of the labeled amount of the drug in the dosage form dissolves within 30 minutes
Bioavailability testing	Establishes the amount of drug available to the systemic circulation when administered by a route other than intravenous	None
Bioequivalence testing	Comparison of the rate and extent of absorption from a test (generic) and reference listed drug (brand)	90% Confidence interval about parameter ratios. Test/RLD must be within ±20%.

Reproduced with permission from USP Convention. United States Pharmacopeia, 34th edn, and National Formulary, 29th edn. Baltimore, MD: United Book Press, 2011.

Please see Chapter 7 for a more extensive discussion of bioavailability and bioequivalence.

Counseling patients

Oral liquids

The patient may be given the following advice.

- Show the patient where the dose will measure on the dosi-spoon or medicine cup.
- Shake suspensions well! If the suspension is particularly difficult to resuspend, the pharmacist should let the patient know how many vigorous shakes the suspension will require to redisperse properly.
- Specify storage conditions and beyond-use date.
- Frequency and onset of dose.

Tablets and capsules

The patient may be given the following advice.

- Take with a glass of water.
- Whether to chew or crush or neither.
- Onset, frequency and storage.
- Store in the least humid environment available at room temperature. Normal humidity is between 40% and 70%.

Orally disintegrating tablets

The patient should be given the following instructions.

- Open the single dose container with dry hands by peeling back the foil layer. Do not push the tablet through the foil.
- As soon as you open the blister, remove the tablet and put it into your mouth.
- The tablet will disintegrate quickly in your saliva so that you can easily swallow it without drinking liquid.

Food advice

The patient may be given the following food advice.

- What to take on an empty stomach:
  - drugs that are not stable to acid
  - drugs that are irritating to the esophagus
  - drugs that are bound by food components (calcium, fiber)
  - if the patient wants rapid results.
- What to take with food:
  - drugs irritating to the stomach
  - drugs absorbed high in the gastrointestinal tract via a special transport mechanism, such as riboflavin and vitamin C
  - poorly water soluble drugs that will be solubilized by bile (best taken with a fatty meal).

Questions and cases

1. The pH of the stomach pouch in patients who have had Roux-en-Y gastric bypass surgery is about 5. Assuming the pH of the normal stomach is 2, which of the following effects is likely to occur with warfarin's solubility? (pK_a 5.05)

Warfarin

(a) The solubility will increase because there will be more uncharged form in a patient with gastric bypass

(b) The solubility will decrease because there will be more uncharged form in a patient with gastric bypass

(c) The solubility will increase because there will be more charged form in a patient with gastric bypass

(d) The solubility will decrease because there will be more charged form in a patient with gastric bypass

2. Amiodarone is a highly lipophilic drug (log P 7.57) with low water solubility but high permeability. What is the likely effect of *faster* gastrointestinal transit time on amiodarone's absorption?

(a) The rate and extent of absorption will increase

(b) The rate and extent of absorption will decrease

(c) The rate may not be affected but the extent is likely to increase

(d) The rate may not be affected but the extent is likely to decrease

3. Research the inactive ingredients in the following dosage forms. Determine the function of the excipients using Appendix Tables A.2 and A.3. Categorize them as contributing to the manageable size of the dosage form, its palatability or comfort, its stability (chemical, microbial, or physical), the convenience of its use or the release of drug from the dosage form. Note that ingredients may be included in many dosage forms for more than one purpose.

Celexa (citalopram) tablets

Ingredient	Category	Quality
Copolyvidone		
Corn starch		
Croscarmellose sodium		
Glycerin		
Lactose		
Magnesium stearate		
Hypromellose		
Microcrystalline cellulose		
Polyethylene glycol		

Ingredient	Category	Quality
Titanium dioxide		

Zyprexa Zydis orally disintegrating tablets

Ingredient	Category	Quality
Gelatin		
Mannitol		
Aspartame		
Methylparaben		
Propyl paraben		

Formulation problems for practice

4. Below is a published formulation for granisetron hydrochloride oral solution (Int J Pharmaceut Compound 2011; 15: 419)

   What is the likely pathway(s) of granisetron's degradation and what functional group(s) is (are) involved? Have the chemical stability issues you identified been addressed by the published formulation?

Granisetron

What is the function of each of the ingredients in the published formulation? What storage and packaging strategies would you recommend be used for the product?

Granisetron HCl 20 mg	
Purified water 50 mL	
Cherry syrup qs 100 mL	

5. You need to make the following suspension for a child with a seizure disorder. What are the likely pathway(s) of the active ingredient's degradation and what functional group(s) is (are) involved? Have the chemical stability issues you identified been addressed by the published formulation? What beyond use date should be used for the product?

Tiagabine

The formulation is presented as a suspension that offers better chemical stability of the active ingredients but the formulation contains no antioxidants or buffers.

Beyond-use date 14 days under refrigeration.

What is the function of each of the ingredients in the formulation? How will you package and store?

Tiagabine hydrochloride 100 mg	
Sodium carboxymethylcellulose 0.25 g	
Methylparaben 200 mg	
Glycerin 10 mL	
Cherry flavor qs	
Syrup 40 mL	
Purified water qs 100 mL	

6. You need to make the following syrup for a child with asthma. (Int J Pharmaceut Compound 2012; 16: 332)
   What is the likely pathway(s) of betamethasone's degradation and what functional group(s) is (are) involved? Have the chemical stability issues you identified been addressed by the published formulation? What beyond use date should be used for the product?

Betamethasone

What is the function of each of the ingredients in the formulation? How will you package and store?

Betamethasone 12 mg		
Alcohol 0.75 mL		
Methylcellulose 1500 300 mg		
Citric acid, anhydrous 500 mg		
Cherry flavor qs		
Orange flavor qs		
Propylene glycol 10 mL		
Sodium benzoate 250 mg		
Sodium chloride 100 mg		
Sorbitol solution 70% 20 mL		
Sucrose 50 g		
Purified water qs 100 mL		

Beyond-use date: 14 days.

7. Rifabutin is an antimycobacterium agent that is commercially available as a capsule.

Rifabutin
$C_{46}H_{62}N_4O_{11}$

Water solubility 0.19 mg/mL.

(a) What are the likely pathways of rifabutin degradation and what functional group(s) are involved?

(b) What dosage form would you choose to make a liquid product with a concentration of 100 mg/5 mL?

(c) What ingredients, storage, packaging and beyond use date would you recommend?

(d) How would you counsel the patient about the product you have prepared?

8. You have a prescription for a metformin oral solution for a diabetic patient. Research the chemical and biological properties of metformin. What challenge and opportunities does this drug present to the formulator in terms of dissolution in body fluids, movement through membranes, stability in dosage forms, loss of drug due to metabolism, access to receptors, or selectivity of the target?

Design issue	Drug property	Analysis
Drug must be dissolved before it can cross membranes		
Drug must have sufficient lipophilicity to cross membranes		
Large molecular weight drugs have difficulty moving across membranes		
Drugs may hydrolyze, oxidize, photolyze or otherwise degrade in solution		
Some drugs cannot be used by mouth because of substantial losses to enzyme degradation		
Intracellular drug receptors or receptors in the brain or posterior eye are challenging to target		
Some drugs cause serious toxicity as a result of poor selectivity of drug action		

9. Because the patient who will receive the metformin oral solution is a diabetic, you will prepare the prescription as a sugar-free product. Make a list of ingredients that you will put in the metformin oral solution. Categorize them as contributing to the manageable size of the dosage form, its palatability or comfort, its stability (chemical, microbial, or physical), the convenience of its use or the release of drug from the dosage form.

Ingredient	Category	Quality
Cherry flavor, other flavor		
Saccharin sodium or sucralose		
Benzoic acid		
Methyl cellulose (or other viscosity agent)		
Purified water		

10. Bariatric surgeries that restrict the mobility of the stomach include gastric banding, vertical banded gastroplasty, gastric bypass, and sleeve gastrectomy. What effect would you predict these surgical procedures would have on dosage form

disintegration and dissolution? What effect would this have on the rate and extent of drug absorption? What can you suggest to the patient to overcome the problem you identify? (possible reference: R Padwal, D Brock, AM Sharma. A systematic review of drug absorption following bariatric surgery and its theoretical implications. Obesity Rev 2010; 11: 41–50.)

11. Jejunal ileal bypass surgery connects the pylorus with the distal small intestine, thus bypassing the mixing of drugs with bile acids that occurs in the upper small intestine. What effect would you expect this surgery to have on the rate and extent of absorption of drugs in Question 1? What can you suggest to the patient's physician to overcome the problem you identify?

Selected reading

1. CH Gu, H Li, J Levons et al. Predicting effect of food on extent of drug absorption based on physicochemical properties. Pharm Res 2007; 24: 1118–1130.
2. MK DeGorter, CG Xia, JJ Yang et al. Drug transporters in drug efficacy and toxicity. Ann Rev Pharmacol Toxicol 2012; 52: 249–273.
3. LV Allen. The Art, Science and Technology of Pharmaceutical Compounding, 4th edn, Washington, DC: American Pharmacists' Association, 2012.
4. JE Thompson, LW Davidow. A practical guide to contemporary pharmacy practice, 3rd edn, Philadephia, PA: Lippincott, Williams & Wilkins, 2009.
5. M Kato. Intestinal first-pass metabolism of CYP3A4 substrates. Drug Metab Pharmacokinet 2008; 23: 87–94.
6. CY Wu, LZ Benet. Predicting drug disposition via application of BCS: transport/absorption/ elimination interplay and development of a biopharmaceutic drug disposition classification system. Pharm Res 2005; 22: 11–23.
7. G Green, C Berg, J Polli et al. Pharmacopeial standards for the subdivision characteristics of scored tablets. Pharmacopeial Forum 2009; 35: 1598–1612.

13

Oral delivery of modified release solid dosage forms

Learning objectives

Upon completion of this chapter, you should be able to:

- Differentiate between the different types of modified release: delayed, sustained, controlled and bimodal.
- Outline the movement of a drug from its site of administration to its target including:
  - release
  - absorption
  - distribution to target.
- Predict the effect of food on drugs in modified release dosage forms.
- Describe how water solubility, permeability, dose and drug metabolism influence modified release design.
- Describe the following dosage forms designs and explain how they modify the release of drug: enteric coats, bacterially cleaved prodrugs, hydrophilic matrix, inert matrix, microparticulate or single-unit reservoir, ion exchange and osmotic tablets.
- Explain how the qualities of the dosage form are evaluated *in vitro* and/or *in vivo*.
- Explain to patients and/or other healthcare providers how to use, store and/or prepare dosage forms for administration via the oral route.

Introduction

Modified release is a general term that refers to any formulation that releases drug other than immediately. The USP differentiates two kinds of modified release dosage forms: delayed and extended. Delayed release dosage forms are designed to disintegrate in a location other than the stomach but when they reach the location of release, they disintegrate rapidly to provide maximal drug concentrations in the area of the gastrointestinal tract where they dissolve. Enteric coatings with pH sensitive solubility are the most common design used to control where a delayed release dosage form disintegrates. These dosage forms protect acid labile drugs and protect the stomach from drugs that can irritate the lining, or can remain intact in the small intestine to release drug in the colon. Extended release dosage forms are those that allow at least a two-fold reduction in dosing frequency as compared to the conventional, immediate release form. There are a number of dosage form designs that produce prolonged release of drug and permit the use of the extended release dosage form once or twice daily. The greater convenience provided by less frequent dosing significantly improves patient compliance with drug therapy. In addition to greater convenience, extended release dosage forms produce more consistent plasma concentrations than their immediate release counterparts resulting in more persistent therapeutic effects and fewer adverse reactions.

Characteristics of modified release oral solids

There are many terms that are used by the pharmaceutical industry to describe products that have delayed or extended release: slow release, prolonged release, slow acting, timed delivery, long acting (Figure 13.1). It is important for the pharmacist to understand the distinction between the release behaviors of the different dosage forms over time.

- *Delayed release* dosage forms are intended to disintegrate in a location other than the stomach but when they do disintegrate and dissolve will produce a rapid rise in blood levels of the drug followed by a decline determined by the ability of the body to eliminate the drug.

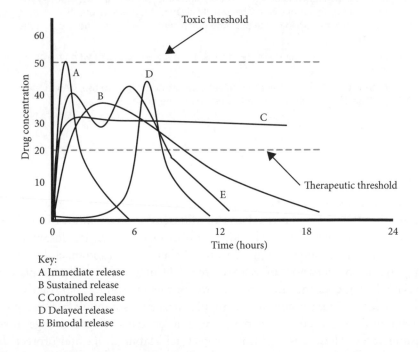

Key:
A Immediate release
B Sustained release
C Controlled release
D Delayed release
E Bimodal release

Figure 13.1 Drug release and absorption patterns

- *Extended release* dosage forms are either sustained or controlled release depending on whether the release of drug occurs in a concentration dependent (first order) or concentration independent (zero order) manner. Often the terms sustained and controlled release are used interchangeably, however, there is a distinction.
  - — *Sustained* or first order drug release is characterized by a slow rise to an initial peak and even slower decline in release rate, the rate slowing as the amount of drug in the dosage form declines.
  - — A *controlled release* dosage form is able to provide a constant rate of release for a predictable period of time, a model of release referred to as zero order because the rate of drug release is independent of the concentration of drug in the dosage form.
- Some modified release dosage forms display bimodal release in which part of the dose is released in an immediate fashion and part is released in a delayed or extended fashion.

Theory behind dissolution controlled and diffusion controlled release

Key Points

- Modified release dosage forms are designed to determine absorption rate by releasing drug more slowly in the small intestine than it will be absorbed.
- Slowing the absorption rate can mean that the time to peak is quite long and the onset of the therapeutic effect after the first dose will be slow compared with an immediate release product.
- Extended release dosage forms ideally release in a constant fashion to provide therapeutic drug concentrations over a convenient time frame.
- Drug is released from the modified release dosage forms by dissolution, diffusion and, for the hydrophilic matrix design, by a combination of dissolution and diffusion.

Inspection of the formulas that describe these processes reveals that diffusion is dependent on the initial dissolution of the drug and dissolution on the diffusion of the drug molecule from the edge of the dissolving particle to the bulk solution.

Diffusion is a process by which molecules *in solution* move from an area of high concentration to an area of low concentration. Fick's Law describes mathematically the amount of drug (moles or grams) transported through a membrane over time (hours, minutes, seconds). In the case of oral extended release products, the membrane is a portion of the dosage form through which the drug must diffuse.

$$\frac{dM}{dt} = \frac{DKA(C_1 - C_2)}{h}$$

Where D is the diffusion coefficient (cm^2/s), K is the partition coefficient between the membrane and the source of drug, A is the surface area of the membrane (cm^2), C_1 is the concentration of drug in the dosage form, C_2 is the concentration of drug in the medium outside the barrier (g/cm^3), and h is the thickness of the barrier (cm).

Dissolution involves diffusion of a drug molecule from the edge of the concentration gradient at the surface of the dissolving particle into the bulk of the dissolution medium, a distance again described as h. Dissolution rate or the amount of drug dissolved per unit time may be described mathematically using the Noyes Whitney equation:

$$\frac{dM}{dt} = \frac{DA(C_s - C)}{h}$$

where D is the diffusion coefficient in the diffusion layer surrounding the dissolving particle, A is the surface area exposed to solvent, h is the thickness of the diffusion layer, C_s is the solubility of the drug and C is the concentration of the drug in the dissolution fluids.

- The design of oral modified dosage forms then is directed at keeping these variables constant as the dosage form moves through the changing environment of the gastrointestinal tract.
  - Many modified release dosage forms are designed to maintain a constant area of drug surface exposed to solvent by using insoluble or selectively soluble coatings on tablets or drug particles.
  - The hydrophilic matrix tablet design produces a viscous diffusion layer of gelatinous polymer that yields a small diffusion coefficient and relatively constant h.
- Other dosage forms contain an excess of drug in the formulation such that the release rate will remain constant as the concentration of drug in solution in the dosage form remains constant at the drug's solubility.

Drug travel from dosage form to target

Release in the small intestine

Key Point

- Undigested solids (intact dosage forms) are moved into the small intestine slowly compared with small particles of disintegrated immediate release dosage form.

- Modified release tablets and capsule granules are designed to remain intact in the stomach.
  - Tablets retain their original surface area.
  - Extended release capsule shells will dissolve and expose slow releasing pellets to the stomach contents.

- — Dosage forms are emptied in order of their size, with the smaller pellets moving into the small intestine sooner than the larger, intact tablets.
- Undigested solids will remain in the stomach after smaller particles such as disintegrated tablets or capsules have been moved into the small intestine for absorption.
- Undigested solids are moved into the small intestine by the migrating myoelectric complex approximately 2 hours after the last small particles of food or disintegrated dosage form have left the stomach.
- Because the emptying of an intact unit by the migrating myoelectric complex can occur any time from a few minutes to 3 hours after administration when the stomach is empty, taking a modified release dosage form with a light meal will produce more predictable, albeit slower, emptying of the dosage form, since it will be synchronized with the meal.
- In general the movement of multiparticulates (pellets and granules) through the gastrointestinal tract is more reproducible than the progression of intact tablets.

Release in the large intestine

Key Point

- Drugs may be targeted to the large intestine by enteric (pH controlled) coatings or as prodrugs that are activated by enzymes found only in the colon.

One of the major functions of the colon is to absorb water and electrolytes, thus the contents of the large intestine become more viscous as they move distally. There are a number of viable microflora that assist in the digestion of polysaccharides and proteins. The large intestine has a smaller surface area than the small intestine and is only moderately permeable.

- Modified release tablets and capsule pellets targeting the colon will remain intact until the terminal ileum or ascending colon.
  - — They reach the proximal (ascending) colon 4–12 hours after they have emptied from the stomach depending on whether they were taken with food or on an empty stomach.
- There is little cytochrome P450 to metabolize drugs and the transit time is slow.
  - — Multiparticulate preparations may offer some advantage over intact units in that intact units move more rapidly than smaller particles in the colon.

Effect of food

- Modified release dosage forms are intended to control the release rate and therefore absorption rate of the drug rather than permitting variables within the gastrointestinal tract such as gastric emptying or permeability to control rate.

- For the most part the rate and extent of drug release and absorption from extended release dosage forms are not affected in a clinically meaningful way by administration with food.
  - A survey of the manufacturer's recommendations for the use of extended and delayed release dosage forms confirms that most of them can be taken with or without food (Table 13.1).

Table 13.1 Modified release dosage forms in relation to meals		
Dosage form	Recommendation	Rationale
Acamprosate DR	With or without food	
Rabeprazole DR (Aciphex)	With or without food	Better bioavailability without food, activity needed during meal
Amphetamine salts ER (Adderall XR)	With or without food	
Albuterol ER (Vospire)	With or without food	
Bupropion ER	With or without food	
Carvedilol ER (Coreg CR)	With or without food	
Dexlansoprazole DR (Dexilant)	With or without food	
Dexmethylphenidate (Focalin XR)	With or without food	Use with or without food may require individualization
Esomeprazole DR (Nexium)	Empty stomach	Faster onset
Glipizide ER	With breakfast	Better bioavailability, reduces hypoglycemia
Glyburide ER	With breakfast	Reduces hypoglycemia
Lamotrigine ER (Lamictal ER)	With or without food	With or without food
Memantine (Namenda ER)	With or without food	
Mesalamine DR (Asacol)	With or without food	
Mesalamine DR (Lialda)	With food	Food increases absorption
Mesalamine ER (Pentasa)	With or without food	Higher extent of absorption with food
Methylphenidate ER (Concerta)	With or without food	
Methylphenidate (Metadate)	With or without food	
Metoprolol succinate (Toprol)	With or without food	
Minocycline ER (Solodyn)	With or without food	

(continues)

Table 13.1 (*continued*)

Dosage form	Recommendation	Rationale
Naproxen DR	With or without food	Better tolerated with food, faster onset without food
Naproxen ER (Naprelan)	With or without food	Better tolerated with food, faster onset without food
Nicotinamide (Niaspan)	With food	Better tolerated and better bioavailability with food
Omeprazole DR	Empty stomach	Faster onset without food
Oxycodone ER (Oxycontin)	With or without food	
Oxymorphone ER (Opana)	Empty stomach	Slow onset, high bioavailability with food
Paliperidone ER (Invega)	With or without food	
Paroxetine ER	With or without food	
Potassium chloride ER	With food	Better tolerated with food
Ranolazine ER (Ranexa)	With or without food	
Quetiapine (Seroquel XR)	Empty stomach	High fat meal significantly increases extent of absorption
Tramadol ER (Ryzolt or ConZip)	With or without food	
Tramadol ER (Ultram ER)	With or without food	Use consistently with or without food
Trazodone ER (Oleptro)	Empty stomach	High C_{max} with food, better tolerated with empty stomach
Venlafaxine ER (Effexor XR)	With food	Better tolerated with food
Zolpidem ER (Ambien CR)	Empty stomach	Faster onset with empty stomach, ER tablet has IR and ER

DR delayed release, ER extended release, IR immediate release.
From manufacturers' labeling.

- Taking a modified release dosage form with fruit juice may synchronize stomach emptying with its administration (Table 13.2).
  - This recommendation is based on the observation that a glucose solution containing 15 grams of glucose triggered rapid stomach emptying in healthy human volunteers.
  - Beverages with fat or protein such as milk will slow stomach emptying compared with carbohydrate solutions (Table 13.3).

The FDA requires that manufacturers investigate 'the food effect' on their extended release dosage forms because they contain enough medication for

Table 13.2 Sugar content of selected juice products

Juice product	Sugar content (g/100 g)
Apple juice, canned or bottled	10.90
Apple juice from frozen concentrate	10.93
Apricot nectar	13.79
Cranberry juice, unsweetened	12.10
Grape juice from frozen concentrate	12.65
Grapefruit juice, white from frozen concentrate	9.63
Orange juice canned, unsweetened	8.40
Orange juice from frozen concentrate	8.40
Orange juice, raw	8.40
Orange, strawberry, banana juice	10.42
Papaya nectar, canned	13.91
Pineapple juice, canned, unsweetened	9.98
Pineapple juice from frozen concentrate	12.57
Prune juice	16.45
Tangerine juice, canned, sweetened	11.80

Reproduced with permission from selected reading 5.

Table 13.3 Stomach emptying of solutions with differing caloric content

Solution	Protein (g)	Glucose (g)	Lactose (g)	Fat (g)	Osmolality (mosmol/kg)	Calories (kcal/mL)	Stomach ($T_{1/2}$ emptying minutes)
Pea peptide hydrol	17.9	15	0	0	367	0.22	16.3 ± 5.4
Whey peptide hydrol	18.3	15	0.9	<0.1	348	0.23	17.2 ± 6.1
Milk protein	18.3	15	25.7	17.6	348	0.66	$26.4 \pm 10*$
Glucose	0	15	0	0	418	0.10	$9.4 \pm 1.2*$

*Statistically significant difference $P <0.05$.
Reproduced with permission from selected reading 5.

12 to 24 hours and there have been isolated reports of 'dose dumping' or the release of the dosage form contents in a prompt fashion. Food, in particular a high fat meal that increases the secretion of bile acids, may cause dose dumping from dissolution controlled extended release systems by speeding their dissolution with surfactant action. There is a possibility of defects in the membrane of membrane controlled extended release designs which can release large amounts of drug if the membrane covers a single unit rather than multiple, small pellets. It has also been suggested that the grinding action of the fed stomach may create mechanical stress on extended release tablets that results in release of drug in the stomach. Another mechanism identified for dose dumping is the use of alcohol with extended release dosage forms.

Pharmaceutical example

Coadministration of a hydromorphone extended release (Pallidone SR) capsule with 40% alcohol resulted in plasma concentrations 5.5-fold higher than the levels observed when the dosage form was ingested with water. Another mechanism of dose dumping is the crushing of tablets or capsule contents that are intended to be ingested intact by patients or caregivers.

Drug properties that influence modified release dosage solids

The following properties influence the design of delayed and extended release dosage forms.

- Solubility:
  - For drugs with good solubility there are many formulation options for the design of extended release products.
  - A drug with very low solubility (<0.01 mg/mL) may have inherently sustained release due to its slow dissolution rate.
- The drug must be sufficiently permeable ($\log P$ 1–5) to provide an absorption rate that is faster than drug release from the dosage form.
- Drugs that are absorbed via carrier-mediated transport are likely to have too narrow an absorption window to allow delivery of 12–24 hours medication in one administration since most active transport carriers operate in only one portion of the small intestine.
- Dose:
  - It is essential that the daily dose of drug be sufficiently low that one day's dose will not be too large to swallow, the maximum swallowable dose generally considered to be between 500 mg and 1 g.
- If a drug is inactivated by gut enzymes more rapidly than it is released by the extended release dosage form, presystemic losses of drug can be higher than with conventional release dosage forms.

- If the drug is absorbed from the colon, extended release formulation can increase drug availability because CYP3A4 activity is absent in the distal small intestine and colon.

Dosage forms of modified release oral solids

Oral solid dosage forms provide exceptional chemical and microbial stability, and they can be easily pocketed for use at work or leisure. Tablets and capsules that exceed 500 mg finished weight can be difficult to swallow, and this is a serious limiting factor for extended release products that are designed to contain medication for 12 to 24 hours. The design of modified release oral tablets and capsules requires attention to ingredients that will produce reproducible delayed or prolonged release of drug upon contact with fluids in the gastrointestinal tract. These dosage forms release drug either by disintegration and subsequent dissolution, diffusion or a combination of dissolution and diffusion.

Delayed release designs

Key Point

- Delayed release dosage forms are designed to disintegrate in a location other than the stomach.

- Delayed release dosage forms may be formulated to release their drug contents for absorption in the small intestine or for release on the surface of the colon (Table 13.4).

Table 13.4 Delayed release formulations			
Dosage form	Stability	Release	Release mechanism
Cymbalta (duloxetine HCl) delayed release capsules	Hypromellose, sodium lauryl sulfate, sucrose, sugar spheres, talc, titanium dioxide	Hydroxypropyl methylcellulose acetate succinate, triethyl citrate	pH dependent dissolution (small intestine)
Asacol (mesalamine) delayed release tablet	Colloidal silicon dioxide, lactose monohydrate, magnesium stearate, polyethylene glycol, povidone, talc	Dibutyl phthalate, methacrylic acid copolymer B (Eudragit S), methacrylic acid copolymer A (Eudragit L), sodium starch glycolate	pH dependent dissolution (colon)
Pentasa (mesalamine) delayed release capsules	Colloidal silicon dioxide, hydroxypropyl methylcellulose, starch, stearic acid, sugar, talc	Acetylated monoglyceride, castor oil, ethylcellulose, white wax	Time dependent diffusion (colon)

(continues)

Table 13.4 (*continued*)

Dosage form	Stability	Release	Release mechanism
Entocort EC (budesonide) delayed release capsules	Antifoam M, polysorbate 80, talc, sugar spheres	Acetyltributyl citrate, triethyl citrate, Eudragit L100–55, ethylcellulose	pH dependent dissolution followed by time dependent diffusion (colon)
Azulfidine EN (sulfasalazine) delayed release tablets	Povidone, magnesium stearate, silicon dioxide, propylene glycol, talc	Cellulose acetate phthalate, carnauba wax, white wax, glyceryl monostearate	pH dependent dissolution plus bacterially cleaved prodrug
Colazide (balsalazide) capsules (available in the United Kingdom)	Magnesium stearate, colloidal anhydrous silica	Shellac, 4-aminobenzoyl-beta-alanine	pH dependent dissolution plus bacterially cleaved prodrug

From manufacturers' labeling.

- The most popular approach has been the use of enteric coats, which are composed of acidic polymers that dissolve as the pH of the intestine increases.
  - There are a number of polymers that dissolve at a suitable range of pHs that can be used to coat either a compressed tablet or pellets or beads (multiparticulates) (Table 13.5).

Table 13.5 pH dependent coating agents

pH dependent polymers	Dissolution threshold pH
Eudragit L100 (methacrylic acid copolymers)	6.0
Eudragit S 100	7.0
Eudragit L 30D	6.0
Eudragit FS 30D	6.8
Eudragit L100–55	5.5
Eudragit P-4135F	7.5
Polyvinyl acetate phthalate	5.0
Cellulose acetate phthalate	6.2
Hypromellose phthalate 50, 55	5.2–5.4
Hypromellose acetate succinate	5.8–7.0
Shellac	7.2–7.6

Adapted from selected reading 6 and 7.

— The methacrylic acid copolymers (Eudragit) are widely used to deliver an intact dosage form through the stomach to the small or large intestine.

- Another approach is to build a lag time into the dosage form with coatings that slowly erode to release the drug at the time typical gastrointestinal transit would carry it to the targeted location.

Pharmaceutical example

Pentasa is a delayed release mesalamine capsule containing pellets coated with ethylcellulose, a water insoluble polymer that allows water into the drug core and slow drug diffusion out. The wax overcoat delays the uptake of water into the pellets until it has eroded. Drug release begins in the small intestine and continues throughout the large intestine.

- The third approach, the preparation of prodrugs that are cleaved to the active agent by colonic bacteria, ensures that drug release is confined to the colon.

Pharmaceutical example

Balsalazide is a prodrug that is split in the colon by bacterial azoreductase into mesalamine, which is the active anti-inflammatory agent, and an amino acid derivative that can be considered an inert carrier.

Extended release designs (Table 13.6)

Key Points

- In most cases, the rate and extent of drug release and absorption from extended release dosage forms are not affected in a clinically meaningful way by administration with food.
- Extended release dosage forms are those that allow at least a two-fold reduction in dosing frequency as compared to the conventional, immediate release form.
- Extended release dosage forms may release drug in a sustained fashion or in a controlled (constant) fashion.

The continued release of drug from extended release dosage forms is a result of limiting the surface area exposed to solvent (dissolution control) or drug solution (diffusion control).

Dissolution controlled

- The earliest design for slow release oral dosage forms was drug granules that were covered with varying thicknesses of lipophilic coat. The coated granules were then pressed into tablets or placed in capsules, and drug

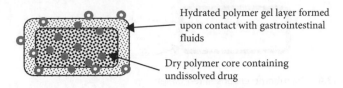

Hydrated polymer gel layer formed upon contact with gastrointestinal fluids

Dry polymer core containing undissolved drug

◉ Dissolved drug

● Undissolved drug

Figure 13.2 Hydrophilic matrix tablets. The gel layer will erode slowly and a layer of the dry polymer core will hydrate to release the drug

release determined by the thickness and the dissolution rate of the coating surrounding the drug core.

Diffusion-dissolution controlled

- The hydrophilic matrix tablet (Figure 13.2) is currently the most common design used for extended release in the industry.
- Drug is homogeneously mixed with a swellable, water soluble polymer and compressed into a single tablet unit.
- The polymer swells on contact with gastrointestinal fluids and forms a gel-like network of polymer fibers through which the drug must diffuse to be released. If the drug is very water soluble, the surface drug dissolves and gives an initial burst of release.
- Drugs with low water solubility are released as the outer hydrated gel layer erodes.
- Coating may be added to hydrophilic matrix dosage forms to reduce the burst effect that can occur with very water soluble drugs. The swelling tablet eventually breaks the coating allowing drug release to proceed.

Diffusion controlled

1. Diffusion through a membrane
   - The membrane enclosed dosage forms are referred to as reservoir systems because they contain a core of drug coated with an insoluble polymer that allows water into the dosage form and, once the drug has dissolved, allows it to diffuse out (Figures 13.3 and 13.4).

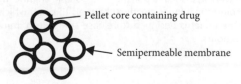

Pellet core containing drug

Semipermeable membrane

Figure 13.3 Membrane encapsulated pellet. The semipermeable membrane allows water to diffuse in and drug to diffuse out

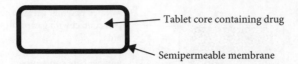

Figure 13.4 Membrane encapsulated tablet

Figure 13.5 Inert matrix tablets. Water enters the tablet through the channels formed in the inert matrix, and dissolved drug diffuses out

- The membrane of reservoir dosage forms may be composed of an insoluble polymer such as ethylcellulose with plasticizers to ensure it is flexible, or the membrane may be a blend of insoluble and soluble polymers with the soluble polymers acting as pore-forming agents.
- The soluble polymer component dissolves on contact with gastro-intestinal fluids producing tiny pores in the membrane to allow the entry of water and the exit of drug molecules. Membrane coatings may be applied to pellets or to single-unit tablet cores containing drug.

2. Diffusion through an insoluble matrix
   - Insoluble polymers such as ethylcellulose, ammonio-methacrylate copolymer and polypropylene may be used to prepare noneroding matrix tablets, which allow drug diffusion out once water has penetrated the matrix and dissolved the drug (Figure 13.5).
   - Because inert matrix tablets do not disintegrate they will appear in the patient's stool.
   - More common than inert matrix tablets are hybrid diffusion systems of membrane coated drug pellets that are embedded in a tablet matrix.

Pharmaceutical example

OxyContin is a microporous matrix tablet containing oxycodone for twice daily use. The drug is mixed with an insoluble polymer and soluble channeling agents that are dissolved by gastrointestinal fluids creating channels through which drug molecules can be released. The tablet is designed to release some drug immediately and the remainder over 12 hours.

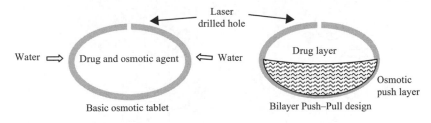

Figure 13.6   Osmotic designs

Table 13.6 Extended release oral solid dosage form designs		
Design	**Release mechanism**	**Examples**
Hydrophilic matrix	Dissolution/diffusion	OraMorph SR, Seroquel XR, Opana ER, Paxil CR, Glucophage XR, TheoChron tablets
Inert matrix	Diffusion	Fero-Gradumet, OxyContin, Slow K
Microparticulate reservoir	Diffusion	Micro-K Extendcaps, Klor-Con tablets, Dilacor XR, Luvox CR, Coreg CR, Amrix ER, Focalin XR, Metadate CD, Cardizem CD, Avinza
Single unit reservoir	Diffusion	Ultram ER
Hybrid: microparticulate reservoir in matrix	Diffusion	Cardizem LA, Toprol XL
Osmotic tablets	Osmotic pressure	Glucotrol XL, Procardia XL, Concerta, Covera HS, Tegretol XR, Invega, Ditropan XL
Ion exchange	pH or ion displacement	Tussionex capsules, Tussionex suspension, Paxil suspension, Delsym suspension

Osmotic pump systems

Figure 13.6 shows osmotic pump designs.

- In the basic osmotic pump tablet (OROS by Alza), drug is loaded into an osmotic core and surrounded by a cellulose acetate membrane that is permeable to water but not to drug or other solutes. As water is drawn across the membrane by the osmotic ingredients, the osmotic pressure pushes the dissolved drug out of the tablet in a zero order fashion through a laser drilled hole.
- The newer push-pull osmotic device has two layers: one containing drug (active layer) and the other containing an osmotically active polymer (push layer). The two layers are surrounded by a semipermeable

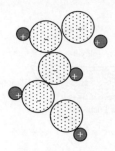
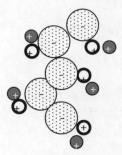

Cationic drug bound to anionic resin in dosage form

Endogenous cations displace the drug from its binding site on the resin

+ Cationic drug

+ Gastrointestinal cation

Figure 13.7 Ion exchange based products

membrane that allows water in to the osmotic layer which swells and pushes on the drug layer, metering the drug through the laser drilled hole.

Ion exchange resins

Ion exchange based products are shown in Figure 13.7.

- Ion exchange technology may be used to prepare oral liquids as well as oral solid dosage forms. Ion exchange resins are polymeric particles (or gels) that contain basic or acidic groups, which can form ionic complexes with oppositely charged drugs.
- The resins are insoluble polymers that are not absorbed by the body. As the pH or ion concentration of the gastrointestinal tract changes, the charged group on the resin will either lose its charge and release the drug, or the drug is released by exchange with a competing ion from gastrointestinal fluids.
- The resin-drug complex is suspended, compressed into tablets or encapsulated.
- Ion exchange resins can mask taste and reduce upper gastrointestinal side effects because of limited release in the stomach.

Gastric retention systems

- Dosage forms designed to be retained in the stomach may be used to treat a condition in the stomach such as *Helicobacter pylori* infection or to provide longer exposure to the upper small intestine for drugs that rely on carrier-mediated transport for absorption.
- Drugs prepared in gastroretentive dosage forms must be acid stable, and the dosage form must be able to withstand the mechanical grinding activity of the stomach.

- The three designs that appear most promising for gastroretention are floating dosage forms, bioadhesive multiparticulates and swelling, single-unit systems.
- Swelling systems appear to be the most promising of the gastroretentive designs. These dosage forms are made with polymers that swell in the stomach, becoming larger than the pyloric opening. They increase the bioavailability of drugs transported by carriers by providing gastric retention for up to 16 hours.

Pharmaceutical examples

DepoMed (now Santarus) has a swellable technology that has been applied to metformin (Glumetza) for once daily use.

Kos Pharmaceuticals, now part of Abbott, has a superporous gel technology that has been marketed as a niacin extended release tablet. Niaspan reduces flushing side effects owing to its consistent release profile.

Quality evaluation and assurance

The quality and performance of oral modified release tablets and capsules are evaluated using many of the same tests that were used for immediate release dosage forms. Drug release testing is required for delayed and extended release tablets and capsules in lieu of dissolution testing to define their release capabilities *in vitro*. The procedure uses the same equipment and procedure as for dissolution. The dosage unit is placed in basket, immersed in dissolution medium at 37°C, and rotated. Samples of the dissolution medium are taken over time and analyzed for drug concentration. Percent of dose released over time is set by individual drug monograph.

Bioavailability and bioequivalence

Bioavailability testing is necessary to understand the release characteristics of an innovator extended release product. Generally the C_{max} will be lower than the immediate release product which means that side effects are less likely to occur. Unlike immediate release dosage forms where dosing interval is determined by half life, the dosing interval for extended release formulations is determined by the size of the dose and the rate of release. Bioequivalence testing is required for generic versions of brand name delayed or extended release drugs to establish whether they can be expected to produce the same drug concentration over time curves. It is important to note that bioequivalence of modified release dosage forms does not imply that the release mechanism is the same, that is to say that a hydrophilic matrix tablet may be considered bioequivalent to an osmotic tablet.

Counseling patients

Delayed release solids

The patient should be given the following instructions and advice.

- Take with a glass of water or apple juice.
- Take on an empty stomach.
  - Proton pump inhibitors need to empty from the stomach to be released and produce an effect on stomach acid secretion. They should be taken on an empty stomach 30 minutes before a meal.
- Other than the above example, food does not significantly affect the bioavailability of delayed release tablets and capsules used for maintenance therapy.
- Do not chew or crush.
- Use acidic foods such as apple sauce or acidic liquids for the administration of enteric coated pellets to patients with dysphagia or feeding tubes.
- Confirm onset, frequency and storage.
  - Store in least humid environment available at room temperature. Normal humidity is between 40% and 70%.

Extended release solids

The patient should be given the following instructions and advice.

- Take with a glass of water or apple juice.
- Food does not significantly affect the bioavailability of extended release tablets and capsules.
- State whether the dosage form can be cut.
- Do not chew or crush.
- Nonerodible dosage form 'ghosts' appear in the stool.
- Confirm onset, frequency and storage.
  - It is not a rapid-onset form of the drug.
  - Store in the least humid environment available at room temperature. Normal humidity is between 40% and 70%.

Questions and cases

1. Research these properties for levodopa

Design issue	Drug property	Analysis
Drug must be dissolved before it can cross membranes		
Drug must have sufficient lipophilicity to cross membranes		
Large molecular weight drugs have difficulty moving across membranes		
Drugs may hydrolyze, oxidize, photolyze or otherwise degrade in solution		
Some drugs cannot be used by mouth because of substantial losses to enzyme degradation		
Intracellular drug receptors or receptors in the brain or posterior eye are challenging to target		
Some drugs cause serious toxicity as a result of poor selectivity of drug action		

2. Read about the effect of food on the absorption of levodopa/carbidopa (C_{max}, T_{max}, AUC) from the controlled release dosage form. Given what you know about the absorption pathways of levodopa in the gastrointestinal tract, what do you recommend to your patients about the use of this product with food and why? (All of this information is available in the package insert (*DailyMed*) or various pharmacy references such as *Clinical Pharmacology* and *AHFS Drug Information*.)

3. Use the tables in the Appendix to determine the function of each of these ingredients in the carbidopa/levodopa extended release tablet by Mylan Pharmaceuticals. What design is this tablet?

Ingredient	Category	Quality
FD&C Blue No. 2		
FD&C Red No. 40		
Hydroxypropyl cellulose		
Hypromellose		
Magnesium stearate		

4. Discuss colonic drug delivery via the oral route of administration with a group of classmates. From what you know about anatomy and physiology, determine the answer to the questions below about the route and whether it is a challenge or opportunity.

Questions	Answer	Challenge or opportunity?
What is the membrane that will absorb the drug from this route and how accessible is it?		
Is the absorbing membrane single or multiple layers and how permeable is it?		
How large is its surface?		
What is the nature of the body fluids that bathe the absorbing membrane?		
Will the dosage form remain at the absorbing membrane long enough to release its drug?		
Will the drug encounter enzymes or extremes of pH that can alter its chemical structure before it is distributed to its target?		
What is the blood flow primacy to the absorbing membrane and the distribution time from it?		

5. Search the FDA Orange book for theophylline extended release dosage forms. Which dosage forms are considered interchangeable? Are there any dosage forms that are not rated AB?

6. Explain why the absorption of extended release tablets and capsules is not affected by food.

7. Gastric emptying will affect the rate of absorption of delayed release dosage forms because:

   (a) They must leave the stomach intact

   (b) They are hyperosmolar

   (c) They are room temperature

   (d) They contain acidic coatings

8. Investigate whether the following extended release dosage forms may be cut in half or opened. Can you tell which designs could be cut in half and which cannot?

Dosage form	Design	Cut or opened?
Dilacor XR		
Seroquel XR		
Opana ER		
Oxycontin		
Klor-Con		
Toprol XL		
Ultram ER		
Glucotrol XL		
Procardia XL		

9. A pair of scientists from Eli Lily investigated the practice of mixing tablets or capsules in pudding or apple sauce before administration. This practice is quite common in long term care facilities where residents have difficulty swallowing intact oral solids. They found that the pH of apple sauce is about 3.5 and the pH of pudding is between 5.5 and 6.0. What effect would this practice have on the release of drugs that are prepared with enteric coats designed to dissolve at the pH of the small intestine? What advice should the pharmacist give to the nursing staff at long term care centers?

10. Discuss the following with classmates. When initiating drug therapy why should a patient start on an immediate release product before being switched to an extended release product?

11. Your company would like to design a sustained release suspension for Adderall. Adderall is a mixture of dextroamphetamine and amphetamine in a 3:1 ratio.

Dextroamphetamine

Amphetamine

In uncharged form amphetamine has a solubility of 1 g/50 mL water, while the sulfate salt has a solubility of 1 g/9 mL water, pK_a 9.94. What functional group should the ion exchange resin have?

(a) Quaternary ammonium

(b) Primary amine

(c) Sulfate

(d) Ester

12. At what pH will more amphetamine be uncharged than charged such that it will be released from the resin and where does that pH occur in the gastrointestinal tract?

(a) pH 6.5 in the duodenum

(b) pH 6.9 in the jejunum

(c) pH 7.4 in the ileum

(d) pH 9.95 which does not occur in the gastrointestinal tract

13. What ions in the gastrointestinal tract could produce competitive exchange with amphetamine?

(a) Chloride

(b) Sodium

(c) Bicarbonate

(d) Hydroxyl

14. For a peptide drug to be bioavailable via the oral route, the dosage form is *not* likely to require:

(a) Enteric coating

(b) Absorption enhancers

(c) Mucoadhesive agents

(d) Metabolic inhibitors

Selected reading

1. KA Wells, WG Losin. In vitro stability, potency, and dissolution of duloxetine enteric-coated pellets after exposure to applesauce, apple juice and chocolate pudding. Clin Ther 2008; 30: 1300–1308.
2. MK Chourasia, SK Jain. Pharmaceutical approaches to colon targeted drug delivery systems. J Pharm Pharmaceut Sci 2003; 6: 33–66.
3. H Lennernas. Ethanol-drug absorption interaction: potential for a significant effect on the plasma pharmacokinetics of ethanol vulnerable formulations. Mol Pharmaceut 2009; 6: 1429–1440.
4. H Wen, K Park (eds). Oral Controlled Release Formulations Design and Drug Delivery: Theory to practice. Hoboken, NJ: Wiley, 2010.

14

Rectal and vaginal drug delivery

Are you ready for this unit?
● Review dispersed systems.

Learning objectives

Upon completion of this chapter, you should be able to:

● Identify the challenges and opportunities presented by each of the following route related characteristics to the rate and extent of drug travel to its target: accessibility, surface area, permeability and type of absorbing membrane, nature of the body fluids that bathe the absorbing membrane, retention of the dosage form at the site of absorption, location and number of drug metabolizing enzymes or extremes of pH and blood flow primacy of the rectum or vagina.

● Predict the effect of each of the following drug properties on the rate and extent of drug travel to its target from the rectum or vagina: water solubility, log *P*, molecular weight, stability in solution, enzymatic degradation and location of drug receptors.

● List the categories of ingredients that are added to vaginal and rectal dosage forms, and identify how the ingredient contributes to the manageable size of the dosage form, its palatability or comfort, its stability (chemical, microbial, physical), the convenience of its use and the release of drug from the dosage form.

● Describe the dosage forms/drug delivery systems used by this route and whether immediate or extended release.

● Outline the movement of a drug from its site of administration to its target including:

   — release
   — absorption
   — distribution to target.

● Explain to patients and/or other healthcare providers how to use, store and/or prepare dosage forms for administration via the rectum or vagina.

Introduction

Key Points

- Rectal and vaginal dosage forms are generally used to treat a local condition and reduce systemic effects.
- These routes may also provide systemic treatment for a patient who cannot take drugs by mouth, or for drugs that are subject to high presystemic losses via the oral route.
- Both the rectum and vagina have a small volume of fluid available for drug dissolution and a small surface area resulting in absorption rate similar to swallowing medications.

Rectal and vaginal dosage forms may be used to treat a local condition and reduce systemic effects, or to provide systemic treatment for a patient who cannot take drugs by mouth because of vomiting or impaired consciousness, or with drugs that are less suitable for the oral route owing to instability in acid or high presystemic metabolism. The rectal route shares the unpredictability and transient contact with the absorbing membrane of the other gastrointestinal routes. However, the accessibility of the rectal membrane and the ease of administration by this route make it attractive for the administration of emergency or comfort care medications in the home setting. The vaginal route offers the opportunity to provide long term drug delivery locally to the reproductive tract and vaginal tissues and systemically with minimal exposure to first pass enzyme activity.

Characteristics of the rectal route

- In the adult the rectum is the final 12 cm of the large intestine.
- The rectal cavity is accessed through the anal canal, a channel between 4 and 5 cm long, where the membrane transitions from columnar to stratified squamous epithelia (Figure 14.1).
- Below the pectinate line, the anal canal is supplied with sensory pain fibers but the rectum lacks specific sensory innervation and requires careful approach for drug administration.
- The rectal columnar epithelium secretes mucus but, lacking villi and physiological absorptive function, has a small surface.
- The fluid available in the rectum for drug dissolution is between 2 and 3 mL, has a pH of 7.0 to 7.4 and few enzymes.
- Suppositories are generally retained at the absorbing membrane long enough to release drug, but solutions and suspensions may leak or be expelled before absorption is complete.
- Although there are few drug metabolizing enzymes in the rectal cavity, the venous drainage of the upper third of the rectum is provided by the superior rectal vein which empties into the portal vein.

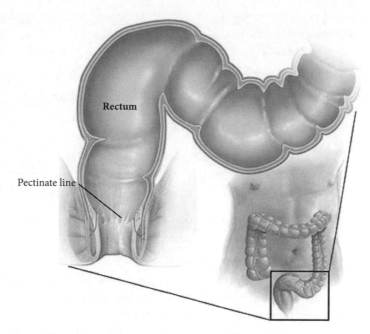

Figure 14.1 The rectum. (Reproduced with permission from http://www.mayoclinic.com/images/image_popup/mcdc7_rectum.jpg)

— There is a modest first pass effect for drugs absorbed rectally.
— The lower two-thirds are drained by the middle and inferior rectal veins which empty into the hemorrhoidal plexuses and the internal iliac veins to the vena cava.

Drug travel from rectal dosage form to target

- Drugs are administered rectally as solids (suppositories), liquids (solutions and suspensions) and semisolid dispersions (creams, gels and foams).
    - All of these dosage forms may be considered immediate release. However, the release rate is more rapid and complete from liquids than from suppositories.
- Rectal solutions or suspensions are intended for local absorption (retention enemas) or to create osmotic or irritant effects that result in the evacuation of lower bowel contents (evacuation enemas).
    - Some extemporaneous solutions or suspensions may be prepared for systemic effect but none are currently commercially available.
- Foams, creams and gels release drug via a combination of diffusion and dissolution. In many cases these semisolids are intended to spread towards the descending colon to provide local anti-inflammatory actions although the success of their migration is variable.

Table 14.1 Release of drugs by suppository base composition and drug solubility

Base	Release	Drug	Release rate
Lipid base	Melts in 3–7 minutes	Lipid soluble drug	Slow
		Water soluble drug	Rapid
Glycerinated gelatin	Dissolves in 30–40 minutes	Not applicable	
Polyethylene glycol	Dissolves in 30–50 minutes	Lipid soluble drug	Moderate
		Water soluble drug	Moderate

Reproduced with permission from USP Pharmacists' Pharmacopeia, Rockville, MD: US Pharmacopeial Convention, 2005.

- — Of the semisolid products only diazepam rectal gel is intended for systemic absorption. Peak diazepam concentrations occur on average at 1.5 hours after rectal gel administration.
- Suppositories are either solid dispersions or solid solutions of drug in a base.
  - — The fat soluble bases such as cocoa butter and hydrogenated vegetable oils melt within 3–7 minutes, releasing drug for dissolution.
  - — Suppositories made from water soluble bases such as polyethylene glycol dissolve very slowly in rectal fluids to release drug, representing the rate limiting step for many drugs (Table 14.1).
- Absorption occurs via passive diffusion through the rectal mucosa; carrier-mediated drug transport has not been demonstrated.
  - — The typical onset for drugs for local action is 15–20 minutes.
  - — Several drugs are available as suppositories for systemic effect. Absorption is difficult to predict, sometimes incomplete, dependent on formulation, drug and the patient's ability to retain the medication (Table 14.2).
- Blood from the rectal region is rapidly returned to the heart for distribution with about 50% of the dose bypassing the liver on absorption.

Table 14.2 Time course of pain relief with morphine

Route	Onset	Peak	Duration
Oral tablet	30–40 minutes	60 minutes	3–6 hours
Rectal suppository	15–20 minutes	20–60 minutes	3–6 hours

Drug properties and the rectal route

- Because of the small volume of fluid in the rectal cavity, it is preferred to use the more water soluble form of an ionizable drug in rectal formulations.
- The neutral pH of the rectal fluids favors the equilibrium ratio of acidic drugs for dissolution and weakly basic drugs for absorption.
- Absorption favors lipophilic drugs (log P of 0–5) with relatively small molecular weights (300–600 Da).
- Drugs formulated in gels, creams or suspensions containing water must be stable to hydrolysis and stabilized to oxidation.

Characteristics of the vaginal route

- The vagina is a muscular organ lined with nonkeratinized, stratified squamous epithelium of modest permeability (Figure 14.2).

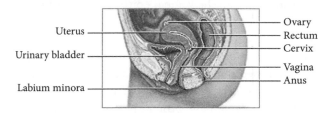

Figure 14.2 Vaginal anatomy. (Reproduced with permission from http://www.nlm.nih.gov/medlineplus/ency/imagepages/1112.htm)

- The vagina is easily, though not conveniently, accessible to drug administration.
- The vaginal fluids have a pH between 3.5 and 4.9 during child bearing years that will vary with age and stage of the menstrual cycle.
- The thickness of the vaginal epithelium is highly sensitive to estrogen levels and thus will vary over the menstrual cycle.
- Premenopausal women produce about 1 mL per hour vaginal fluid while postmenopausal women produce about half that amount.
- As they disintegrate, semisolid and solid dosage forms leak from the vagina while the patient is upright, reducing contact time between the medication and the vaginal tissue. Many of these medications are used at night to provide longer residence time.
  - Some newer formulations may contain bioadhesive polymers.
- A small number of esterases and amidases have been identified in the vaginal fluids but no cytochrome P450 drug metabolizing enzymes have been reported.

Drug travel from vaginal dosage form to target

- The immediate release dosage forms for vaginal use include solids (suppositories and tablets), liquids (solutions) and semisolid dispersions (creams and gels).
  - Vaginal tablets must dissolve to release drug while creams and gels release drug by diffusion and dissolution.
  - The vaginal inserts are extended release dosage forms used for local or systemic drug therapy. They contain a reservoir of drug with an outer, rate controlling polymeric covering.

Pharmaceutical example

The Estring vaginal ring contains a reservoir of 2 mg of estradiol and releases 7.5 µg/day over 90 days. NuvaRing (Organon) is a combination oral contraceptive ring that releases ethinyl estradiol and etonogestrel at a constant rate over 21 days.

- The main route of absorption for drugs absorbed systemically is by passive diffusion through the vaginal epithelium; no vaginal drug transporters have been reported.
- Most drugs formulated for vaginal use are for treatment of local conditions and are absorbed systemically only to a small extent (1–10%).
- The blood is returned to the circulation from the vaginal tissues through the vaginal plexus to the internal iliac veins. The blood supply is somewhat leaky allowing large protein molecules such as the immunoglobulins into the vagina creating some interest in the delivery of proteins or vaccines via this route.

Drug properties and the vaginal route

- Because of the small volume of fluid in the vagina, the water solubility of the drug is particularly significant in determining the rate and extent of release from vaginal formulations.
- Systemic absorption from the vagina favors lipophilic drugs demonstrating a similar dependence on log P to rectal absorption.
- Small molecular weight compounds such as progesterone and ethinyl estradiol are well absorbed systemically. However, the molecular weight characteristics of vaginal absorption have not been extensively studied.
- Drugs formulated in gels, creams or solutions containing water must be stable to hydrolysis and stabilized to oxidation.

Dosage forms for the rectal and vaginal routes

See Tables 14.3 and 14.4.

Table 14.3 Example rectal dosage forms

Dosage form	Stability	Release	Other function	Target
Anusol-HC Rectal ointment	White wax	White petrolatum		Intracellular induction of anti-inflammatory peptide
Diazepam rectal gel (Diastat)	Hydroxypropyl methylcellulose, sodium benzoate, benzoic acid	Propylene glycol, ethyl alcohol, water	Benzyl alcohol (comfort)	GABA$_A$ receptors, CNS
Mesalamine rectal suspension (Rowasa) 4 g/60 mL	Carbomer 934P, xanthan gum, edetate disodium, potassium acetate, sodium benzoate	Purified water		Cyclooxygenase inhibitor in colonic, rectal epithelium
Sodium phosphates rectal solution, USP		Purified water		Draws water into the feces resulting in bowel evacuation
Pramoxine aerosol foam (Proctofoam)	Emulsifiers: glyceryl monostearate, PEG-100 stearate blend, polyoxyl 40 stearate, trolamine, polyoxyethylene 23 lauryl ether, viscosity enhancer: cetyl alcohol, preservatives: methylparaben, propylparaben	Propylene glycol, purified water, isobutane and propane (propellants)		Local sensory neurons
Indomethacin rectal suppositories	Butylated hydroxyanisole, butylated hydroxytoluene, edetic acid	Glycerin, purified water and sodium chloride	Size: polyethylene glycol 3350, polyethylene glycol 8000	Cyclooxygenase inhibitor in inflamed or painful body part
Hydrocortisone rectal cream (Anusol HC)	Antimicrobial: benzyl alcohol; emulsifiers: sodium lauryl sulfate, isopropyl myristate, polyoxyl 40 stearate	Water phase: propylene glycol, carbomer 934, edetate disodium, purified water	Oil phase: petrolatum, stearyl alcohol	Intracellular induction of anti-inflammatory peptide

Table 14.4 Example vaginal products

Dosage form	Stability	Release	Other function	Target
Clindamycin phosphate cream (Cleocin)	Benzyl alcohol, cetostearyl alcohol, cetyl palmitate, mineral oil, polysorbate 60, sorbitan monostearate, stearic acid	Propylene glycol, purified water		Susceptible microorganisms in vaginal tissues
Teraconazole (Terazole 3) suppositories	Butylated hydroxyanisole	Triglycerides derived from coconut and/or palm kernel oil (a base of hydrogenated vegetable oils)		Susceptible microorganisms in vaginal tissues
Metronidazole (MetroGel) vaginal gel	Edetate disodium, methyl paraben, propyl paraben,	Propylene glycol, carbomer 934P, purified water, sodium hydroxide		Susceptible microorganisms in vaginal tissues
Progesterone (Endometrin) vaginal effervescent tablet	Magnesium stearate, pregelatinized starch, polyvinylpyrrolidone, colloidal silicone dioxide	Adipic acid, sodium bicarbonate, sodium lauryl sulfate	Size: lactose	Intracellular receptor in the uterine endometrium
Estradiol (Estring) vaginal ring	Silicone polymers and barium sulfate	Silicone polymers and barium sulfate		Vaginal epithelial cells
Gentian violet solution		Ethanol, purified water		Vaginal candida

Design of rectal and vaginal liquids

Key Points

- Solid dosage forms (suppositories and tablets) provide more exact doses.
- Semisolid or liquid dosage forms are messy but more soothing to irritated tissues and spread more easily.

- There are relatively few liquid dosage forms available specifically for vaginal or rectal administration.
- Solutions and suspensions applied to the rectal and vaginal cavities will require viscosity enhancing agents to increase contact time and antimicrobial preservatives to prevent microbial growth in the water vehicle.

Pharmaceutical example

Hydrocortisone is available as a unit of use rectal suspension for ulcerative colitis. The structured vehicle consists of carbomer 934P as a viscosity enhancing agent, polysorbate 80 as a wetting/dispersing aid, methylparaben as a preservative and purified water as the vehicle. It is administered nightly using a disposable applicator with the patient positioned on the left side during administration and for 30 minutes after, so that the fluid will distribute throughout the left colon. If possible the enema should be retained all night.

Design of gels, creams and foams

- Gels, creams and foams are usually packaged with an applicator to facilitate measurement of the dose. They are messy but soothing to irritated tissues.
- Gels and creams are dispersed systems that can overlap in their composition.
  - Aqueous gels are semisolid colloidal dispersions of hydrophilic polymers that are popular because they do not contain oils that can stain clothing.
  - Organogel and cream formulations contain lipid and water phases and must be stabilized with emulsifying agents to maintain their uniformity.

Pharmaceutical example

Crinone is a progesterone oil-in-water organogel for vaginal use. Its water phase contains glycerin, sorbic acid as a preservative and water. The oil phase consists of hydrogenated palm oil glyceride and mineral oil. The homogeneity of the mixture is maintained by two polymers, polycarbophil and carbomer 934P. In addition, polycarbophil is bioadhesive and enhances the retention of the gel in the vaginal cavity. The progesterone is partially dissolved and partially suspended in the gel creating a reservoir that produces sustained release. The T_{max} of progesterone released from Crinone gel is between 5.4 and 6.8 hours.

- Foams are dispersions of a gas in a liquid that are stabilized by surface active agents

Pharmaceutical example

Proctofoam HC is rectal foam composed of an oil-in-water liquid emulsion that is expanded into a foam by two propellants, isobutane and propane. The liquid emulsion contains cetyl alcohol and emulsifying wax in the oil phase, and methyl and propylparaben as preservatives, propylene glycol and purified water in the water phase. Trolamine and polyoxyethylene-10-stearyl ether are used as emulsifying agents to stabilize the foam, which is formed by the conversion of the propellants to gases as they are released into atmospheric pressure.

Design of suppositories and tablets

- Suppositories and tablets provide a means for administering a precise dose to the rectal or vaginal cavity.
- Suppositories for the rectal cavity are 2.5–4 cm long and weigh roughly 2 g.
- Vaginal suppositories have traditionally been somewhat larger (3–5 g) but the extra weight is generally inactive base, and not needed to prepare a functional unit.
- Urethral suppositories are very thin, usually 5 mm in diameter and 50 mm long for females, 125 mm long for males. Suppositories are usually dispersions of drug in a solid medium but may be solid solutions.

Ingredients

- The bases used for the preparation of suppositories are presented in Table 14.5.
  - The fat soluble bases are bland and nonirritating and melt rapidly to release drug. Cocoa butter can be difficult to work with because if

Table 14.5 Suppository bases

Base	Description	Release
Lipid bases		
Cocoa butter	Fat derived from cocoa bean composed of mixed triglycerides of oleic, palmitic and stearic acids, specific gravity 0.858–0.864	Melts in 3–7 minutes between 31°C and 34°C
Witepsol H15	Waxy brittle solid composed of triglycerides of saturated fatty acids with emulsifier, density 0.95 to 0.98 at RT	Melts in 3–7 minutes between 33.5°C and 35.5°C
Fattibase	Triglycerides derived from palm, palm kernel and coconut oil with glyceryl monostearate and polyoxyl stearate as emulsifying and suspending agents, specific gravity 0.89	Melts in 3–7 minutes between 32°C and 36.5°C
Water soluble bases		
PEG 8000 50%, PEG 1540 30%, PEG 400 20%	Good general purpose, water soluble base	Dissolves in 30–50 minutes
PEG 3350 60%, PEG 1000 30%, PEG 400 10%	Good general purpose base that is softer than the base above and dissolves more rapidly	Dissolves in 30–50 minutes

(continues)

Table 14.5 (*continued*)

Base	Description	Release
PEG 8000 30%, PEG 1540 70%	Has a higher melting point than other PEG bases, which is usually sufficient to compensate for melting point lowering effects of such drugs as chloral hydrate and camphor	Dissolves in 30–50 minutes
Polybase	Preblended suppository base consisting of PEGs and polysorbate 80. Melts between 55°C and 57°C	Dissolves in 30–50 minutes

Adapted from JE Thompson and LW Davidow. A Practical Guide to Contemporary Pharmacy Practice, 3rd edn. Lippincott, Williams and Wilkins, Philadelphia, 2009 and RE King (ed) Dispensing of Medications, 9th edn. Easton, PA: Mack Publishing, 1984.

    melted too rapidly it will not solidify properly and is difficult to release from the mold. It is rarely used for commercially manufactured suppositories.

— Fatty bases must be refrigerated or stored at room temperature below 25°C. They may begin to melt in the patient's hand so it can be helpful to rinse them with cold water before insertion.

— The polyethylene glycol (PEG) bases dissolve rather than melt to release drug and, because they are solids with a higher melting point, can be stored at room temperature.

— The PEG bases draw body fluids into the base and will sting upon insertion so the pharmacist should recommend that they be rinsed with cool tap water after unwrapping and before insertion. This may help them dissolve more rapidly as this process can be quite slow in the presence of the small amount of fluid in the rectum or vagina.

— Drug release rate from suppositories can be difficult to predict.

- The choice of a base should be made based on the solubility of the drug and its intended use (see Table 14.1).

— The drug solubility in body fluids is the major driving force for absorption. However, if a water soluble drug is formulated in a water soluble base, the tendency of the drug to leave the base and partition into body fluids can be low.

— Lipid soluble drugs partition better from water soluble bases, however, both the base and the drug will dissolve very slowly.

— Water soluble drugs will partition well from fat soluble bases and dissolve relatively rapidly but will require some lipid solubility usually as the unionized form of the drug, to move across the absorbing membrane.

Preparation

- Suppositories are prepared by melting the base, incorporating the drug and pouring into a mold. See selected reading 4 and 5 for more information about compounding suppositories.
- Vaginal tablets contain diluents, binders and disintegrants and are prepared using the same processes as tablets for oral use.
  - Because there is little fluid for the dissolution of vaginal tablets, they disintegrate and release drug slowly.
  - Tablets are placed in the upper third of the vagina using an applicator before bedtime to prolong the contact of the drug with the vaginal tissues.

Design of ring inserts

Key Point

- Extended release dosage forms for vaginal administration include the ring with drug contained in an internal reservoir and the drug matrix insert.

- The vaginal inserts are extended release devices for local or systemic drug release.
- Vaginal ring inserts are constructed of silicone rubber measuring between 5 and 9 cm in diameter.
- The drug is contained in a reservoir within the ring and released in a zero order fashion over a defined period.
- The flexible ring is wedged in the upper third of the vagina and intended to be worn continuously for the specified period of treatment.

Pharmaceutical example

The dinoprostone vaginal insert is a cross-linked polymer matrix containing drug that forms a slowly eroding hydrogel to release the medication at a constant rate. The polymer matrix is enclosed in a polyester knit retrieval system that allows the patient to remove the system after the dosing interval is over.

Over-the-counter anorectal products

There are a number of nonprescription products available for anorectal discomfort with brand names that imply their use for anorectal disorders. The pharmacist must read the labeling of these products with a discriminating eye to determine which products can be used in the rectum and which are intended to be applied to the external perineal area. Please see selected reading 6 for a more detailed discussion of these products.

Quality evaluation and assurance

The uniformity of drug content in the individual suppositories must be determined using either the weight variation test or the content uniformity test (see Table 12.7).

- The weight variation test requires manufacturers to weigh a specified number of units individually and calculate average weight and relative standard deviation of active substance based on their formulation.
- Suppositories with very low doses of drug must be tested using the content uniformity test. This test requires that manufacturers assay a specified number of individual units for drug content. Content must be within the limit specified in the USP monograph (i.e. ±10% of the label claim).
- Compounding pharmacists may also record the time to dissolution for PEG or glycerin suppositories, or time to melting for cocoa butter or other lipid based suppositories.

Counseling patients

Rectal dosage forms

Suppositories

The patient should be given the following instructions.

- Wash your hands and rectal area and pat dry.
- Remove the wrapper.
- A suppository can be moistened before insertion. For fat soluble bases, cool water can firm them and make them easier to handle. For water soluble bases, water treatment will reduce their tendency to cause stinging.
- Lie on your side with your lower leg straight and your upper leg bent.
- Gently insert the suppository pointed end first, with the index finger into the rectum. For an adult patient or child older than 12 years the suppository should be inserted a finger length from the anal opening to ensure it is placed beyond the anal canal and into the rectum. For younger children the suppository should be inserted to a length proportionate to their smaller size (Figure 14.3).
- Wash and dry your hands again.

The patient may be given the following advice.

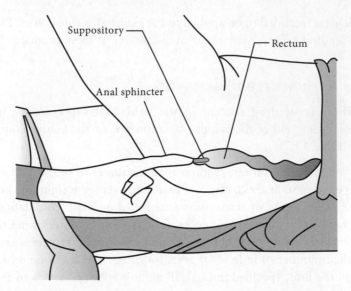

Figure 14.3 Positions for administration of rectal suppositories and enemas. (Reproduced with permission from http://www.ncbi.nlm.nih.gov/pubmedhealth/PMH0000146/bin/a601174g1.jpg)

- Onset of action is fairly rapid for local treatments, similar to oral medications for systemic treatments (30 minutes to 2 hours).
- Some suppositories require refrigeration but many fewer than most people realize because commercially prepared suppositories are now rarely made with cocoa butter.

Enemas

The patient should be given the following instructions.

- Wash your hands and rectal area and pat dry.
- Some enemas require dilution. If so, dilute according to the product instructions.
- Shake the liquid well.
- Assemble or uncap the applicator tip and lubricate with petrolatum or KY jelly.
- Lie on your left side with your left leg extended and your right leg bent for balance. This position facilitates migration of the liquid into the sigmoid colon.
- Gently insert the applicator tip fully into your rectum with the tip aimed towards the navel.
- To administer the dose tilt the bottle so the applicator nozzle is aimed towards your back and instill slowly. If you experience discomfort, slow down the instillation of the liquid.

- Retain the enema for the recommended length of time. The parent or caregiver may hold a child's buttocks closed to prevent the solution from being expelled too soon.
- Wash and dry your hands again.

Creams and foams

The patient should be given the following instructions.

- Wash your hands and rectal area and pat dry.
- Remove the cap of the ointment and attach the applicator to the tip or shake the foam canister and attach the applicator to its opening.
- Fill the applicator to the designated dose.
- The applicator can be lubricated with a small amount of the ointment, KY Jelly or petrolatum before insertion.
- Lie on your left side with your left leg extended and your right leg bent.
- Gently insert the applicator and administer the dose into your rectum.
- Wash and dry your hands again.

   The patient may be given the following advice.

- Onset of action is fairly rapid for local treatments, similar to oral medications for systemic treatments (30 minutes to 2 hours).

Vaginal dosage forms

The patient may be given the following advice.

- A sanitary pad is recommended to keep the medication away from clothing.
- Administration at bedtime is preferred.
- Vaginal medications can be used during menstruation. An exception would be the vaginal rings that are designed for birth control, which are inserted for 21 days and removed for 7 days.

Suppositories or tablets

The patient should be given the following instructions.

- Wash and dry hands and vaginal (perineal) area.
- Vaginal suppositories and tablets are packaged with applicators. Place the unit into the open end of the applicator.
- Stand or lie on your back and insert the tablet or suppository high in the upper part of your vagina. Use the plunger to push the unit out of the applicator (Figure 14.4).
- Wash and dry hands again.

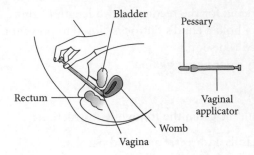

Figure 14.4 Administration of vaginal medications. (Reproduced with permission from CA Langley, D Belcher, Applied Pharmaceutical Practice, p. 245; published by Pharmaceutical Press, 2009.)

Creams

The patient should be given the following instructions.

- Wash and dry hands and vaginal (perineal) area.
- Vaginal creams and gels come with applicators. The applicator is intended to screw onto the threads of the medication tube. The seal on the tube of medication is broken with a cone shaped point on the outside of the cap.
- The applicator is generally marked with a line to indicate where it should be filled to provide a dose.
- Stand or lie on your back and insert the tablet or suppository high in the upper part of your vagina. Use the plunger to push the cream or gel out of the applicator.

Ring inserts

The patient should be given the following instructions.

- Wash and dry your hands and vaginal (perineal) area.
- Stand or lie on your back to place the insert.
- Remove the vaginal ring from its pouch.
- Hold the vaginal ring between your thumb and index finger and press the opposite sides of the ring together.
- Gently push the compressed ring into your vagina as far as possible. The exact position of the vaginal ring is not critical, as long as it is placed in the upper third of the vagina.
- When the ring is in place, you should not feel anything.
- Wash and dry your hands again.

  The patient may be given the following advice.

- The patient may bathe or participate in other activities while using the vaginal ring, including sex.

- Estring stays in place for 90 days, and is then replaced immediately with a new ring.
- NuvaRing is like an oral contraceptive, in that it is kept in place for 3 weeks and then is removed for 1 week before inserting a new one.
- To remove the vaginal ring, wash and dry hands thoroughly. Assume a comfortable position, either standing with one leg up, or lying down. Loop a finger through the ring and gently pull it out. Discard the used ring in a waste receptacle in an area where children or pets cannot reach it; do not flush it.
- Missed doses for NuvaRing: If the ring has been out of the vagina for more than 3 hours, reduced contraceptive effectiveness may occur and an additional method of contraception (e.g., male condoms or spermicide) must be used until NuvaRing has been used continuously for 7 days.
- If NuvaRing has been left in place for up to 1 extra week (i.e., up to 4 weeks total), it should be removed and the patient should insert a new ring after a 1-week, ring-free interval.

Questions and cases

1. The anesthesiology department at your hospital has asked the pharmacy to prepare a 100 mg/mL thiopental sodium solution for rectal administration as a pediatric sedative. Research the chemical and biological properties of the following drug. What challenges does this drug present to the formulator in terms of dissolution in body fluids, movement through membranes, stability in dosage forms, loss of drug due to metabolism, access to receptors or selectivity of the target?

Thiopental prepared extemporaneously for rectal administration

Design issue	Drug property	Analysis
Drug must be dissolved before it can cross membranes		
Drug must have sufficient lipophilicity to cross membranes		
Large molecular weight drugs have difficulty moving across membranes		
Drugs may hydrolyze, oxidize, photolyze, or otherwise degrade in solution		
Some drugs cannot be used by mouth because of substantial losses to enzyme degradation		

Design issue	Drug property	Analysis
Intracellular drug receptors or receptors in the brain or posterior eye are challenging to target		
Some drugs cause serious toxicity as a result of poor selectivity of drug action		

2. What pharmaceutical necessities would you put into the thiopental rectal suspension? Categorize them as contributing to the manageable size of the dosage form, its palatability or comfort, its stability (chemical, microbial, or physical), the convenience of its use or the release of drug from the dosage form. Note that ingredients may be included in many dosage forms for more than one purpose.

Ingredient	Category	Quality
Methylcellulose (or other viscosity enhancing agent for oral use)		
Glycerin		
Purified or sterile water		

3. Discuss the rectal route of administration with a group of classmates. From what you know about anatomy and physiology, determine the answer to the questions below about the route.

Questions	Answer	Challenge or opportunity?
What is the membrane that will absorb the drug from this route and how accessible is it?		
Is the absorbing membrane single or multiple layers and how permeable is it?		
How large is its surface?		
What is the nature of the body fluids that bathe the absorbing membrane?		
Will the dosage form remain at the absorbing membrane long enough to release its drug?		
Will the drug encounter enzymes or extremes of pH that can alter its chemical structure before it is distributed to its target?		

Questions	Answer	Challenge or opportunity?
What is the blood flow primacy to the absorbing membrane and the distribution time from it?		

4. You are advising a physician about what drug to use to treat a yeast infection in a pregnant patient. What are the factors that determine whether a drug is absorbed systemically?

5. You are advising a patient about what drug to use to treat a yeast infection in a pregnant patient. Which of the following factors would *reduce* the amount of drug absorbed systemically?

   (a) Hydrophilic, basic, larger molecular weight

   (b) Lipophilic, basic, larger molecular weight

   (c) Hydrophilic, acidic, larger molecular weight

   (d) Lipophilic, acidic, larger molecular weight

Diazepam

6. Diazepam has a water solubility of 50 mg/L, pK_a 3.4, and a log P of 2.82. At pH 7.2 is there more water soluble diazepam or membrane soluble diazepam, and what is likely to limit the amount of diazepam that can be absorbed in the rectum?

   (a) Amount of diazepam that dissolves in rectal fluids

   (b) Amount of diazepam that moves through membranes

   (c) The amount of both forms of diazepam limit its absorption since their concentrations are approximately equal

7. You need to make a rectal suppository containing 4 mg of hydromorphone, pK_a 8.2. Should you use hydromorphone or hydromorphone hydrochloride?

Hydromorphone

(a) Hydromorphone to provide a higher concentration of membrane soluble drug for absorption

(b) Hydromorphone hydrochloride to provide a higher concentration of water soluble form for dissolution

(c) Hydromorphone to provide a higher concentration of water soluble drug for dissolution

(d) Hydromorphone hydrochloride to provide a higher concentration of membrane soluble form for absorption

8. What suppository base would provide the most rapid release of hydromorphone hydrochloride?

(a) Polyethylene glycol to provide better solubility in the base

(b) Polyethylene glycol to provide better partitioning out of the base

(c) A fat soluble base to provide better solubility in the base

(d) A fat soluble base to provide better partitioning out of the base

Selected reading

1. GD Anderson, RP Saneto. Current oral and non-oral antiepileptic drug delivery. Adv Drug Del Rev 2012; 10: 911–918.
2. DC Corbo, JC Liu, YW Chen. Drug absorption through mucosal membranes: effect of mucosal route and penetrant hydrophilicity. Pharm Res 1989; 6: 848–852.
3. E Baloglu, ZA Senyigit, SY Karavana et al. Strategies to prolong the intravaginal residence time of drug delivery systems. J Pharm Pharmaceut Sci 2009; 12: 312–336.
4. LV Allen. The Art, Science and Technology of Pharmaceutical Compounding, 4th edn, Washington, DC: American Pharmacists' Association, 2012.
5. USDA Database for the added sugars content of selected foods, Release 1, United States Department of Agriculture, Beltsville, MD, 2006, accessed 8/23/2012 at http://www.nal.usda.gov/fnic/foodcomp/Data/addsug/addsug01.pdf.
6. MK Chourasia, SK Jain. Pharmaceutical approaches to colon targeted drug delivery systems. J Pharm Pharmaceut Sci 2003; 6: 33–66.
7. LF Ali Asghar, S Chandran. Multiparticulate formulation approach to colon specific drug delivery: current perspectives. J Pharm Pharmaceut Sci 2006; 9: 327–338.

15

Nasal drug delivery

Learning objectives

Upon completion of this chapter, you should be able to:

- Identify the challenges and opportunities presented by each of the following route related characteristics to the rate and extent of drug travel to its target: accessibility, surface area, permeability and type of absorbing membrane, nature of the body fluids that bathe the absorbing membrane, retention of the dosage form at the site of absorption, location and number of drug metabolizing enzymes or extremes of pH and blood flow primacy of the nasal mucosa.

- Predict the effect of each of the following drug properties on the rate and extent of drug travel to its target from the nasal route: water solubility, log P, molecular weight, stability in solution, enzymatic degradation and location of drug receptors.

- List the categories of ingredients that are added to nasal dosage forms, and identify how the ingredient contributes to the manageable size of the dosage form, its palatability or comfort, its stability (chemical, microbial and physical), the convenience of its use and the release of drug from the dosage form.

- Describe the dosage forms/drug delivery systems used by this route.

- Outline the movement of a drug from its site of administration to its target including:
  - release
  - absorption
  - distribution to target.

- Explain to patients and/or other healthcare providers how to use, store and/or prepare dosage forms for administration via the nasal route.

Introduction

Most of the products available for administration in the nose are used to treat local conditions, however, there is an increasing number of drugs applied to the nose for systemic activity. The route provides faster systemic drug levels and fewer enzymes than the oral route and better permeability than the oral cavity. These qualities make the nasal route attractive for the administration of macromolecules such as proteins and oligonucleotide drugs. In addition the olfactory epithelium at the apex of the nasal cavity provides the possibility of direct access to the central nervous system that bypasses the traditional blood–brain barrier. Thus several drugs with receptors in the brain have been formulated as nasal dosage forms. However, the respiratory and olfactory epithelia are difficult to access reproducibly, and the ciliated cells of the respiratory epithelium carry applied drug rapidly to the nasopharynx. These design challenges are currently addressed with devices such as metered dose pumps and the use of bioadhesive gelling agents. Proper use of nasal drops and sprays may also enhance the placement and persistence of drug at the absorbing membrane.

Characteristics of the nasal route

See Appendix Table A.4.

- The function of the nose is to filter, warm and moisten air on its way to the lungs. It also serves us with the sense of smell.
- The nose is well suited for its warming and moistening functions in that the mucous membrane that lines the nose is well supplied with blood vessels, and while the nose has no natural absorptive function, the three turbinates and other nasal crevices provide about $160\,\text{cm}^2$ surface for drug absorption.
- The surface of the nasal vestibule, the area just inside the nostrils, is lined with skin punctuated with coarse hairs to filter out particulate from inspired air (Figure 15.1).
- Moving internally, the lining of the atrium transitions from keratinized stratified squamous cells to a single layer of ciliated, pseudostratified columnar cells.

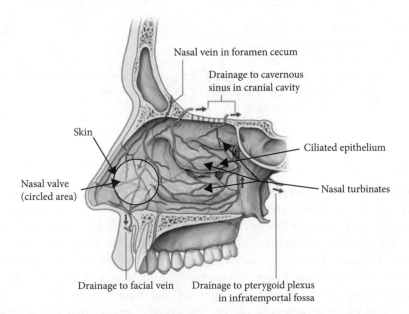

Nasal vein in foramen cecum

Drainage to cavernous sinus in cranial cavity

Skin

Ciliated epithelium

Nasal valve (circled area)

Nasal turbinates

Drainage to facial vein

Drainage to pterygoid plexus in infratemportal fossa

Figure 15.1 Veins on the lateral surface of the nose. (Modified from StudyBlue.com.)

- The ciliated columnar cells are found throughout the respiratory and olfactory areas until the nasal cavity merges with the pharynx and the lining transitions back to stratified squamous cells.
- The olfactory area is relatively small, 1–5 cm^2, but of significant interest because drugs may be absorbed directly into the central nervous system through its surface.
- There is a relatively small volume of fluid in the nose at any one time, from 0.2 to 1.1 mL. The pH of the nasal epithelial cells is 7.4, but the mucus that drug particles encounter first is acidic, pH 5.5–6.5.
- Drugs applied to the nose are absorbed by passive diffusion through the epithelium of the respiratory or olfactory areas, which comprise the posterior two-thirds of the nasal cavity.
- The membrane in these two areas is moistened by the secretions of interspersed serous glands, which produce a lower 'sol' layer of fluidized proteins, and the goblet cells (unicellular mucous glands), which produce the sticky upper layer of glycoproteins.
- Foreign particles including drugs inhaled into the nose become trapped in the sticky top layer of mucus and are carried by the movement of cilia under the mucus to the nasopharynx where they are swallowed and their activity in the nose terminated.
- Liquid formulations are generally cleared from the nose entirely in 30 minutes.
- Thus if the drug is not dissolved and absorbed within half an hour, what absorption is observed may occur from drug delivered to the nasopharynx and swallowed.
- Drug metabolizing enzymes in the nose include a number of cytochrome P450 isozymes, peptidases and esterases.

Drug travel from dosage form to target

Release

- Nasal solutions and suspensions may be applied to the nose as drops or sprays.
- Powders as well as solutions and suspensions may be administered with metered dose devices that provide a more precise dose than a multidose squeeze bottle.
  - Drugs applied as powders or suspensions require dissolution before drug is available for action.
- A few drugs have been formulated as gels or ointments for nasal administration to provide longer contact time with the membrane.
  - These drugs are released by diffusion from their semisolid vehicle. The volume of the dose administered intranasally should be limited to 200 µL (100 µL per nostril) because the excess is lost either from the front or the back of the nasal cavity.

- Examples of commercially available dosage forms and their ingredients can be found in Tables 15.1 and 15.2.

Absorption

- The nasal respiratory epithelium is difficult to access because it is behind the nasal valve.
- Drug doses are rapidly removed from the nose by mucociliary clearance.

Table 15.1 Local nasal dosage forms

Dosage form	Comfort	Stability	Retention	Release
Equate (phenylephrine) nose drops	Sodium chloride	Benzalkonium chloride, citric acid, sodium citrate		Distilled water
Rhinaris moisturizing (sodium chloride) nasal gel	Sodium chloride, potassium chloride	Benzalkonium chloride	Carbomers, sodium hydroxide (pH sensitive gel), carboxymethyl cellulose sodium, polyethylene glycol	Distilled water, propylene glycol
Afrin (oxymetazoline) spray		Benzalkonium chloride, benzyl alcohol, edetate disodium, sodium phosphate dibasic and monobasic	Polyethylene glycol, povidone	Propylene glycol, purified water
Astelin (azelastine) metered dose pump	Sodium chloride	Benzalkonium chloride, edetate disodium, citric acid, dibasic sodium phosphate	Hypromellose	Purified water
Beconase AQ (beclomethasone dipropionate) metered dose suspension	Dextrose	Benzalkonium chloride, polysorbate 80, phenylethyl alcohol	Microcrystalline cellulose, carboxymethyl cellulose sodium	Purified water
Nasacort HFA (triamcinolone acetonide) metered dose suspension				Tetrafluoroethane (HFA-134a), dehydrated alcohol

Table 15.2 Systemic nasal dosage forms

Dosage form	Comfort	Stability	Retention	Release
FluMist (attenuated influenza virus) single dose sprayer	Sucrose	Monosodium glutamate, arginine, potassium phosphate, mono- and dibasic	Gelatin	Purified water
Miacalcin (calcitonin) nasal solution in mMDP	Sodium chloride	Benzalkonium chloride		Purified water
Butorphanol tartrate nasal solution in mMDP	Sodium chloride	Benzethonium chloride, citric acid		Purified water
Imitrex (sumatriptan) nasal spray units		Potassium phosphate, mono- and dibasic		Distilled Water
Synarel (nafarelin) mMDP		Acetic acid, glacial benzalkonium chloride		Distilled Water
DDAVP (desmopressin) mMDP	Sodium chloride	Benzalkonium chloride, citric acid, disodium phosphate dodecahydrate		Distilled Water

mMDP = mechanical metered dose pump.

- Drugs applied to the nose are absorbed by passive diffusion through the epithelium of the respiratory or olfactory areas which comprise the posterior two-thirds of the nasal cavity.
- While there is general agreement that drugs are absorbed from the nose through the single layer of columnar epithelium in the respiratory region, the best place for deposition of nasal sprays to achieve absorption in this area has not been determined.
  - Evidence gathered to date suggests that drug deposition just anterior to the three turbinates is associated with better bioavailability.
  - From the preturbinate area, applied drug is spread across the absorbing membrane by ciliated cells.
  - Owing to the narrowing of the nasal airway about 2–3 cm inside the nostrils, a location referred to as the nasal valve, this area can be difficult to reach with nasal dosage forms.
- The efflux transporters, p-glycoprotein and others return drug absorbed via the respiratory and olfactory epithelia back into the nose.
- Active influx transport has only been demonstrated for amino acids.

Distribution

- The capillaries of the respiratory epithelium drain into the sphenopalatine, facial and ophthalmic veins and then into the internal jugular.

Drug properties and the nasal route

Key Point

> ● The respiratory epithelium of the nose offers rapid absorption for small molecules and good permeability to large drug molecules.

See Appendix Table A.1.

- Because the removal of drug from the nasal cavity by mucociliary clearance is relatively rapid, drugs applied as suspensions and solids must dissolve rapidly to be absorbed before they are removed. Thus drugs with good water solubility or active at very low doses are required to provide rapid dissolution in the nose.
- Water soluble drugs with small formula weight may be absorbed via paracellular channels while absorption through the lipid bilayer favors drugs with a higher log P.
- The nasal epithelium permits adequate absorption of drugs with molecular weights between 1000 and 6000 Da, although absorption promoting ingredients are required with higher molecular weights.
- Drugs formulated as aqueous solutions will require good stability in solution. However, drugs with poor aqueous stability can be formulated as nonaqueous dispersions in propellant and administered using a metered dose device.
- Nasal 'first pass' has not been characterized for most drugs but given the assortment of enzymes found in the nose, there is potential for some first pass loss to occur.
- Currently most dosage forms for administration to the nose contain drugs that act at local targets, however, the number of drugs applied to the nose for systemic targets increases each year.
- Local drug receptors include the histamine receptors in vascular endothelial cells, mast cells in connective tissue near blood vessels, and alpha adrenergic receptors on vascular smooth muscle. Corticosteroids need to reach intracellular receptors on endothelial cells in nasal vessels, as well as local fibroblasts, basophils, macrophages and lymphocytes. Intranasal influenza vaccine is targeted at nasal-associated lymphoid tissue in the pharynx.
- Drugs applied for systemic effect need to reach opiate and serotonin receptors in the brain, calcitonin receptors in bone and kidney, and vasopressin receptors in renal collecting ducts.

Table 15.3 Ingredients for nasal products

Category	Contribution
Solvents	Dissolution
Cosolvents	Dissolution, manageable size dose, viscosity
Ointment base	Semisolid nonaqueous vehicle with good retention
Wetting agents	Physical stability
Tonicity agents	Comfort. Some drug solutions are hypertonic and will require no tonicity agent
Buffers	Chemical stability (preferred pH between 5 and 8)
Antimicrobial preservatives	Microbial stability for all multi dose products containing water
Antioxidants	Chemical stability for oxidizable drugs
Propellants	Release drug in small droplets to enhance accessibility to respiratory membrane
Mucoadhesive viscosity enhancers	Increase retention at the site of administration. Convenience
Penetration enhancers	Increase permeability of absorbing membrane

Dosage forms for the nasal route

Design

See Table 15.3 and Appendix Table A.2.

1. **One dose in a manageable size unit**
   - Dosage forms designed for nasal administration are mostly aqueous solutions and suspensions that contain the drug dose in a volume between 25 and 100 µL.
2. **Comfort**
   - Nasal liquids are preferably isotonic and will contain tonicity agents such as sodium chloride and dextrose to provide for patient comfort.
3. **Stability**
   - Microbiological: All multiple dose nasal formulations must contain an antimicrobial preservative unless their packaging is able to exclude microbial contamination (propellant driven, metered dose devices).
   - Unit of use nasal products prevent the need for a preservative and eliminate concern about contamination as a result of touching the tip of the application device to the nose.
   - Chemical: Buffers may be included to stabilize or solubilize drug; the preferred pH range for patient comfort is 5.0–8.0.

- Physical: Viscosity enhancing agents are included in suspensions to prevent rapid sedimentation after shaking.

4. Convenience/ease of use

- Devices are used to facilitate placement of drug solutions, suspensions and powders on the preturbinate area of the nasal epithelium.
- Viscosity enhancing agents may be added to solutions and suspensions to provide longer residence time of the drug in the nasal cavity.
  - There are a number of viscosity enhancing agents that are also mucoadhesive and that gel when exposed to body temperature, pH or cations.
- The formulation of drugs as gels or ointments provides longer residence time of the drug in the nasal cavity but their increased viscosity and mucoadhesion does not necessarily provide better bioavailability.
  - The reasons for this are not entirely understood but may include slow release of drug from the dosage form because of its viscosity.

5. Release of drug

- Drugs are released from suspension dosage forms by dissolution; those in solution are ready for absorption.
- The peptide drugs currently available as nasal sprays have formula weights around 1000 Da and have low bioavailabilities, but are formulated without absorption enhancers because they require only a small amount to produce an effect.
- Absorption enhancers such as bile acids, fatty acids and surfactants are not well tolerated in the nose. There are several absorption promoters under development for nasal applications: cyclopentadecalactone (CPE-215), alkylsaccharides (Intravail), chitosan (ChiSys) and Macrogol 15 hydroxystearate (CriticalSorb). Cyclopentadecalactone is the absorption promoter in the nasal insulin product Nasulin that is in Phase II clinical trials.

Dosage forms and devices

Key Points

- Liquid doses applied to the nose should not exceed 100 μL per nostril.
- Multidose products for the nose are easily contaminated and must be preserved.

- Nasal liquids are administered as sprays or drops and packaged in devices that facilitate their use (Figure 15.2).
- Nasal drops are packaged as multidose liquids in dropper bottles or as single dose liquids in nasules, small, disposable plastic droppers that can be easily pocketed.

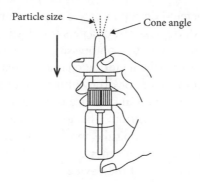

Particle size ⟶   Cone angle

Figure 15.2 Properties of the nasal spray. (Reproduced with permission from http://images. rxlist.com/images/rxlist/calomist4.gif)

- Nasal sprays are packaged in squeeze bottles, rhinal tubes and metered dose devices.
  - A squeeze bottle is a flexible plastic container with a small atomizing tip. When the bottle is squeezed, the air inside the container is pressed out of the nozzle, releasing a small volume of the liquid as a spray. The squeeze bottle is easily contaminated when the pressure on the container is released and air or nasal secretions are drawn back into the bottle. Because the volume of the dose released can vary considerably, squeeze bottles are primarily used for decongestants and humectants.
  - A rhinal tube is a thin catheter that is loaded with a dose of the nasal solution from a squeezable dropper. One end of the tube is placed in the nostril and the other end is placed in the patient's mouth. The patient delivers the dose with a short, strong puff of air through the tube.
  - Metered dose devices include the propellant driven, metered dose inhalers and the mechanical metered dose pumps. Both types of devices deliver a reproducible dose with consistent spray properties. Spray properties that affect the deposition of the drug solution in the nose include particle (droplet) size, cone angle and velocity. Droplet size for nasal sprays should be 10 µm or larger in diameter to prevent them from being drawn into the lower airways with inspiratory flow.
  - Propellant driven, metered dose inhalers are formulated with propellant and, if needed, a cosolvent and surfactant to improve the uniformity of the mixture. A propellant is a gas that is liquefied by a small decrease in temperature or increase in pressure. The propellant produces a constant pressure inside the aerosol container that pushes the liquid drug mixture in the dose chamber out when the device is actuated. The propellant expelled with the dose becomes a gas when exposed to atmospheric pressure and creates small droplets of drug mixture that are carried into the nose.

— Mechanical metered dose pumps are more commonly used for nasal liquids. These devices consist of a glass or plastic bottle capped with a small displacement pump. Compression of the actuator moves a piston downward in the metering chamber, creating pressure that forces air (if not primed) or drug solution out through the spray orifice. When the actuation pressure is released, a spring returns the piston to its original position. This generates a vacuum that draws another dose through a dip tube and into the chamber. If the pump is not used for several days, air may enter the chamber and the dip tube, necessitating repriming of the device.

Quality evaluation and assurance

The United States Pharmacopeia in its general chapter <601> Aerosols, and individual monographs for official nasal preparations provide the following requirements:

- Delivered dose may not be outside the range 75–125% of the label claim and no more than 2 out of 20 may be outside the range 80–120% of label claim.
- Drug concentration between a specified range of the labeled concentration.
- Osmolality between a specified range.
- pH between a specified range.
- The preparation should be tested to determine the total aerobic microbial count and for the presence of specific excluded microorganisms.

The FDA in its *Guidance for Industry* document, a nonbinding description of the data that should be provided by manufacturers for approval of their products, provides the following additional recommended qualities for nasal sprays:

- Impurities exceeding 1 mg per day must be identified and tested for toxicity.
- The weight of the volume delivered is reproducible (±15% of target weight).

And for comparisons between test and reference nasal product, the FDA recommends that:

- the percent of particle or droplet size less than 9 μm be comparable
- the spray pattern and angle be comparable.

Counseling patients

The proper administration of nasal liquids is complicated by the rapid movement of a liquid bolus in response to gravity and fundamental nasal anatomy.

- The goal is to deposit the nasal medication in the area of the nose just posterior to the nasal valve (anterior turbinate region) where it can be spread across the respiratory mucosa by the cilia.
- Nose drops that are administered onto the floor of the nasal cavity as a small bolus of medication will flow out either the front or the back of the nose without contacting much surface unless the patient is careful about the placement and movement of their head.

Key Point

- Nose drops should be used in the lying-head-back position or the lateral-head-low position.

- The recommended head positions for the administration of nasal drops are shown in Figures 15.3 and 15.4.
- Patients should be discouraged from using the sitting-head-back posture that is shown in many instructions to patients from pharmaceutical companies and older reference books. This posture allows the drop of medication to roll immediately to the back of the nasopharynx and into the gastrointestinal tract.

Key Points

- There is no consensus on the proper administration of nasal sprays but the evidence suggests holding the head upright facilitates placement of the dose in the preturbinate area.
- Nasal sprays should be applied to the preturbinate area where they are spread across the absorptive membrane by the cilia.

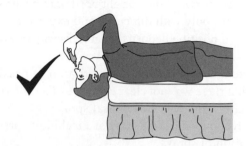

Figure 15.3 Lying-head-back position for administration of nasal drops. (Reproduced with permission from CA Langley, D Belcher, Applied Pharmaceutical Practice, p. 240; published by Pharmaceutical Press, 2009.)

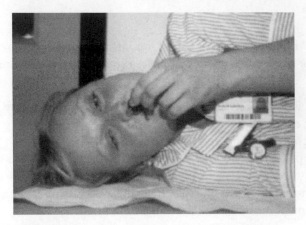

Figure 15.4 Lateral-head-low position for administering nose drops. (Reproduced with permission from U Raghavan et al. J Laryngol Otol 2000; 114: 456–459.)

- The ability of the patient to reach the anterior turbinate area with a nasal spray from squeeze bottles or metered dose devices is affected by the angle at which the patient directs the device into the nose.
- Studies in humans and in human nasal models suggest that the spray bottle should be tilted 30–45° from the horizontal to deposit the drug just beyond the nasal valve. The angle at which the spray is aimed into the nose is complicated by differing head positions advised by the various manufacturers of nasal liquids, and the need to ensure that the dip tube of the mechanical metered dose pumps is immersed in liquid to deliver a reproducible dose (Table 15.4).
- There appear to be two explanations for the head-forward position recommended in several patient package inserts (Figure 15.5).
  - The most common is that the dose delivered by mechanical dose pumps is more consistent if the bottle is held upright.
  - Another reason is to prevent the nasal liquid from flowing into the nasopharynx.
- When the head is tilted forward the upright bottle can be directed towards the anterior turbinates only with difficulty, and experiments in models suggest that this head position deposits the nasal spray mostly anterior to the nasal valve.
- A review of the literature on proper administration of nasal sprays in 2004 concluded that there was not clear evidence of a preferred technique to maximize the safety and efficacy of nasal sprays.
- The expert panel recommended the standard technique presented below for the administration of sprays.
- The panel advocated the use of the right hand to spray the left nostril and the left hand to spray the right nostril to reduce the frequency of nose bleeds (Figure 15.6).

Table 15.4 Head position recommended in product labeling

Product	Device	Head position
Beconase (beclomethasone)	pMDI	Tilt your head forward slightly
Nasacort AQ (triamcinolone)	mMDP	Tip your head back a little and aim the spray toward the back of your nose
Nasacort HFA	pMDI	Tilt your head back slightly (side rather than septum)
FluMist (atten influenza virus)	Single dose sprayer	Head upright
Miacalcin (calcitonin) nasal solution	mMDP	Head upright
Synarel (naferlin)	mMDP	Tilt head forward during administration of the dose, then tilt it back to spread the dose
Atrovent nasal solution	mMDP	Tilt head forward, bottle upright and pointed back towards the outer side of the nose
Astelin nasal solution	mMDP	Tilt head forward to keep the medicine from going down throat
Imitrex nasal solution	Single dose spray	Head upright
Desmopressin acetate nasal solution	mMDP	Tilt head forward so that the mMDP can remain upright
Afrin nasal solution	Squeeze bottle	Head upright

pMDI = propelled metered dose inhaler; mMDP = mechanical metered dose pump.

- The studies published subsequent to this review have been primarily in nasal models. Their findings do not contradict the presented technique.
- Experiments examining the practice of sniffing as the dose is applied do not find any benefit to either deposition in the target area or bioavailability.

Figure 15.5 Head-forward position for administration of drugs from nasal metered dose pumps. (Reproduced with permission from Synarel packaging.)

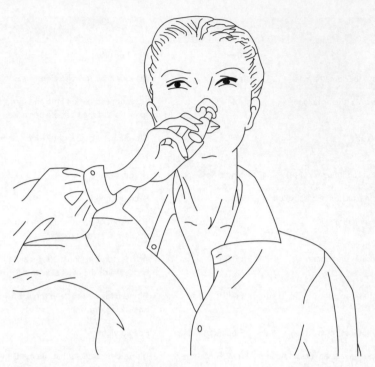

Figure 15.6 Recommended head position for the administration of nasal sprays. (Reproduced with permission from MS Benninger et al. Otolaryngol Head Neck Surg 2004; 130: 5–24.)

Nose drops

The patient may be given the following advice.

- The nose generally responds to local treatment within an hour, usually earlier.
- If you miss a dose of this medicine, use it as soon as possible. But if it is almost time for your next dose, skip the missed dose and go back to your regular schedule. Do not double the dose.
- Nose drops generally cause local side effects only, such as burning or itching. However, if side effects are very bothersome or persist, consult the pharmacist or prescriber.

Lying-head-back position

The patient should be given the following instructions.

- Blow your nose gently before using the medication if either nostril is blocked.
- Wash your hands with soap and water before and after using the nose drops. Avoid touching the dropper against the nose or any other surface to prevent contamination of the nasal solution.

- Squeeze the rubber bulb to draw up the dose of the medicine or remove the cap from the dropper bottle.
- Lie down on your back with your head dropped over the side of the bed.
- Insert the dropper into one nostril and squeeze the dose into the nose.
- Turn your head from side to side to spread the liquid across the nasal surface.
- Repeat this process with the other nostril and remain still for a couple of minutes to permit spreading.

Lateral-head-low position

- The patient should be given the following instructions. Blow your nose gently before using the medication if either nostril is blocked.
- Wash your hands with soap and water before and after using the nose drops. Avoid touching the dropper against the nose or any other surface to prevent contamination of the nasal solution.
- Squeeze the rubber bulb to draw up the dose of the medicine or remove the cap of the dropper bottle.
- Lie on your side with your head at the same level as your shoulder (no pillow).
- Insert the dropper into the lower nostril and squeeze the dose into the nose.
- Maintain the side-lying position for 30 seconds to allow spreading.
- Turn to the other side and repeat the administration of the other drop in the second nostril.

Nasal metered dose pumps

The patient should be given the following instructions.

- Gently blow your nose to clear any thick or excessive mucus, if present, before using the medication.
- Wash your hands with soap and water before and after using the nasal spray.
- Remove the cap. Shake the bottle. The first time you use the pump spray, you will have to 'prime' it by squirting a few times into the air until a fine mist comes out. You will need to prime it again if you don't use it for 3 to 7 days.
- Hold your head in a neutral upright position. Breathe out slowly.
- Hold the pump bottle with your thumb at the bottom and your index and middle fingers on top using the hand on the side opposite the nostril you want to treat.
- Use a finger on your other hand to close your nostril on the side not receiving the medication.
- Insert tip of pump into the nostril directing the spray laterally towards the outer portion of the eye or the top of the ear.

- Activate the pump by pressing down with your middle and forefingers.
- If this drug is for treatment of a condition outside of your nose, you may need to administer the dose to only one nostril. For a local treatment, if you are using more than one spray in each nostril, follow all these steps again.
- Try not to sneeze or blow your nose just after using the spray.
- Nasal aerosols for local effect will begin to work almost immediately (within 1 minute). However, it may take up to 2 weeks of using a nasal steroid spray before you notice the full effects.
- The nasal applicator can be freed from the pump for washing.
- If the pump spray is used correctly, the spray should not drip from your nose or down the back of your throat.
- If your nose hurts, or you begin to have nosebleeds or if the inside of your nose stings, stop using the spray for 1 or 2 days. If you have nosebleeds, stop using the medicine for a few days and use a saline nasal spray instead. You can also use a cotton swab to spread a thin layer of petroleum jelly inside your nose right after using the saline spray. If the bleeding or irritation continues, talk to your doctor.
- Most nasal sprays work best when used regularly and consistently.

Questions and cases

1. Drugs administered to the nasal vestibule will:
   (a) Be carried to the nasopharynx by columnated cilia
   (b) Reach the central nervous system before they reach the circulation
   (c) Be poorly absorbed due to the keratinized squamous cell lining
   (d) Be completely absorbed due to lack of metabolizing enzymes

2. To reach their drug target, antihistamine drugs must:
   (a) Be absorbed systemically
   (b) Be carried to the nasopharynx
   (c) Avoid the soluble layer of the nasal mucosa
   (d) Reach the nasal vasculature

3. Which of the following drugs will be absorbed better systemically in the nose?
   (a) Candesartan (FW 440)
   (b) Angiotensin (FW 1046)

4. Which of the following drugs is more likely to remain locally in the nose?
   (a) Xylometazoline (log P 5.3)
   (b) Phenylephrine (log P −0.31)

5. The preferred head position for the administration of nasal sprays is:
   (a) Head upright
   (b) Head down
   (c) Head back
   (d) Lying head lateral

6. An elderly man who has a new prescription for intranasal vasopressin should be advised to direct the metered dose pump to:

   (a) The vestibule

   (b) The anterior turbinate region

   (c) The olfactory epithelium

   (d) The nasopharynx

7. Research the inactive ingredients in the following dosage forms. Determine the function of the excipients using Appendix Tables A.2 and A.3. Categorize them as contributing to the manageable size of the dosage form, its palatability or comfort, its stability (chemical, microbial, or physical), the convenience of its use or the release of drug from the dosage form. Note that ingredients may be included in many dosage forms for more than one purpose.

Momentasone furoate nasal suspension

Ingredient	Category	Quality
Benzalkonium chloride		
Carboxymethylcellulose		
Citric acid		
Sodium citrate		
Glycerin		
Microcrystalline cellulose		
Polysorbate 80		
Purified water		

8. You have a prescription for nasal terbinafine 1% liquid to treat a patient who has a nasal fungal infection. Research the chemical and biological properties of terbinafine. What challenges does this drug present to the formulator in terms of dissolution in body fluids, movement through membranes, stability in dosage forms, loss of drug due to metabolism, access to receptors, or selectivity of the target.

Terbinafine HCl nasal liquid

Design issue	Drug property	Analysis
Drug must be dissolved before it can cross membranes		
Drug must have sufficient lipophilicity to cross membranes		
Large molecular weight drugs have difficulty moving across membranes		

Design issue	Drug property	Analysis
Drugs may hydrolyze, oxidize, photolyze, or otherwise degrade in solution		
Some drugs cannot be used by mouth because of substantial losses to enzyme degradation		
Intracellular drug receptors or receptors in the brain or posterior eye are challenging to target		
Some drugs cause serious toxicity as a result of poor selectivity of drug action		

9. Make a list of ingredients that you will put in the terbinafine nasal product. Categorize them as contributing to the manageable size of the dosage form, its palatability or comfort, its stability (chemical, microbial, or physical), the convenience of its use or the release of drug from the dosage form.

Ingredient	Category	Quality
Propylene glycol or glycerin		
Benzalkonium chloride or other antimicrobial preservative		
Sodium chloride		
Methyl cellulose (or other viscosity agent)		
Purified water		

10. Discuss the nasal route of administration with a group of classmates. From what you know about anatomy and physiology, determine the answer to the questions below about the route and whether the characteristic represents a challenge or an opportunity for drug delivery.

Nasal route

Questions	Answer	Challenge or opportunity?
What is the membrane that will absorb the drug from this route and how accessible is it?		
Is the absorbing membrane single or multiple layers and how permeable is it?		

Questions	Answer	Challenge or opportunity?
How large is its surface?		
What is the nature of the body fluids that bathe the absorbing membrane?		
Will the dosage form remain at the absorbing membrane long enough to release its drug?		
Will the drug encounter enzymes or extremes of pH that can alter its chemical structure before it is distributed to its target?		
What is the blood flow primacy to the absorbing membrane and the distribution time from it?		

11. You have a patient who complains of a bad taste after she uses her nasal spray. What can you advise the patient about reducing this problem?

Selected reading

1. F Ishikawa, M Murano, M Hiraishi et al. Insoluble powder formulation as an effective nasal drug delivery system. Pharm Res 2002; 19: 1097–1104.
2. X Ding, LS Kaminsky. Human extrahepatic cytochromes P450 function in xenobiotic metabolism and tissue-selective chemical toxicity in the respiratory and gastrointestinal tracts. Ann Rev Pharmacol Toxicol 2003; 43: 149–173
3. L Illum. Nasal drug delivery – recent developments and future prospects. J Control Rel 2012; 161: 254–263.
4. P Merkus, FA Ebbens, B Muller et al. The 'best method' of topical nasal drug delivery: comparison of seven techniques. Rhinology 2006; 44: 102–107.
5. U Raghavan, BM Logan. New method for the effective instillation of nasal drops. J Laryngol Otol 2000; 114: 456–459.
6. MS Benninger, JA Hadley, JD Osuthorpe et al. Techniques of intranasal steroid use. Otolaryngol Head Neck Surg 2004; 130: 5–24.
7. SP Newman, KP Steed, JG Hardy et al. The distribution of an intranasal insulin formulation in healthy volunteers: effect of different administration techniques. J Pharm Pharmacol 1994; 46: 657–660.

16

Drug delivery to the lung and from the lung

Learning objectives

Upon completion of this chapter, you should be able to:

- Define aerodynamic diameter, inspiratory flow rate and fine particle fraction.

- Identify the challenges and opportunities presented by each of the following route related characteristics to the rate and extent of drug travel to its target: accessibility, surface area, permeability and type of absorbing membrane, nature of the body fluids that bathe the absorbing membrane, retention of the dosage form at the site of absorption, location and number of drug metabolizing enzymes or extremes of pH and blood flow primacy of the lung.

- Predict the effect of each of the following drug properties on the rate and extent of drug travel to its target from the lung: water solubility, log P, molecular weight, stability in solution, enzymatic degradation and location of drug receptors.

- List the categories of ingredients that are added to respiratory dosage forms, and identify how the ingredient contributes to the manageable size of the dosage form, its palatability or comfort, its stability (chemical, microbial and physical), the convenience of its use and the release of drug from the dosage form.

- Describe the dosage forms/drug delivery systems used by this route and how they generate aerosols.

- Outline the movement of a drug from its site of administration to its target including:
  - release
  - absorption
  - distribution to target.

- Explain to patients and/or other healthcare providers how to use, store and/or prepare dosage forms for administration via the nasal route.

- Explain how the qualities of the dosage form are evaluated *in vitro* and/or *in vivo*.

- Explain to patients and/or other healthcare providers how to use, store and/or prepare dosage forms for administration via the oral inhalation route.

Introduction

The respiratory epithelium of the small airways and alveoli presents a very large surface with excellent blood flow designed for the absorption of gases. The surface has tremendous potential for drug absorption but is only accessible to particles in the respirable range, less than 5 μm in diameter. The delivery of these small particles requires the use of devices that generate aerosols such as nebulizers or portable inhalers. When used properly, drugs administered as aerosols by oral inhalation provide good outcomes in patients with asthma and chronic obstructive pulmonary disease with minimal side effects. However, a number of studies have demonstrated that improper use of inhalation devices is common. While device manufacturers continue to improve the portability and ease of use of these devices, pharmacists must take the time to educate their patients in the proper use of their inhaler devices and reinforce appropriate technique at regular intervals. Currently drugs available for oral inhalation are used to treat local conditions. The brief availability of insulin for oral inhalation has stimulated further interest in delivering proteins and other large molecules via the respiratory epithelium.

Characteristics of the respiratory route

(Figure 16.1, Appendix Tables A.1 and A.4)

- The function of the respiratory tract is the exchange of gases from blood to air and air to blood. Its upper airways include the nose, nasopharynx, trachea and bronchi that branch several times before reaching the smallest conducting airways, the terminal bronchioles.
- The airways are lined with a single layer of ciliated columnar epithelium that contains mucous and serous glands.
  - Cartilage rings prevent the collapse of the larger conducting airways, however, these are incomplete in the smaller bronchi and disappear in the bronchioles both of which can be narrowed by contraction of smooth muscle (Figure 16.2).

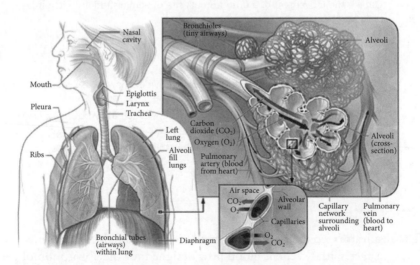

Figure 16.1 The respiratory tract and its blood supply to the alveolar surface. (Reproduced with permission from http://www.nhlbi.nih.gov/health/health-topics/topics/hlw/system.html)

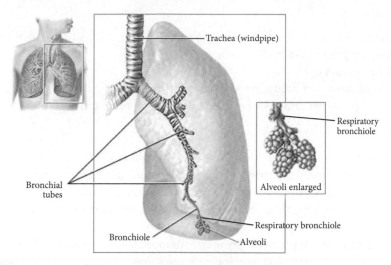

Figure 16.2 Airways of the respiratory tract. (Reproduced with permission from http://www.nlm.nih.gov/medlineplus/ency/imagepages/1103.htm)

- At the respiratory bronchioles the function of the tissue transitions from conducting airway to gas exchange, and the epithelium changes to ciliated cuboidal cells without mucous or serous glands.
- At the alveolar ducts the epithelium becomes a single layer of simple squamous cells, and at their ends the alveoli form clusters of small sacs lined with a film of phospholipid. The alveoli expand as they fill with air but do not contain smooth muscle.
- The respiratory tract has a very large surface area with an extensive capillary system at the alveolar surface.
- There are several different drug metabolizing enzyme categories found in the lung including cytochrome P450 isozymes, esterases, proteases and peptidases.
  - Cytochrome P450 isozymes can be found from the bronchi to the alveoli but their activity is 5–20 times lower than in the liver.
  - Lung esterases cleave the ester bonds of steroid prodrugs. Another mechanism responsible for the removal of drug from the airways is mucociliary clearance.

Drug travel from dosage form to target

Key Points

- The lung is better able to absorb polar drugs and macromolecules than the small intestine.
- Systemic absorption of inhaled drugs is small because the doses delivered are small.
- The lung's large absorptive surface is only accessible to drug aerosols ≤5 μm in diameter.

Release

- The respiratory epithelium is only accessible to particles in the respirable range, 5 μm or less in diameter.
- Drug aerosols:
  - Solutions are ready for absorption.
  - Suspensions and powders must dissolve to release drug.

Absorption

- Most drugs are absorbed via passive diffusion, and the alveolar epithelium creates a barrier to hydrophilic molecules greater than 0.6 nm.
- Proteins and peptides are transported across the alveolar membrane at a rate dependent on their molecular weight via mechanisms that remain unclear.
  - It is known that several proteins: albumin, insulin, immunoglobulin G and transferrin have specific receptors that mediate their transport across the alveolar epithelium via endocytosis/transcytosis.
- The respiratory epithelium is the rate limiting barrier for the movement of drugs into the blood and not the alveolar capillary membrane.
- Drugs are used by inhalation to reduce systemic absorption. However, given the permeability and surface area of the lower respiratory tract, there is some systemic drug absorption and the amount of drug that is found in the systemic circulation is related either to the inhaled dose or to the drug deposited on the mouth and swallowed.
  - The systemic absorption of inhaled drugs is relatively small because the doses delivered are small, but the potential for systemic absorption with higher than recommended doses is great.
- Transporter gene expression in the lung is quite high, ranking third after the liver and kidneys but because of the rapid pulmonary clearance of inhaled drugs, drug absorption from the lungs is less likely to be significantly influenced by transporters as is absorption in the small intestine.
  - p-Glycoprotein and a few other ATP binding cassette (ABC) efflux transporters have been detected at the various levels of the airways and lung.
  - The organic cation transporters, organic anion transporting poly-peptides and peptide transporters of the solute carrier family are also present.
  - Several drugs interact with these carriers *in vitro* but it is not known whether the carriers contribute to or detract from their absorption in humans.

Distribution

- Most drug targets are within the lung.
- The alveoli and respiratory bronchioles receive the entire cardiac output (from the right heart). However, blood flow to the larger airways (trachea

to terminal bronchioles) comes from the systemic circulation. Blood from the lung is returned to the heart within a few seconds and rapidly distributed out to the systemic circulation.

Elimination

- The ciliated epithelial cells from the trachea to the terminal bronchioles beat upwards to drive foreign particles including drug aerosols trapped in the respiratory mucus towards the esophageal opening (Figure 16.3). The activity of the cilia is reduced by cigarette smoking and some respiratory disease states.
- The alveoli have phagocytic cells that can remove foreign particles but have no cilia.

Respiratory function tests

Respiratory function can be objectively measured in terms of flow rates and volumes of expired or inspired air. Because most of the drugs used by the respiratory route are used to treat a respiratory condition and would not be expected to produce measureable drug concentrations in the peripheral veins, respiratory function tests are useful for the evaluation of aerosol devices. The descriptions, normal and abnormal values for some of the respiratory function tests are provided in Table 16.1.

Drug properties and the respiratory route

- Adequate water solubility is a desirable property for drugs applied to the lung. However, since they are delivered as very small droplets or particles and encounter sufficient airway surface fluid to dissolve them, water solubility is rarely the limiting factor for absorption.

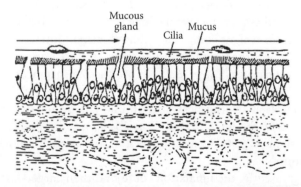

Figure 16.3 Mucociliary clearance. Cilia propel mucus mouthward at 15–20 mm/minute. (Reproduced with permission from University of Detroit Mercy, College of Sciences, Detroit, MI, 2004.)

Table 16.1 Respiratory function tests

Parameter	Description	Normal	Asthmatic or COPD values
Respiratory rate	Respirations per minute	12–16 per minute	12–16 (40–50) per minute
Tidal volume	Volume of a normal breath	0.5 L	0.5 L
Oxygen saturation	Oxygen carried by hemoglobin (arterial)	95–100%	30–100%
Forced vital capacity	Volume exhaled forcefully and completely after maximal inspiration	3.8–5.5 L	1.9–4 L
Forced expiratory volume	Volume exhaled forcefully and completely after maximal inspiration in 1 s	3.2–4.5 L	0.9 to 3.2 L
Peak expiratory flow rate	Maximum airflow during a forced expiration	300–550 L/minute	180–400 L/minute
Peak inspiratory flow rate	Maximum airflow during a forced inspiration	225–330 L/minute (elderly)	40–110 L/minute

COPD = chronic obstructive pulmonary disease.

- The solubility of the drug in propellant, its particle size and its surface characteristics will affect the ease with which it may be aerosolized in the respirable range.
- Most drugs are absorbed via passive diffusion, and the alveolar epithelium creates a barrier to hydrophilic molecules greater than 0.6 nm, but animal studies indicate that the lung is better able to absorb polar and large molecular weight compounds than the small intestine.
  - The lung absorbs lipophilic drugs faster and to a greater extent than hydrophilic drugs (Table 16.2) but produces measurable levels of very hydrophilic drugs such as cromolyn sodium.
  - Macromolecular drugs less than 40 000 Da are relatively rapidly absorbed by the lung (insulin T_{max} 15–60 minutes) while molecules larger than 40 000 Da are absorbed more slowly (albumin T_{max} 20 hours).
- Most drugs used by the inhalation route find their targets in the respiratory tract (Table 16.3).
  - It is desirable to deposit drugs such as the bronchodilators to the airways where there is responsive smooth muscle and the

Table 16.2 Effect of physicochemical properties on absorption rate and extent in the rat lung

Drug	Bioavailability to the lung (%)	T_{max} (minutes)	Water solubility	log P	log D	Molecular weight
Cromolyn	36	3.5	210 mg/mL	1.9	−8.4	468
Cyanocobalamin	84	13.3	12.5 mg/mL	−0.9	na	1356
Formoterol	101	5.3	0.66 mg/mL	1.1	0.5	330
Imipramine	98	2.4	0.0182 mg/mL	4.4	2.4	280
Losartan	92	4.6	0.048 mg/mL	3.5	0.3	423
Metoprolol	138	2.3	1000 mg/mL	1.9	0.0	267
Talinolol	81	10.9	1.23 mg/mL	3.2	1.4	364
Terbutaline	74	3.6	213 mg/mL	0.1	−1.3	225

Reproduced with permission from A Tronde, et al. J Pharm Sci 2003; 92: 1216–1233.

Table 16.3 Drug receptors in the respiratory tract

Drug class	Location of receptors
Histamine blockers	Vascular endothelial cells, smooth muscle of bronchi
Beta agonists	Beta-2 receptors in tracheal, bronchial and bronchiolar smooth muscle (relax), serous glands (increase secretions)
Mast cell stabilizers	In the connective tissue near blood vessels, especially alveoli
Anticholinergics	Cholinergic receptors in tracheal, bronchial smooth muscle (relax), serous glands (decrease secretions)
Corticosteroids	Intracellular receptors on epithelial cells, fibroblasts, basophils, macrophages, and lymphocytes. These are located in connective tissue at all levels of the respiratory tree

    corticosteroids to the respiratory epithelium from the trachea to the alveolar surface.
— Systemic therapies should be deposited on the alveolar epithelium for the best bioavailability.
• Because most drugs administered by oral inhalation are for treatment of lung and airway conditions, drug metabolism at the lung tends to reduce systemic absorption and side effects but does not apparently limit the therapeutic effect.

- Pulmonary first pass has not been characterized for most drugs but given the variety of enzymes found in the lung, there is potential for some first pass loss to occur.
  - Large proteins such as albumin move across the lung epithelium intact, whereas smaller proteins and peptides may undergo significant degradation.
- Most products for use in nebulizers are aqueous solutions and require adequate stability of the drug in water.
- Drugs may be aerosolized as powders in dry powder inhalers, or as nonaqueous solutions or suspensions in metered dose inhalers providing options for the formulation of drugs with poor aqueous stability.

Dosage forms for the respiratory route

Design

Aerosol products may be applied as dry powder inhalers (DPIs), nonaqueous solutions or suspensions in metered dose inhalers (MDIs) or aqueous solutions and suspensions for nebulization. The newer metered dose liquid inhalers are formulated as aqueous solutions.

1. One dose in a manageable size unit
  - Presently available inhalation devices cannot conveniently deliver more than approximately 5–10 mg of drug. Thus a drug administered to the lungs must be active in very small doses.
  - To reach the large surface of the lower airways and alveoli, a delivery device must produce drug particles small enough to be carried with the airstream without depositing on the progressive turns and narrowing of the conducting airway walls.

Definition

The *aerodynamic diameter* is a parameter that describes the ability of a particle to move with the airstream.

- The parameter used to describe particle sizes released from an aerosol dosage form is the aerodynamic diameter. If one considers the properties of a particle that would determine its ability to move with the airstream, those properties would include the diameter (size) of the particle, its shape and its density. The aerodynamic diameter combines these properties by acknowledging that two particles with the same diameter but very different densities or shapes will not flow with the airstream into the lungs in the same fashion.

$$D_{ae} = D\sqrt{\frac{particle\ density}{unit\ density}}$$

where D_{ae} is the aerodynamic diameter, D is the measured (geometric) diameter and unit density is 1 g/mL.

- The use of aerodynamic diameter classifies two particles with different diameters and densities that both behave like a particle with a diameter of 2 μm and a density of 1 g/mL (unit density) in an airstream as if they had the same diameter.
- Respirable drug particles or droplets must have an aerodynamic diameter less than 5 μm. This is called the *fine particle fraction* of the aerosol particles and may be used to predict the lung deposition of a product. These particles are more likely to deposit by sedimentation or diffusion than inertial impaction (Table 16.4).
- Particles less than 3 μm aerodynamic diameter are more likely to deposit in the peripheral airways of the lungs rather than in the large airways. Drug particles larger than 5 μm are deposited in the oropharynx, swallowed and absorbed by the gastrointestinal tract. If the drug is not extensively metabolized by the liver, it can cause systemic side effects. Drug deposited in the oropharynx can also have local side effects. The inhaled corticosteroids increase patient susceptibility to oral yeast infections.
- Drug particles or droplets released from respiratory dosage forms can range in size from 0.5 to 80 μm.

Table 16.4 Mechanisms of deposition

Aerodynamic diameter (μm)	Mechanism of deposition	Location
≥10	Inertial impaction. Particles this size have high momentum and will hit the airway walls rather than turning corners with the airstream	Mouth
5–10	Inertial impaction	Mouth and large central airways (first 10 generation of airways)
1–5	Sedimentation – settling out of the airstream owing to gravity	Smaller bronchi, bronchioles and alveoli where air velocity is low or zero
0.5–1	Sedimentation and diffusion	Alveoli
<0.5	Diffuse – move down a concentration gradient	Alveoli. May be expired before they deposit

Reproduced with permission from NR Labris, et al. Br J Clin Pharmacol 2003; 56: 588–599.

- The particle size of an aerosol is determined by the physico-chemical properties of the drug, whether the drug is prepared as a solution, suspension or dry powder and the device used to create the aerosol.
- In addition, the patient's ability to use the device properly can affect both the particle size emitted and the location in which the drug deposits.

2. Comfort
   - Nebulizer solutions should be isotonic or slightly hypotonic.
3. Stability
   - Microbial:
     - Multidose, nebulizer solution should contain a compatible antimicrobial preservative.
     - The pressurized container of MDIs protects the drug from oxidation and microbial contamination.
     - DPIs exhibit good chemical and microbial stability in solid state.
   - Physical:
     - Suspension aerosols require surfactants to prevent aggregation of drug particles
     - DPIs must be protected from moisture to flow properly.
4. Convenience/ease of use
   - Requires the use of an aerosol device to produce particles <5 μm aerodynamic diameter.
   - Proper use of device is essential to good therapeutic outcomes.
   - MDIs are more difficult to use properly than DPIs, which are more difficult to use than nebulizers.
5. Release of drug
   - Release from device.
     - MDI: expelled by pressure created by propellant.
     - DPI: patient draws powder into air stream. Requires good powder flow.
     - Nebulizer: aerosol droplets produced by drawing air across the surface of the drug solution or by vibrating the drug solution using ultrasound energy.
   - Release from formulation.
     - Drugs are released from powder or suspension dosage forms by dissolution; those in solution are ready for absorption.

Aerosol ingredients and devices

Table 16.5 details the design and ingredients of selected nebulizer, MDI and DPI products.

Table 16.5 Formulation of commercially available aerosol products

Aerosol product	Other	Stability	Release
Combivent Respimat MDLI		Benzalkonium chloride, edetate disodium	Water for injection
QVAR MDI solution			Alcohol, norflurane
Symbicort MDI		Povidone, polyethylene glycol 1000	HFA 227
Pulmicort Flexhaler DPI			Lactose
Foradil Aerolizer DPI	Gelatin capsule		Lactose
Atrovent MDI solution		Citric acid	Alcohol, distilled water, norflurane
Formoterol nebulizer solution	Sodium chloride	Citric acid, sodium citrate	Water
Pulmicort Respules nebulizer suspension	Sodium chloride	Citric acid, edetate disodium, polysorbate 80, sodium citrate	Sterile water for injection

MDI = metered dose inhaler; DPI = dry powder inhaler.

Metered dose inhalers

Key Point

- Pressurized MDIs produce aerosols using propellants that vaporize at atmospheric pressure to disperse drug as small particles or droplets.

- MDIs (Figure 16.4) contain drug dissolved or suspended in a propellant that is liquefied by a relatively small decrease in temperature or increase in pressure.
- This permits the drug mixture to be contained in a thin walled, aluminum or plastic coated, glass canister that is light weight and portable.
- When released into atmospheric pressure the propellant becomes a gas and the drug is dispersed as droplets or small particles.
- The dose chamber and valve of a metered dose inhaler reproducibly release volumes from 25 to 100 µL when the device is properly primed.
- Depression of the valve stem causes the contents of the chamber to be expelled. When the valve is released (closed) it refills with the predetermined volume.
- Drug exits an MDI at high velocity, between 2 and 8 m/s, making it challenging for the patient to coordinate the breath inward at the moment of actuation.
- If properly used, the newer MDIs deliver between 30% and 60% of the dose to the lung.

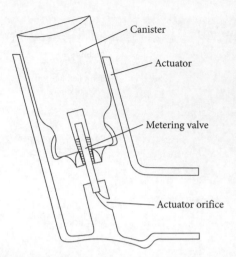

Canister

Actuator

Metering valve

Actuator orifice

Figure 16.4 Metered dose inhaler. (Reproduced with permission from L Felton (ed.), Remington: Essentials of Pharmaceutics, p. 646; published by Pharmaceutical Press, 2012.)

- A pressurized MDI is unable to provide a reproducible dose after the labeled number of doses has been actuated. However, when shaken it will contain audible liquid, thus the 'shake test' is not a good way to determine if doses are left.

Metered dose inhalers with spacers and valved holding chambers

- A spacer is a simple tube that provides extra distance between the released aerosol and the patient's mouth.
- A valved holding chamber is a device that is designed to collect the aerosol released from an MDI until the patient is able to start the inhalation that will carry the aerosol into the lungs.
- Spacers and valved holding chambers:
    - provide more time for the patient to coordinate the breath with the actuation of the pressurized MDI.
    - collect the larger particles on their sides that would otherwise deposit in the patient's mouth and be swallowed.
- While the benefit of a spacer or holding chamber may be small for patients with good inhaler technique, it is considered clinically significant for patients with poor technique.
- Spacers and valved holding chambers were designed to be used with chlorofluorocarbon (CFC) inhalers, thus their effect on the delivered dose of the newer hydrofluoroalkane (HFA) inhalers is uncertain.
    - The pharmacist should monitor the patient's response to inhaled medication when initiating the use of a spacer or valved holding chamber to determine whether it has the desired effect.

- Plastic spacers and valved holding chambers develop static charge that attract drug particles to the surface of the spacer and can decrease aerosol delivery to the lung. It is recommended that patients wash these devices in detergent, rinse well and allow to air dry to reduce static charge.

Ingredients in MDI aerosols

See Appendix Table A.2.

- Propellants produce a constant pressure inside the MDI container by vaporizing into the available head space.
- The pressure pushes the liquid drug mixture out of the dose chamber when the MDI is actuated, and as the propellant becomes a gas at atmospheric pressure, it disperses the drug in the airstream.
- The CFC propellants used before their mandated phase-out in December 2008 photolyzed in the ozone layer to chloride radicals that react with and damage the ozone.
- The HFA propellants that replaced CFCs are:
  — nonflammable and do not react with ozone
  — liquefied by a relatively small decrease in temperature or increase in pressure, producing a slower forward velocity of spray and smaller aerosol droplet sizes than CFCs
  — miscible with water and alcohol so that many of the newer products contain drug in solution
  — are not good solvents for surfactants, which may be required for suspended drug mixtures. Surfactants are used to prevent the growth of suspended particles in MDI mixtures.

Metered dose liquid inhalers

Key Point

- Metered dose liquid inhalers disperse an aqueous solution of drug by generating two rapidly moving jets of liquid that collide.

Respimat Soft Mist inhaler is the first metered dose liquid inhaler to be marketed. The device delivers a dose of medication from a solution in a collapsible bag contained in a replaceable plastic cartridge. Twisting the inhaler's base compresses a spring that when released will disperse the dose by generating two rapidly moving jets of the formulation that collide. The aerosol produced is a slow velocity mist with a fine particle fraction of 65–80%. The slow velocity (0.8 m/s) of the aerosol and longer duration of the plume emitted allows the patient more time to coordinate inspiration with the activation of the device. Lung deposition studies comparing the device with an MDI demonstrate that with Respimat, the amount of drug deposited in the mouth is reduced and the amount deposited in the lower respiratory tract is doubled.

Dry powder inhalers

Key Point

- The patient generates drug aerosol by drawing his breath through loose powder in a DPI.

- DPIs include devices that contain drug powder:
  - in a reservoir
  - on a strip of individual blisters
  - in unit dose capsules that are loaded into the device before use.
- The blisters or capsules are broken or the dose of powder from the reservoir is dropped into the inhalation channel of the inhaler, and the drug aerosol is created by directing air through the loose powder.
- The patient actuates a dry powder inhaler with his own inspiration.
  - A relatively fast inspiratory rate is required to produce a particle cloud in the airstream with a sufficiently small aerodynamic diameter to reach the lower lung.
- Drug particles micronized for aerosolization (<5 μm in diameter) tend to be highly cohesive and flow poorly.
  - For this reason dry powder inhaler formulations consist of micronized drug particles that adhere to larger (50–200 μm diameter) carrier powders or low density particles that form larger, spherical agglomerates.
  - Lactose has been the most commonly used carrier powder.
  - Because carrier powders are large and deposited in the mouth, they may be tasted.
  - The larger carrier particles and spherical agglomerates overcome the poor flow properties of the smaller drug particles but in order for the drug to reach the lower respiratory tract, these assemblies must disassociate as they leave the device.
  - Detachment occurs when the combination of shear forces resulting from rapid inspiratory flow and inertial forces resulting from impact with other particles or with the inhaler walls exceed the adhesive forces.
  - Each DPI is designed to capture the energy of the inspiratory flow in a different fashion to generate shear or inertial forces.
  - The inspiratory flow requirements for different DPIs are generally presented in the product labeling.
- Powders for aerosolization that absorb water will be too heavy or not deaggregate properly; thus a variety of packaging strategies are used to protect powders from humidity.

Dry powder inhaler types

- The first DPIs packaged the drug dose in a gelatin capsule that is broken prior to administration (Foradil Aerolizer, Spiriva Handihaler; Figure 16.5). The gelatin capsules are individually packaged in blisters to be opened just before loading. The capsule is loaded into a well in the inhaler and pierced prior to inhaling the dose.
- The reservoir devices (Easyhaler, Flexhaler, Asmanex) are extremely compact; however, dose reproducibility may be suboptimal. Drug powder is loaded into the inhalation channel by shaking and clicking or twisting the device. The Flexhaler contains a drying agent in its base while the other two reservoir devices have an outer foil wrap and limited shelf life once opened.
- Blister packaged powders (Diskus) contain the dose preloaded in foil covered blisters. The blister is broken by sliding a lever to expose drug powder just before inhalation.

Small volume nebulizers

Key Points

- The aerosol generated by a nebulizer can be inhaled during normal breathing.
- Nebulizer solutions must be sterile.

- Small volume nebulizers (SVNs) are table top devices that deliver larger volumes of drug (1–3 mL) in small droplets dispersed in room air.
- Drug is delivered via a mouthpiece or mask during normal breathing.

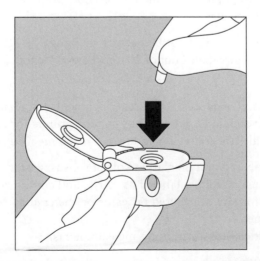

Figure 16.5 Spiriva Handihaler dry powder inhaler. (Reproduced with permission from drugs.com.)

- SVNs are useful for children and elderly patients who typically have difficulty with inhalers.
- A nebulizer will require 10–15 minutes to deliver a dose, and has a significant dead volume (0.5–1 mL).
- Use of a mouthpiece provides a larger dose delivered to the lung than a mask because nasal inhalation reduces the number of droplets that reach the lungs.
  - If a mask is used with a small child or infant, care must be taken to prevent leaks along the mask seal to minimize deposition in the eyes and on the child's face.
- Nebulizers generate drug aerosols either by drawing air across the surface of the drug solution or by vibrating the drug solution using ultrasound energy.
- The aerodynamic diameter of aerosols delivered by different nebulizers can vary significantly.
- Nebulizers are easily contaminated by microorganisms, and while they are rarely a source of infection for patients, they should be carefully cleaned after use.
- Pharmacists occasionally prepare compounded inhalation solutions.
  - Sterile, isotonic or slightly hypotonic aqueous solutions with a minimum of ingredients ensure stability and solubility of the active drug.
  - If antioxidant ingredients are required, generally the sulfites are avoided because some asthmatic patients have airway reactions to these compounds.

Quality evaluation and assurance

The United States Pharmacopeia in its general chapter <601> Aerosols, and individual monographs for official aerosol preparations provide the following requirements:

- The delivered dose may not be outside the range 75–125% of the label claim and no more than 2 out of 20 may be outside the range 80–120% of label claim.

The FDA in its Guidance for Industry document, a nonbinding description of the data that should be provided by manufacturers for approval of their products, provides the following additional recommended qualities:

- all aqueous-based oral inhalation solutions, suspensions and spray drug products must be sterile, labeled as sterile and confirmed by testing
- content uniformity of metered doses or premetered blisters
- plume geometry
- *in vitro* particle size distribution measurements.

Because of the different mechanisms by which each device generates aerosols, the same dose of the same drug from an MDI, DPI or nebulizer would produce different particle size ranges and deposit differently in the lung. As a result, from a regulatory standpoint, the aerosol devices are not considered interchangeable.

Counseling patients

Choosing an aerosol generating device

The efficacy the different devices available for the treatment of respiratory disease states with aerosolized drugs has been the subject of several reviews. The evidence reveals that nebulizers, MDIs and DPIs work equally well in a variety of clinical settings in patients who use them properly. In the absence of a clear advantage for any one device, the choice of an appropriate device for treatment of a particular patient will require a number of considerations (Table 16.6).

- Clinical: For patients in acute distress with a significant degree of breathlessness, the use of a nebulizer or pressurized MDI with holding

Table 16.6 Advantages and disadvantages of aerosol devices

Advantages	Disadvantages
Metered dose inhalers	
Portable and compact	Coordination of breathing and actuation needed
Treatment time is short	Not all medications available as MDI
No drug preparation required	High pharyngeal deposition
No contamination of contents	Not suitable for children ≤5 years
Dose–dose reproducibility high	Deliver relatively small drug doses
MDI plus spacer or valved holding chamber	
Reduces need for patient coordination	Less portable than MDI alone
Reduces pharyngeal deposition	Can reduce dose available if not used properly
Suitable for children >4 years and impaired adults	May develop static charge that reduces the dose
Dry powder inhalers	
Breath actuated	Requires moderate to high inspiratory flow
Less patient coordination required	Some units are single dose
Propellant not required	Can result in high pharyngeal deposition
Small and portable	Not all medications available

(continues)

Table 16.6 (*continued*)	
Advantages	**Disadvantages**
Short treatment time	Not suitable for children <5 years
Good chemical and microbial stability	Deliver relatively small drug doses
Small volume nebulizers	
Patient coordination not required	Lack of portability
Effective with tidal breathing (easy to use!)	Pressurized gas or power source required
High dose possible	Lengthy treatment time
Suitable for children ≤2 years and elderly	Contamination possible
	Large dead volume
	Do not aerosolize suspensions well
	Device and drug preparation required

MDI = metered dose inhaler.

chamber is recommended. In patients requiring chronic maintenance therapy, more portable devices are preferred unless the patient has severely diminished respiratory or cognitive function (Table 16.7).

- Age of patient: infants and very young children have difficulty cooperating with the aerosol devices that require training.
- Patient's ability to use the device: Patients need initial training to use aerosol devices properly and benefit from periodic retraining to assure that appropriate technique is retained. The use of a training device such as an inspiratory flow meter may be helpful to provide feedback to the patient on their inspiratory flow rate and coordination of the actuation of an MDI with inspiration.
- Availability of the formulation.

Table 16.7 Respiratory function indicators for nebulizer use outside of the hospital
Vital capacity <1.5 times the predicted tidal volume (7 mL/kg body weight)
Inspiratory flow rate less than 30 L/minutes
Breath holding capacity less than 4 s
MMSE <23/30 (this number is for propelled metered dose inhalers)

Reproduced with permission from National Association for Medical Direction of Respiratory Care (NAMDRC) Consensus Group. Guidelines for the use of nebulizers in the home and at domiciliary sites. Report of a Consensus Conference. Chest 1996; 109: 814–20. http://dx.doi.org/10.1378/chest.109.3.814.

- Patient preferences for device: Patients show a strong preference for portability and ease of use. However, once properly trained to use one type of device, a particular patient may prefer to continue the use of that type of device rather than use both an MDI and a DPI.
- Healthcare professional's lack of time or skills to instruct the patient in proper use.
- Cost of therapy and potential for reimbursement.

Use of aerosol devices

Key Point

- Proper use of an aerosol device is essential to achieve good patient outcomes.

- Use of proper technique by patients is *essential* to the effectiveness of inhaler devices.
- Studies with patients who use MDIs estimate that up to 59% of users have poor or inadequate technique. The most commonly identified problems are presented in Table 16.8.
  — The evidence reveals that MDI aerosols should be inhaled after a complete exhalation, the device should be actuated at the start of a slow inhalation (inspiratory flow rate of less than 60 L/minute) followed by a 10 second breathhold at the end of inspiration.
  — Spacers provide the user more time to coordinate their inspiration with the actuation of the device and reduce side effects, because most of the dose that would have deposited in the mouth is deposited in the spacer.

Table 16.8 Common errors in patient technique	
Device	**Errors**
pMDI	Difficulty coordinating pMDI actuation with inhalation
	No exhalation before inhalation
	No breath hold
DPI	Failure to achieve a forceful and rapid flow at the beginning of inspiration
	No exhalation before inhalation or exhalation into the device
	No breath hold

pMDI = propelled metered dose inhalers; DPI = dry powder inhalers.
Reproduced with permission from selected reading 6.

- Despite their breath actuated design, about one-third of DPI users make serious errors in their inhalation technique.
  - DPI users need a deep and forceful inspiration with rapid onset to produce the necessary fine particle fraction for good lung deposition.
  - Because the objective is to empty the dose from the device, the patient should continue the inhalation as long as possible.
  - The deep inhalation can be facilitated by a complete exhalation before an inhalation through an inhaler to maximize the inhalation volume.
  - Patients are advised to exhale away from the dry powder inhaler. This reduces the amount of humidity in the DPI and prevents the patient from blowing out the dose loaded into the inhalation channel.
- Most patients can attain sufficient inspiratory flow rates to use a DPI; however, pressurized MDIs with spacer or nebulizers are recommended in acute care situations.
- An inspiratory flow meter is a device that can be used to provide the patient with feedback about technique for both DPIs and MDIs.
- The pharmacist should take the time to review the instructions and have the patient demonstrate proper use on initial dispensing and follow up with regular review of technique as needed. This is particularly important if the patient's asthma or chronic obstructive pulmonary disease symptoms are not well controlled.

Use of metered dose inhalers

The patient should be given the following instructions.

- Shake the inhaler and remove the mouthpiece cap.
- Exhale fully and place the mouthpiece into your mouth. Some clinicians recommend that the inhaler be held 2.5–5 cm (1 to 2 inches) from the open mouth or that a spacer device be used. Both of these suggestions reduce the amount of drug solution deposited on the sides of the mouth and swallowed (Figure 16.6).
- As you take a deep breath, squeeze the canister and mouthpiece together at the same time.

Figure 16.6 Appropriate positioning of a MDI. (Reproduced with permission from http://minorityhealth/hhs.gov/assets/pdf/checked/1/factscontrolling%20asthma.pdf)

- Draw the spray from the inhaler into your lungs with a slow, deep breath. Hold your breath as long as comfortable.
- Exhale slowly. If you did not use a space or a spacer, rinsing your mouth before you swallow drug solution that may be on the sides of your mouth can reduce the chances of side effects. If you are using a corticosteroid inhaler, rinsing your mouth is necessary to prevent development of mouth infections. The rinse water should be spat into the sink.

Caring for MDIs:

- Wipe or rinse the mouthpiece clean. Rinse the plastic actuator once a week to prevent clogging. Take out the drug filled canister and run the plastic actuator under warm water and allow to air dry before reassembling.

The patient may be given the following advice.

- Inhaled medications begin to work almost immediately (within 1 minute). If more than one puff is to be given, wait at least 1 minute between each administration.
- If several different medicines from different inhalers are to be given, it is best to give the bronchodilator before any other.

The following are instructions for use of MDIs with spacers or holding chambers in children 3 years old or under (Figure 16.7).

- Shake the inhaler and place the mouthpiece into the open end of the chamber ensuring a tight fit.
- Place the facemask over the nose and mouth.
- Actuate one dose into the chamber.
- Allow the child to inhale and exhale normally into the chamber at least 10 times.
- Clean the face after the mask is removed in case medication was deposited on the skin.

The following are instructions for use of MDIs with spacers or holding chambers in children over 3 years old.

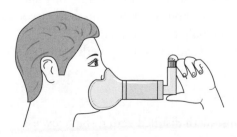

Figure 16.7   Aerochamber with mask. (Reproduced with permission from CA Langley, D Belcher, Applied Pharmaceutical Practice, p. 233; published by Pharmaceutical Press, 2009.)

- Shake the inhaler and place the mouthpiece into the open end of the chamber ensuring a tight fit.
- Instruct the child to exhale to empty their lungs.
- Place the mouthpiece of the chamber into the mouth with lips sealed around it.
- Actuate one dose into the chamber.
- Ask the child to inhale slowly through the mouthpiece.
- Ask the child to hold their breath as long as possible.

Use of dry powder inhalers

The following are instructions for the patient.

- Prepare the dose and the mouthpiece of the inhaler according to the Patient Information Leaflet.
- Keeping the inhaler level, turn your head away from the DPI and exhale completely to empty your lungs.
- Begin your inhalation rapidly and forcefully into the mouthpiece.
- Continue inhaling as long as you can.
- Hold your breath about 10 seconds.
- If you are using a corticosteroid medication, rinse your mouth.

Caring for DPIs:

- Keep your inhaler in a dry place at room temperature.
- Wipe it out once a week with a dry cloth.

Use of a nebulizer

The following are instructions for the patient.

- Assemble the nebulizer according to its instructions.
- Fill the medicine cup or reservoir with your prescription, according to the instructions.
- Connect to the power source or compressor.
- Place the mouthpiece in your mouth. Breathe through your mouth until all the medicine is used, about 10–15 minutes. Some people use a nose clip to help them breathe only through the mouth.
- Wash the medicine cup and mouthpiece with water, and air dry until your next treatment.

Caring for a nebulizer:

- After each use, rinse with distilled water, shake excess moisture from the cup and leave to air dry on a clean towel.

- Once or twice a week follow the manufacturer's instructions on disinfection. If the manufacturer's instructions allow, disinfect the nebulizer by soaking in benzalkonium chloride solution 30 mL to 3.8 L (1 gallon) of water for 10 minutes.

Questions and cases

1. Drug residence time in the alveoli is longer than in the bronchi because:
    (a) The alveoli have no mucociliary clearance
    (b) The alveoli are expandable
    (c) The alveoli have no metabolic enzymes
    (d) Drug absorption is rapid from the bronchi

2. A study found that a bronchodilator beta agonist drug with a mass median aerodynamic diameter (MMAD) of 1.03 μm did not produce good bronchodilation. The likely explanation of this finding is:
    (a) The aerosol deposits mostly in the mouth
    (b) The aerosol deposits in the trachea where the cartilage rings prevent changes in the diameter of the airway
    (c) The aerosol deposits in the alveoli where there is no smooth muscle
    (d) The aerosol is absorbed too quickly into the blood to have an effect on the smooth muscle

3. Drugs used to treat pneumonia need to reach the:
    (a) Trachea and bronchi
    (b) Bronchi and bronchioles
    (c) Trachea and bronchioles
    (d) Bronchioles and alveoli

4. What biological barrier reduces systemic absorption of inhaled drugs?
    (a) Endothelial tight junctions in lung capillaries
    (b) Drug metabolizing enzymes in lung tissue
    (c) Small absorptive surface in the lungs
    (d) Drug induced reductions in blood flow to the lung

5. Which steroid drug would theoretically be better absorbed systemically?
    (a) Fluticasone propionate log P 3.46
    (b) Beclomethasone dipropionate log P 4.07
    (c) Budesonide log P 2.81
    (d) Flunisolide log P 2.66
    (e) Triamcinolone log P 1.16

6. Propellants produce small aerosol particles by:
    (a) Reacting with atmospheric oxygen to form ozone
    (b) Dispersing as small droplets in the airstream when the patient inhales
    (c) Dispersing as small droplets as the drug solution moves through a mesh in the metered dose chamber
    (d) Becoming a gas as they are released into atmospheric pressure

7. Drug powders are able to flow easily out of a dry powder inhaler because:

    (a) The drug powder exits the device at 30 m/s

    (b) The drug powder is μmized (<5 μm)

    (c) The drug powder is carried on a larger, inert solid

    (d) The patient does not have compromised pulmonary function

8. Budesonide suspension for nebulization contains citric acid monohydrate, edetate disodium, polysorbate 80, sodium chloride, sodium citrate and sterile water for injection. The presence of edetate disodium suggests that this drug is susceptible to:

    (a) Hydrolysis

    (b) Oxidation

    (c) Photolysis

    (d) Racemization

9. Chronic obstructive pulmonary disease (COPD) predominantly affects the terminal bronchi, respiratory bronchioles and alveoli. The preferred particle size range for the treatment of patients with COPD would be:

    (a) 0.5–1 μm

    (b) 1–2 μm

    (c) 2–3 μm

    (d) 3–5 μm

    (e) Answers (a) and (b)

10. One of the approaches to the preparation of dry powder inhaler particles that flow well has been to make porous drug particles with low density and larger geometric diameter. The effect of reducing the density of a particle on the particle's aerodynamic diameter would be:

    (a) To increase the aerodynamic diameter

    (b) To decrease the aerodynamic diameter

    (c) To keep the aerodynamic diameter constant

    (d) Answers (b) and (c)

11. You have a pregnant asthmatic patient whose physician has advised her to continue using her inhaled steroids and beta agonist medications. Which of the following counseling points will reduce systemic absorption of drugs applied to the lung?

    (a) Use the bronchodilator medication first

    (b) Breathe out all the way

    (c) Hold your breath as long as you can

    (d) Rinse your mouth after using your inhaler

    (e) All of the above

12. A group of pharmaceutical scientists demonstrated that the bronchodilator response to albuterol was significantly greater when nebulized using a nebulizer that produced an aerosol with an aerodynamic diameter of 3.3 μm compared with one that generated a larger size aerosol (7.7 μm). Explain this finding.

13. Flunisolide is available as a metered dose inhaler with built in spacer. What impact would this have on the particle size emitted, and the fraction of dose deposited in the mouth and lungs compared with a regular metered dose inhaler (MDI)?

14. In 2008 the use of chlorofluorohydrocarbon propellants was phased out in favor of hydrofluroalkane (HFA) propellants. The MMAD of a newer HFA beclomethasone dipropionate inhaler is 1 μm (range 0.4–6 μm) compared with its older formulation using CFC propellant, which was 3 μm (range 0.4–12 μm). Compare the deposition of these two inhalers in the respiratory tract. How would these two products differ in their side effects?

15. BC, a 3 year old boy with a 1½ year history of daily recurrent wheezing, is referred to your clinic for difficult to control wheezing. Current medications:

- Slo-Phyllin Syrup 80 mg tid prn wheezing

- Prednisolone oral liquid 1 tsp bid prn severe wheezing.

He has been taking both medications maximally lately. The physician would like him to have daily corticosteroids and as needed beta agonists but she is concerned about slowed bone growth with the corticosteroid therapy. What delivery device do you recommend for this patient? Explain how the family should use and care for this device.

16. MT is a 25 year old woman who presents to the emergency department with complaints of dyspnea and coughing that have progressively worsened over the past 2 days. Physical examination reveals that she is wheezing, RR 30 breaths/minutes, BP 110/83, HR 130 bpm, temperature 37.8°C, O_2Sat 72%. The physician would like to provide beta agonist therapy to relieve the dyspnea before performing respiratory function tests. What do you recommend?

17. An elderly patient who is hard of hearing presents to the emergency department with severe difficulty breathing. The physician would like to give him short-acting beta agonists. What delivery device do you recommend for this patient?

18. The elderly patient above will be discharged home with a diagnosis of COPD on albuterol and ipratropium every 4 hours as needed. What delivery device do you recommend for this patient? Explain how the patient should use and care for this device.

Selected reading

1. NR Labris, MB Dolovich. Pulmonary drug delivery. Part 1: Physiological factors affecting therapeutic effectiveness of aerosolized medications. Br J Clin Pharmacol 2003; 56: 588–599.
2. NR Labris, MB Dolovich. Pulmonary drug delivery. Part II: The role of inhalant delivery devices and drug formulations in therapeutic effectiveness of aerosolized medications. Br J Clin Pharmacol 2003; 56: 600–612.
3. DR Hess. Aerosol delivery devices in the treatment of asthma. Resp Care 2008; 53: 699–725.
4. AARC Clinical Practice Guideline. Selection of a device for delivery of aerosol to the lung parenchyma. Resp Care 1996; 41: 647–653.
5. TC Carvalho, JL Peters, RO Williams. Influence of particle size on regional lung deposition - What evidence is there? Int J Pharmaceut 2011; 406: 1–10.
6. F Lavorini, A Magnan, JC Dubus et al. Effect of incorrect use of DPIs on management of patients with asthma and COPD. Resp Med 2009; 102: 593–604.
7. MB Dolovich, RC Ahrens, DR Hess et al. Device selection and outcomes of aerosol therapy: Evidence-based guidelines, American College of Chest Physicians/American College of Asthma, Allergy, and Immunology. Chest 2005; 127: 335–371.
8. MG Cochrane, MV Bala, KE Downs et al. Inhaled corticosteroids for asthma therapy: Patient compliance, devices and inhalation technique, Chest 2000; 117: 542–550.

17

Drug delivery to the skin

Are you ready for this unit?

● Review dispersed systems.

Learning objectives

Upon completion of this chapter, you should be able to:

● Define thermodynamic activity, penetrant polarity gap and fingertip unit.

● Identify the challenges and opportunities presented by each of the following route related characteristics to the rate and extent of drug travel to its target: accessibility, surface area, permeability and type of absorbing membrane, nature of the body fluids that bathe the absorbing membrane, retention of the dosage form at the site of absorption, location and number of drug metabolizing enzymes or extremes of pH and blood flow primacy of the skin.

● Outline the movement of a drug from its site of administration to its target including:
  − release
  − absorption
  − distribution to target.

● Predict the effect of each of the following drug properties on the rate and extent of drug travel to its target in the skin: water solubility, log P, molecular weight, stability in solution, enzymatic degradation and location of drug receptors.

● Describe the dosage forms/drug delivery systems used by this route and their effect on hydration of the stratum corneum and drug absorption.

● List the categories of ingredients that are added to products for the skin and identify how the ingredient contributes to the manageable size of the dosage form, its palatability or comfort, its stability (chemical, microbial and physical), the convenience of its use and the release of drug from the dosage form.

● Given a drug with targets in the skin and adequate references, determine its relevant physicochemical properties and use the penetrant polarity gap method to design a formulation to maximize thermodynamic activity of the drug.

● Given a specific skin condition, choose a base appropriate for that type of lesion and determine whether the selected base and form of drug maximize concentration of drug in the vehicle, diffusion coefficient or partition coefficient.

● Explain how the qualities of the dosage form are evaluated *in vitro* and/or *in vivo*.

● Explain to patients and/or other healthcare providers how to use, store and/or prepare dosage forms for administration via the skin.

Introduction

The skin is a major barrier to the entry of microorganisms, chemical and physical agents into the body. It regulates water and electrolyte loss and serves immunological and sensory functions. The permeation of drugs into or through the skin is controlled by the dense layer of protein and lipids at the surface called the stratum corneum. As a result of the barrier properties of this layer, most drugs applied to the skin are confined to acting on receptors in the skin and do not produce significant systemic absorption or effects. The dosage forms that are designed to reach drug targets in one or more layers of the skin (local treatment) are considered in this chapter. The transdermal products designed to reach the systemic circulation and regional transdermal products which penetrate the tissues beneath the skin where they are applied are considered in Chapter 18.

Characteristics of the dermal route

Key Points

- The skin is the most accessible of the absorbing membranes we have studied but also the least permeable.
- Most drug absorption occurs through the intercellular *(paracellular)* regions of the stratum corneum.

See Appendix Table A.1.

- The skin is the most accessible of the absorbing membranes we have studied but also the least permeable.
- The surface of the skin is large, on average 1.73 m^2, and the surface exposed to the drug is a major approach to control the dose applied.
- The skin's accessibility permits the easy removal of the dosage form for termination of drug effects if that is desired.
- Topical dosage forms may be lost onto clothing, or other surfaces in contact with the skin, resulting in loss of part of the applied dose.
- The epithelial portion of the skin consists of 15–20 layers of stratified squamous, keratinized cells cemented together by desmosomes and dense lipid lamellae (Figure 17.1).
  - The *stratum corneum* is the outermost 10 μm of skin and its cells, keratinocytes, in their last stage of differentiation, are considered to be nonviable.
  - The cells are keratinized, meaning that they are filled with the highly cross-linked protein keratin, rather than the cell organelles immersed in intracellular fluid found in the viable epidermal cells.
  - The structure of the stratum corneum may be characterized as 'bricks and mortar' with the keratinocytes representing the impermeable 'bricks' and the intercellular lipids a hydrophobic 'mortar'.

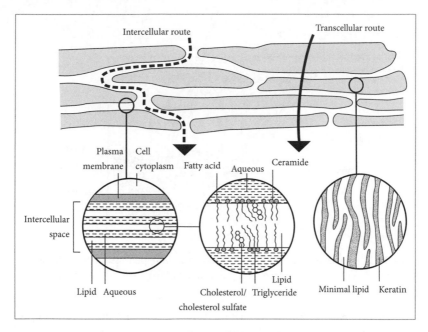

Figure 17.1 Intracellular and transcellular diffusion pathways in the stratum corneum. (Reproduced with permission from Elsevier, 1996. HR Moghimi, AC Williams and BW Barry. A Lamellar Matrix Model of Stratum Corneum, Intercellular Lipids.)

- — The barrier properties of the stratum corneum are a result of the thick, cross-linked keratinocyte envelope and the tightly packed, cohesive lipids that fill the space between the epithelial cells.
- — The intercellular lipids form lamellae (Figure 17.1) that provide a pathway for lipophilic drug diffusion through their fatty acid chains and a pathway for hydrophilic drugs between the polar heads of two lipid arrays.
- — The lipids found between the keratinocytes are more lipophilic and pack with less fluidity than the phospholipids in cell membranes.
- The permeability of the stratum corneum differs from site to site with the face more permeable than the trunk or arms.
- Skin fluids:
  - — Sebum is a lipid secretion that enhances the skin's elasticity and reduces water loss. The lipid secretions maintain a pH of about 5.0 on the skin surface.
  - — Eccrine sweat glands secrete a fluid that is mostly water onto the skin surface to produce cooling.
  - — The stratum corneum layer provides little water for drug dissolution being composed of 10–20% water and the remainder lipid and protein.
- The viable portion of the epidermis consists of layers of metabolically active cells approximately 100 µm thick (Figure 17.2).

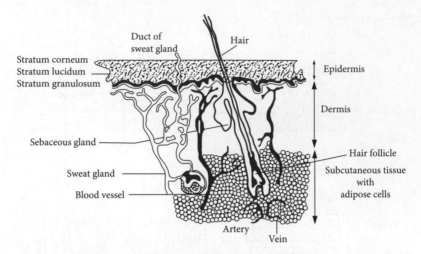

Figure 17.2 The layers of the skin. (Reproduced with permission from L Felton (ed.), Remington: Essentials of Pharmaceutics; published by Pharmaceutical Press, 2012.)

- — It consists of mostly keratinocytes, but also includes melanocytes, Merkel cells, and the Langerhans, antigen-presenting cells.
- — All of these cell types have the typical watery cytoplasm and lipoprotein membranes characteristic of viable cells.
- — The density and consistency of the viable epidermis has been likened to 0.9% sodium chloride.
- There is some drug metabolizing activity in the viable epidermis, dermis and skin appendages including esterases, peptidases, cytochrome P450 isozymes and others, although the effect on drugs applied to the skin has not been well characterized.
  - — Despite moderate enzyme activity, drug metabolism does not appear to limit the use of drugs by the dermal route.
- The cells of the stratum corneum turn over once every 2–3 weeks being replaced by cells from the viable epidermis that become more keratinized as they move away from the blood supply in the dermis.
- The blood flow to the dermis is between 2 and 2.5 mL/minute per 100 g tissue, in the intermediate range.

Drug travel from dosage form to target

Drug release

- Drugs are applied to the skin for treatment of local conditions as liquid and semisolid preparations. These products vary in their polarity and include the ointments, liquid and semisolid emulsions (creams), gels and pastes.
- With the exception of the pastes, these dosage forms contain dissolved drug that diffuses from the dosage form into the top layer of the skin.

— Pastes contain large quantities of undissolved active ingredient that is intended to protect the skin surface and therefore is not released from the dosage form.

Absorption

- Most drug absorption occurs through the 0.1 µm wide intercellular *(paracellular)* regions of the stratum corneum.
  - The cells of the stratum corneum, while occupying a greater surface than the intercellular space, are too tightly packed with protein to provide a suitable medium for drug diffusion.
  - Hydration of the stratum corneum significantly increases drug absorption by swelling the 'bricks' (keratinocytes) of the membrane, which loosens the mortar (the lipid lamellae) between the cells.
- The skin appendages include the hair follicles, the sebaceous glands and the eccrine glands, extending from the skin surface to the subcutaneous tissue.
  - The skin appendages represent about 0.1% of the total skin surface, and their role in the transport of drugs remains unsettled, but it seems likely that these structures make some contribution despite their small surface area due to their thinner layer of stratum corneum. Their role may be limited to transport of ions, high molecular weight and hydrophilic drugs.
- It is estimated that between 0% and 15% of most drugs applied topically are absorbed systemically.

Distribution

- Drugs with targets in the viable epidermis or dermis diffuse through the stratum corneum and then partition into the viable epidermis.
- Drugs move through the viable epidermis primarily by passive diffusion. Normal human keratinocytes have been found to express organic transporting polypeptide carriers and p-glycoprotein. None of the transporters have been characterized in detail.
  - There is evidence that p-glycoprotein has *influx* transporter activity in the skin, a surprising finding given the skin's barrier function.
- Drug next partitions from the viable epidermis into the dermis. The dermis is composed of the structural proteins, collagen and elastin, suspended in a watery gel of complex carbohydrates. In addition there are scattered cells in the dermal layer including:
  - the fibroblasts that synthesize collagen and elastin
  - macrophages and mast cells that produce allergic and inflammatory reactions
  - sensory neurons.

- The dermis contains the skin's vasculature with capillaries around the hair follicles and glands and just below the epidermis (see Figure 17.2).

Drug properties and the dermal route

- Movement through the stratum corneum as for any membrane is determined by solubility and molecular weight.
- The solubility preference of a drug for water and lipid phases determines its concentration in its vehicle and tendency to partition out of the vehicle into the stratum corneum lipids.
  - Drugs salts will be soluble in the aqueous phase of creams and aqueous gels, while the uncharged form will partition into the lipids of the stratum corneum.
  - Drug derivatives may be prepared that are more soluble in polar bases and partition well into skin lipids.
  - Uncharged forms of drugs will be highly soluble in lipophilic bases but may need the addition of cosolvents to encourage partitioning out of the vehicle into skin lipids.
- The molecular weight of a drug along with the viscosity of the barrier will determine its mobility in the barrier. For a drug molecule to penetrate the skin it should be small (<500 Da) with a log P between 1 and 4.
- Many corticosteroid drugs are formulated in dermatological products as derivatives to alter their lipid solubility and enhance their ability to partition into the watery layers of skin.
- Very lipophilic solutes are rate limited by the viable epidermis and dermis layers, while the polar solutes are rate limited by the small dimensions of the polar channels in the paracellular lipids of the stratum corneum.
- Drug targets found in the various layers of skin are presented in Table 17.1.

Table 17.1 Drug targets in the skin	
Skin layer	Drugs with targets in this layer
Stratum corneum	Sunscreens, protectants, emollients, keratolytics
Viable epidermis	Antibiotics, antifungals, antivials, keratoplastics, depigmenting and pigmenting agents, retinoids, immunosuppressants
Dermis	Corticosteroids, antihistamines, local anesthetics, immunosuppressants, drugs applied for systemic action
Skin appendages	Antibiotics, antifungals, antiproliferative agents, antiperspirants

Dosage forms for the dermal route

Design

1. One dose in a manageable size unit
   - The size of a dose of topical drug will depend on the surface area to be treated.
   - The bulk of the dose of a medicated cream or ointment is created by the vehicle or base.
2. Comfort
   - The oil-in-water creams and gels are favored by patients because they are not greasy and are easily removed. However, these qualities also reduce their contact time and ability to hydrate the skin.
   - The bases available to the pharmacist and their descriptions are detailed in Table 17.2.
3. Stability
   - Signs of physical instability of dermatological products that may be detected by inspection are presented in Table 17.3.
   - Chemical: Hydrolysis, oxidation and photolysis can occur in topical products.
     - The hydrocarbon bases are water free and may be used for drugs that are unstable to water, assuming the base provides appropriate release characteristics.
     - Drugs that are susceptible to oxidation may be stabilized with an antioxidant with appropriate solubility characteristics to place it in the same phase as the drug.
   - Dermatological products containing water must be preserved against antimicrobial growth.
4. Convenience/ease of use
   - Dermatological dosage forms are easy for patients to use but may require frequent application.
   - Semisolid bases such as ointments, creams and gels are better retained on the skin than liquid dosage forms, the latter requiring more frequent application.
5. Release of drug
   - The greatest challenge to the formulator is to choose the appropriate base and additives to maximize the release of the drug from the dosage form into the stratum corneum.
   - For the pharmacist who lacks the analytical equipment of the pharmaceutical scientist, the choice of a base must be based on a solid understanding of the determinants of drug release and penetration, and grounded in subsequent patient monitoring to ensure a therapeutic response.

Table 17.2 Characteristics of dermatological bases and dressings

Class	Examples	Formulation	Characteristics	Effect on skin hydration	Effect on skin permeability
Occlusive dressings	Transdermal patches, Tegoderm, Tegasorb, DuoDerm, Op-Site, Bioclusive, Blisterfilm	Polmeric hydrocolloids, polyurethane, copolyester films	Occlusive, nongreasy, protectant	Prevent water loss	Marked increase
Oleaginous vehicles	Petrolatum, white ointment	Petrolatum, oils, stiffeners	Stable, greasy, emollient, occlusive	Prevent water loss	Marked increase
Absorption bases	Aquaphor, Aquabase, lanolin, hydrophilic petrolatum, Polysorb	w/o Emulsifiers, petrolatum, oil, stiffeners	Stable, absorb water, greasy, emollient, occlusive	Prevent water loss	Marked increase
Water-in-oil creams	Cold cream, hydrous lanolin, Eucerin, hydrocream	w/o Emulsifiers, petrolatum, oil, stiffeners, water, antimicrobial preservatives	Susceptible to coalescence, not water washable, greasy, emollient, occlusive	Slow water loss, raise hydration	Increase
Oil-in-water creams	Dermabase, hydrophilic ointment, vanishing cream	o/w Emulsifiers, petrolatum, oil, thickeners, cosolvents, water, antimicrobial preservatives	Susceptible to coalescence, water washable, non-occlusive, non-greasy	Donate water to skin	Owing to evaporation, produce only slight increase
Water soluble vehicles	Polyethylene glycol ointment	PEG 3350, PEG 400	Water washable, non-occlusive, non-greasy	May withdraw water from skin	Negligible
Water based gels	Carbomer gel, Liqua-Gel (HPMC gel), methylcellulose gel, hydroxypropyl cellulose gel, polaxamer gel	Thickeners, cosolvents, water, antimicrobial preservatives	Water washable, non-occlusive, non-greasy	Donate water to skin	Owing to evaporation, produce only slight increase
Powder	Dusting powder, talcum powder	Magnesium stearate, zinc stearate, talc, starch, zinc oxide	Stable, non-occlusive, non-greasy	May adsorb water to their surface	Negligible, may reduce

Adapted from R Daniels. Strategies for skin penetration enhancement. BASF Skin Care Forum http://www.skin-care-forum.basf.com/en/articles/regulatory-toxicology/strategies-for-skin-penetration-enhancement/2004/08/12?id=5b9a9164-6148-4d66-bd84-6df76bd6d111&mode=Detail accessed October 21, 2012, and JE Thompson, LW Davidow. A Practical Guide to Contemporary Pharmacy Practice, 3rd edn, Philadephia, PA: Lippincott, Williams & Wilkins, 2009.

Table 17.3 Physical instability of dermatological bases

Instability	Observation
Coalescence	Noticeable layering in emulsion dosage forms. Warm storage temperatures can accelerate because it lowers the viscosity of the base
Evaporation	The pharmacist will notice puffiness in the formulation, or it may pull away from the sides of the container or stiffen. Volatile components such as alcohol and water may evaporate and change the solubility characteristics of the vehicle causing precipitation of drug in the vehicle or on the skin (and decreased penetration)
Phase separation	Formation of visible droplets of the dispersed phase on the surface of the product

- The discussion that follows is intended to provide a theoretical framework for the choice of dermatological bases for drugs with known physicochemical characteristics.

Fick's law

For a drug applied to the skin to produce a response at a target in the dermis, the following steps must occur. The drug must:

- be dissolved in the vehicle
- diffuse through the vehicle to the skin
- partition into the stratum corneum
- diffuse through the stratum corneum
- partition into viable epidermis
- diffuse through viable epidermis
- partition into the dermis
- diffuse through the dermis.

Any of these steps is potentially rate limiting. For many drugs the slowest diffusion step is diffusion through the stratum corneum, although very hydrophobic drugs may find the diffusion through the aqueous cytoplasm of the viable epidermal cells or the watery matrix of the dermis rate limiting. Diffusion is the movement of a drug from an area of high concentration to an area of low concentration by random molecular motion.

Definition

Partitioning is the movement of a drug to divide its mass between two media.

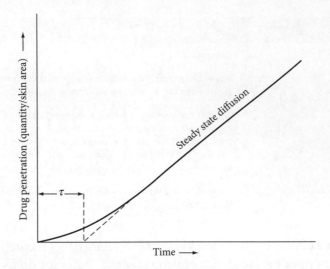

Figure 17.3 Steady state flux through a membrane. (Reproduced with permission from AT Florence, D Attwood, Physicochemical Principles of Pharmacy, 5th rev. edn, p. 379; published by Pharmaceutical Press, 2011)

The partition coefficient of a drug between dosage form (vehicle) and stratum corneum may be expressed as:

$$K_{sc/vehicle} = \frac{C_s}{C_v}$$

Fick's law describes mathematically the amount of drug (moles or grams) transported through a membrane over time (hours, minutes, seconds). In the case of products applied to the skin, the membrane is the skin through which the drug must diffuse (Figure 17.3).

$$\text{Steady state flux} = J = \frac{dM}{dt} = \frac{DSK(C_v - C_r)}{h}$$

Where D is the diffusion coefficient (cm^2/s), K is the partition coefficient between the membrane and the vehicle, S is the surface area of the membrane to which the drug has been applied (cm^2), C_v is the concentration of drug in the vehicle, and h is the thickness of the membrane barrier (cm). The second concentration in the gradient, C_r, is near zero because the blood carries the drug away as soon as it is absorbed so it is removed from the equation entirely. This gives the following equation for flux:

$$J = \frac{DSKC_v}{h}$$

Influence of formulation on flux

Key Points

- Choose a base that will dissolve the drug and choose a drug form that will dissolve in the base. In the case of two phase systems (water-in-oil and oil-in-water emulsions) there must be an appreciable amount dissolved in the continuous phase.
- Use of a cosolvent ingredient with polarity similar to the stratum corneum enhances partitioning of the drug and cosolvent into the stratum corneum.
- Occlusive properties increase the hydration and permeability of the stratum corneum.
- Water from the formulation will be absorbed into the skin but will only have transient effects on hydration.

Concentration of drug in vehicle (C_v) (Figure 17.4)

- In order to have the maximum concentration gradient from the vehicle to layers beneath the stratum corneum, the vehicle must be saturated with dissolved drug.
- The first formulation consideration is the choice of a base that will maximize the solubility of the drug.
  - For emulsion dosage forms the drug must be dissolved in the continuous phase. Suspended drug can act as a reservoir but does not increase the thermodynamic activity of the drug in the vehicle or the response.
- If the choice of base is dictated by the condition to be treated, the form of the drug should be chosen to be soluble in the base, drug salts being more soluble in gels and uncharged forms in oleaginous bases or oil phases of emulsions.
  - A drug derivative or amorphous form can provide more suitable solubility characteristics in some cases.

Partition coefficient K $K = \dfrac{C_s}{C_v}$ (Figure 17.5)

- If we examine the diffusion equation, we can recognize that when we maximize the concentration of drug in the vehicle, C_v, it will be difficult to have a high partition coefficient, K.

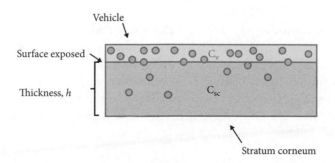

Figure 17.4 Concentration of the drug in the vehicle, C_v.

- Formulation scientists refer to the combination of saturation drug solubility in the vehicle and the 'escaping tendency' of the drug from the vehicle into skin as the *thermodynamic activity* of the drug.
- Conceptually thermodynamic activity of a drug in dermatological formulation may be estimated from $C_v \times K$. The higher the value of the product of $C_v \times K$, the higher the thermodynamic activity and the higher the drug permeation. Thus, rather than attempting to maximize one of these variables at the expense of the other, we would strive to maximize the product of $C_v \times K$.
- Wiechers *et al.* (selected reading 3) presented a rational approach for optimizing the diffusion rate of drugs from a vehicle based on the relative polarity index or log P of the drug in relation to the log P of the stratum corneum, a value called the penetrant polarity gap (PPG). They estimate the log P of the stratum corneum to be 0.8 and use this value along with the log P of the drug to calculate the PPG:

Penetrant polarity gap = PPG = $|\log P_{\text{penetrant}} - \log P_{\text{stratum corneum}}|$

- The relative polarity of the phase of the formulation in which the active ingredient is dissolved should be the magnitude of the PPG greater or less than the log P of the active ingredient itself.

Pharmaceutical example

- For hydrocortisone with a log P of 1.53, the PPG would be:

$$\text{PPG}_{\text{hydrocortisone}} = |1.53 - 0.8| = 0.73$$

The log P of the vehicle chosen for hydrocortisone should be either $1.53 + 0.73 = 2.26$ or higher OR $1.53 - 0.73 = 0.8$ or lower.

- The log partition coefficients for complex mixtures are not available but it may be determined that the log P of a vehicle would be higher or lower than a certain value from the log P available for pure solvents and emollients in Table 17.4.
- A vehicle consisting of water and alcohol would have a log P of 0.8 or lower, while a vehicle in which hydrocortisone is dissolved in the oil phase would have a log P of 2.26 or greater.
- If the drug's polarity differs greatly from the polarity of stratum corneum lipids, its skin penetration can be enhanced by the addition of a cosolvent that dissolves the drug and which also has a high affinity for stratum corneum lipids.

Table 17.4 Estimated log *P* for selected dermatological ingredients

Ingredient	Log *P*
Water	−4.00
Ethylene glycol	−1.93
Glycerin	−1.76
Dipropylene glycol	−1.20
N-N-dimethylformamide	−1.01
Propylene glycol	−0.92
Polyethylene glycol	−0.88
Methanol	−0.77
DMSO	−0.6
Ethanol	−0.32
Isopropanol	0.3
Glyceryl isostearate	4.76
Isopropyl myristate	5.41
Mineral oil	>6.0
Petrolatum	>6.0
Propylene glycol isostearate	6.08
Cetyl alcohol	7.30
Isopropyl isostearate	7.40
Stearic acid	7.40
Isopropyl palmitate	8.20
Stearyl alcohol	8.40
Ethylhexyl palmitate	9.12
Lecithin (range of components)	9.1–14.9
Ethylhexyl isostearate	10.05
Vegetable squalene	14.93
Lanolin	15.60
Triisostearin	18.60
Trimethylolpropane triisostearate	20.27
Pentaerythrithyl tetraisostearate	25.34
Isostearyl isostearate	26.98

- The adjuvants that facilitate the uptake of drugs into the skin are those taken up into skin lipids because they are small (low molecular weight) and have a similar polarity to skin lipids.
- They include propylene glycol, ethanol, isopropanol and others.
- The inclusion of a cosolvent in the product produces an effective reduction in the relative polarity gap between the drug and the skin lipids.

- In selecting a base for a drug to be applied topically, it is best to strike a balance between maximizing the concentration of the drug in the vehicle, C_v, and maximizing the partition coefficient, K.
  - The pharmacist should begin with a base that has a similar polarity to the drug to maximize its solubility in the formulation, and if necessary include a cosolvent with a different polarity from the drug to maximize its thermodynamic activity and facilitate its partitioning into the skin.

Diffusion coefficient (D) (Figure 17.5)

- Diffusion coefficient is a measure of the mobility of a drug in a medium and will vary with the nature of the medium and with the size of the drug molecule diffusing through the medium.
  - Dermatological formulations that increase the water content (hydration) of the stratum corneum can substantially increase diffusion coefficient, D.
  - The use of occlusive dressings over topical corticosteroids increases penetration 10-fold.
  - Increasing the hydration of keratin in corneocytes induces swelling that loosens the intercellular lipid and polar channels of the stratum corneum.

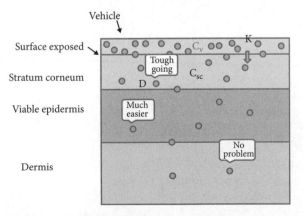

Figure 17.5 Partition coefficient, K, and diffusion coefficient, D.

— See Table 17.2 regarding the occlusiveness of different categories of dermatological bases.

Patient factors

The patient's age, skin condition, location and surface area affected can influence the process of diffusion out of the product and into the target site. Many of these factors affect:

Depth of the stratum corneum barrier (h)

- Variability in skin permeability among individuals: For a given skin site and skin condition, there is an approximately 10-fold variation in response.
- Condition of the skin: Skin conditions that cause inflammation, produce defective stratum corneum (psoriasis) or damage skin (burns, eczema) will have a greater permeability than intact skin. Pharmacists should monitor burn patients for excessive use of topical antiseptics which are not intended to be absorbed systemically. Some diseases (ichthyosis) thicken skin.
- Age of the patient: Preterm infants have no barrier properties in their skin. These develop about 9 months after conception. In older patients the stratum corneum is thickened and less hydrated.
- Location of the treatment or device: For any given individual there can be an approximately 50-fold variation in the dose absorbed between skin sites. Treatments for the skin must be applied to the affected area but when a drug targets the systemic circulation, the patient should be advised to use the sites specified in the product labeling for placement of patches.

Moist or dry lesions

- An important consideration in the choice of a vehicle for the skin is the nature of the skin lesion to be treated (Table 17.5).

Table 17.5 Moist and dry skin conditions

Moist skin lesions	Dry skin lesions	Either
Acne	Dry skin	Atopic dermatitis
Bacterial infections	Psoriasis	Seborrheic dermatitis
Fungal Infections	Ichthyosis	Contact dermatitis
Diaper rash		
Burns		
Wounds		

- Pastes are used to create a barrier between skin and sunlight or irritating body fluids like urine or feces.
- Gels or oil-in-water creams are used to treat moist, weeping lesions because their nonocclusive properties prevent further maceration of the affected area.
- The occlusive bases are used for dry lesions to enhance skin hydration and penetration of the medication. In addition the emollient properties of these bases will reduce dryness and scaling.

Surface area exposed to drug(s)

Key Points

- Surface area equals dose.

- The surface area exposed is an important consideration in the use of products applied to the skin for treatment of skin conditions because the affected surface area can vary considerably and the barrier properties of the skin may be compromised.
- This is an important factor to keep in mind as you monitor patients for signs of toxicity resulting from extensive skin exposure to drugs or chemicals.
- There have been a number of case reports of serious drug toxicity resulting from the exposure of a large surface area to medicated ointments.
- When an area greater than 20% of the body surface area is exposed to the drug, systemic dose can be significant.
- Figure 17.6 provides the student with a perspective on the surface area of different body parts that is useful for estimating the amount of dermatological product the patient will require to complete a course of treatment.
- Remember that surface area equals dose!

Dosage forms

Ointments

- The term 'ointment' is often used synonymously with semisolid dosage forms for the skin but the pharmaceutical industry uses the term to describe the hydrocarbon-based products.
- Most ointments are composed of hydrophobic vehicles such as mineral oil, petrolatum, paraffin, silicones and waxes.
  - The hydrophobic ointments are very occlusive, emollient meaning they soften skin, and are preferred for dry lesions but their greasiness can stain clothing.

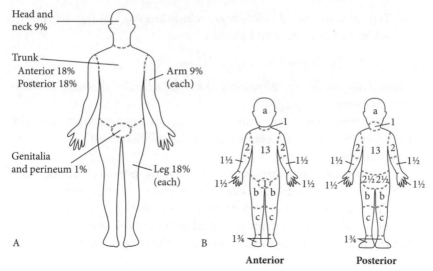

Relative percentage of body surface ara (% BSA) affected by growth

Body Part	0 yr	1 yr	5 yr	10 yr	15 yr
a = 1/2 of head	9 1/2	8 1/2	6 1/2	5 1/2	4 1/2
b = 1/2 of 1 thigh	2 3/4	3 1/4	4	4 1/4	4 1/2
c = 1/2 of 1 lower leg	2 1/2	2 1/2	2 3/4	3	3 1/4

Figure 17.6 Lund Browder chart: Percent of body surface for different body parts. (Reproduced with permission from http://www.merckmanuals.com/media/professional/figures/PHY_rule_of_nines.gif)

- Polyethylene glycol (PEG) ointment is made with polyethylene glycols of different molecular weights and is water soluble.
  - PEG ointment is nonocclusive and preferred for moist lesions.
- The ointment bases make extremely stable medicated products because they contain no water.

Absorption bases and water-in-oil creams

- Absorption bases are formulations that, in addition to hydrophobic ingredients, contain water-in-oil emulsifying agents in significant amounts.
  - When an aqueous drug solution is added, they form creamy, water-in-oil emulsions but because the continuous phase is oil based, they will behave more like ointments.
- Water-in-oil emulsion bases are often referred to as creams because of their appearance but the USP uses the term 'cream' to refer to oil-in-water emulsion bases only. This book uses the term 'cream' more loosely to include the water-in-oil and oil-in-water emulsion bases.
  - Because the continuous phase of water-in-oil creams is oil based, they are relatively occlusive and used for dry lesions.

— Typical water-in-oil emulsifiers include lanolin, cholesterol, fatty acids, sorbitol esters and glycol esters.

Emulsifying bases and oil-in-water creams

- Emulsifying bases are anhydrous, semisolid vehicles that contain oil-in-water emulsifying agents.
  - They form oil-in-water emulsions when an aqueous solution is added.
- Oil-in-water creams are emulsions with an oil dispersed phase and a water continuous phase.
  - Oil-in-water emulsifiers include fatty acids, polyoxyethylene esters, sodium lauryl sulfate, sodium borate and triethanolamine.
  - The water phase will require an antimicrobial preservative.
- When an oil-in-water cream is rubbed into the skin, the water (continuous phase) evaporates and leaves the drug concentrated in the remaining film.
  - This can cause the drug to precipitate on the skin and the use of the less volatile cosolvents can provide continued solubility on the skin surface.
- The oil-in-water creams are not occlusive, and may be used for moist lesions.
- Both water-in-oil and oil-in-water emulsion bases can lose water, and the emulsion crack or cream.

Gels

- Gels are semisolid products that contain a liquid (usually water or alcohol) bound by a polymeric matrix.
- Most gels are transparent while others contain polymers in colloidal aggregates that disperse light and are opaque.
- Polymers used as gelling agents are presented in Table 17.6.
- The physical stability of gels is good, but they must contain an antimicrobial preservative.
- The gels are not occlusive, and they are useful for areas of the body where ointments would be too greasy, or for moist lesions.

Other topical formulations

- Pastes are ointments containing a large percentage of insoluble particles, as high as 50% of the finished weight.
  - The insoluble particles include protectant ingredients such as starch, zinc oxide, talc and calcium carbonate.
  - They are used to provide protective barriers and not to produce drug absorption.
- Medicated powders contain medications mixed with inert, lubricating solids such as starch or talc.

Table 17.6 Excipients in dermatological bases

Excipient class	Function	Examples
Antioxidant	Used to prevent degradation of preparations by oxidation (chemical stability)	Ascorbic acid, ascorbyl palmitate, butylated hydroxyanisole, butylated hydroxytoluene, hypophosphorous acid, monothioglycerol, propyl gallate, sodium ascorbate, sodium bisulfite, sodium formaldehyde, sodium metabisulfite
Aqueous phase	In addition to water, this phase will contain antimicrobial preservative	Purified water, benzoic acid, sorbic acid, ethyl, methyl and propyl paraben
Cosurfactant cosolvents	Paired with surfactant emulsifier to enhance stability of emulsions	Alcohol, butanol, glycerin, isopropanol, polyethylene glycol 300, 400, propylene glycol, Transcutol (diethylene glycol monoethyl ether)
Emollient	Used in topical preparations, ease spreading and soften skin. Prevent loss of water from skin increasing its permeability	Cetyl alcohol, cetyl esters wax, isopropyl myristate, lanolin, mineral oil, olive oil, petrolatum
Emulsifying agent	Used to promote and maintain fine dispersions of one immiscible liquid in another (water in oil or oil in water). These ingredients maintain the physical stability of dispersed systems	Acacia (oral), carbomer, cetyl alcohol, cremophor, glyceryl monostearate, mono-oleate, labrasol (caprylocaproyl, olyoxylglycerides), lanolin alcohols, lecithin (oral, parenteral), polaxamer, polyoxyl ethers, polyoxyethylene fatty acid esters, polysorbates, stearic acid, trolamine
Humectant	Used to prevent drying of preparations, particularly ointments and creams	Glycerin, propylene glycol, sorbitol
Levigating agent	Liquid or semisolid, used to facilitate the reduction of the particle size of a powder by grinding on an ointment slab	Mineral oil, glycerin, propylene glycol
Oil phase	Contributes to the emollient and occlusive effects of a topical product	Corn oil, cottonseed oil, decane, decanol dodecane, dodecanol, ethyl oleate, isopropyl myristate, isopropyl palmitate, isostearyl isostearate, jojoba oil, lauryl alcohol, medium chain triglycerides, mineral oil, petrolatum, oleic acid, silicon oil, soybean oil

(continues)

Excipient class	Function	Examples
Stiffening agent	Used to increase thickness or hardness of a semisolid preparation, and prevent coalescence in w/o creams	Cetyl alcohol, cetyl esters wax, microcrystalline wax, paraffin, stearyl alcohol, white wax
Viscosity increasing agent	Thickening agent used to prepare water-based gels and prevent coalescence in o/w emulsion bases	Acacia, carbomer, carboxymethylcellulose, colloidal silicon dioxide, guar gum, hydroxyethyl cellulose, hydroxypropyl cellulose, hypromellose (hydroxypropylmethylcellulose), methylcellulose, polaxamer, povidone, pregelatinized starch, xanthan gum

Table 17.6 (*continued*)

- — They are useful to prevent skin irritation and chafing or for treatment of moist lesions such as athlete's foot.
- Aerosols contain drug dissolved or suspended in propellant or a propellant/solvent mixture.
  - — They are particularly useful for applying to large areas or to lesions that are tender to rubbing.

Microemulsions

- Microemulsions have generated particular excitement in the formulation of dermatological products because they demonstrate enhanced penetration properties compared with conventional vehicles.
- They differ from conventional emulsions in that microemulsions are thermodynamically stable, clear or translucent and require little input of energy (mixing) to prepare.
- There are some 'microemulsions' that are actually molecular dispersions because there is enough cosolvent to make the water and oil phases miscible.
- Microemulsions can increase both extent and rate of drug absorption but it is unclear whether they have any effect on the stratum corneum or if they produce enhanced absorption entirely through their high solubilization capacity for both hydrophilic and lipophilic drugs and high drug mobility.

Preparation

- Most extemporaneously prepared dermatologicals are made by adding a smooth paste of the drug to a commercially available base.

- The selection of the base to use in the formulation of a dermatological preparation should be based on the nature of the patient's lesion, the physicochemical properties of the drug and the ability of the base to dissolve and release the drug for absorption.
- The use of the PPG method described above is recommended for selection of a base for its biopharmaceutical properties.
- Please see selected reading 9 and 10 for detailed information about the preparation of compounded dermatologicals.

Quality evaluation and assurance

- The USP specifies that products to be applied to the skin should contain no more than 10^2 bacterial colony forming units (CFUs) and no more than 10 yeast or mold CFUs per gram or milliliter of product.
  - In addition they must be free of *Pseudomonas aeruginosa* and *Staphylococcus aureus*. The individual monographs of official dermatological products specify the permissible potency range for the active ingredient.
- The development of *in vitro* and *in vivo* correlations to simplify the evaluation of the bioequivalence of dermatological products has not evolved to the point that the FDA permits the use of *in vitro* release testing as a surrogate for bioequivalence or bioavailability.
- Bioequivalence of topical corticosteroid products is presently demonstrated *in vivo* using a pharmacodynamic study of the vasoconstrictor effect between test and reference products, while other product categories use clinical endpoint studies.

Doses of dermatological products

- Doses of ointments, creams or gels are expressed to the patient as a specific length of ointment ribbon or as a *fingertip unit* (FTU).
  - One fingertip unit is the ribbon of ointment expressed from a tube with a standard 5 mm tip applied along the length of the index fingertip.
  - One FTU is about 0.5 cm × 2.5 cm and can be considered equal to 0.5 g.
- This method applied as a guide for dosing of ointments and creams may be used to estimate the amount of product required for the patient to complete a specific course of therapy.
- Some estimates of the number of fingertip units required to cover different body parts are presented in Table 17.7.
- The hand may be used as a gauge to estimate the amount of skin surface affected and the dose of dermatological required to treat that surface.
  - One fingertip unit provides enough medication to cover both sides of one hand. Thus if the affected surface can be covered by an adult's

Table 17.7 Fingertip units for adult dermatologic treatment

Area of skin	% BSA	No. FTUs	Weight (g)
Hands (both)	4	2	0.5
Face + neck	5	2.5	1
Arm + hand	9	4	2
Chest + abdomen	18	7	3.5
Back + buttocks	18	7	3.5
Leg + foot	18	8	4

% BSA = percentage of body surface area; FTU = fingertip unit.
Adapted from J Dermatolog Treat 2007; 18: 319–320.

hand (palmar surface), it will require one-half of a fingertip unit to cover the affected area.
— The palm (side) represents 1% of the total body surface in an adult.

Counseling patients

The patient should be given the following instructions.

- Apply to a clean surface with clean hands.
- Apply a thin layer. A thick layer will only rub off on the clothing and will not provide more medication to the skin.

The patient may be given the following advice.

- Doses of ointment, creams or gels are expressed as a specific length of ointment ribbon (when dispensed in tubes). Patients who should use 1 g can be instructed to use 2 FTUs.
- Duration of treatment: In acute conditions, patients should be advised to continue use for 2 days after the symptoms have resolved. In chronic conditions, they should be advised to apply daily.
- Response: In acute conditions, at least 24 hours may elapse before improvement can be noted. In chronic conditions it may require 4–6 weeks to observe maximum benefit.
- Covering: May be covered with a breathable dressing to prevent contact with clothes. Only cover with occlusive dressing (plastic or plastic/gauze combinations) if directed by doctor.
- Cosmetics: Most acne medications can be used with cosmetics. Advise waiting 30 minutes after applying the medication.
- Side effects: When talking to patients about side effects, bear in mind that somewhere between 0% and 15% of the drug in a topical product is

actually absorbed systemically and capable of causing side effects other than local ones. Patients who use corticosteroids or other drugs with many potential side effects topically generally do *not* receive a dose that would cause side effects.

Questions and cases

1. You have a prescription for nifedipine 2% topical product for wound healing. Research the chemical and biological properties of nifedipine. What challenges does this drugs present to the formulator in terms of dissolution in body fluids, movement through membranes, stability in dosage forms, loss of drug owing to metabolism, access to receptors, or selectivity of the target.

Nifedipine topical

Design issue	Drug property	Analysis
Drug must be dissolved before it can cross membranes		
Drug must have sufficient lipophilicity to cross membranes		
Large molecular weight drugs have difficulty moving across membranes		
Drugs may hydrolyze, oxidize, photolyze or otherwise degrade in solution		
Some drugs cannot be used by mouth because of substantial losses to enzyme degradation		
Intracellular drug receptors or receptors in the brain or posterior eye are challenging to target		
Some drugs cause serious toxicity as a result of poor selectivity of drug action		

2. Make a list of ingredients that you will put in the nifedipine 2% topical product for wound healing. Categorize them as contributing to the manageable size of the dosage form, its palatability or comfort, its stability (chemical, microbial, or physical), the convenience of its use, or the release of drug from the dosage form.

Ingredient	Category	Quality
Propylene glycol or glycerin		
Polyethylene glycol ointment		

3. A dermatologist would like you to combine 15 g of betamethasone 0.05% cream and 15 g of Fototar cream (2% coal tar in water-oil base) for a patient with psoriasis. Which of the following betamethasone products should be selected to combine with Fototar?

(a) Betamethasone 0.05% ointment by AlphaPharm

(b) Betamethasone 0.05% cream (w/o) by Schering

(c) Betamethasone 0.05% cream (o/w) by Fougera

(d) Betamethasone 0.05% gel by Taro

(e) All of the above

4. You have a prescription for cyproheptadine 5% topical product for itching associated with chicken pox. The drug has a molecular weight of 287, log P 4.682 and pK_a 9.3. The hydrochloride salt has a water solubility of 5 mg/mL.

Cyproheptadine

(a) Are there any stability concerns with this molecule?

(b) Where are the target receptors for this drug and what layer(s) will the drug need to penetrate in order to reach them?

(c) Would an occlusive or nonocclusive base be preferred for this indication?

(d) Calculate the penetrant polarity gap (PPG) for this compound. What should be the log P of the vehicle you choose to maximize thermodynamic activity?

(e) Will you use cyproheptadine or its hydrochloride salt to prepare this prescription and why?

(f) What specific base would you choose for this prescription and why?

Case studies 1 and 2: Background information

- Commercially available topical lidocaine products contain between 1% and 5%. We will assume if the patient is using a commercially available product, the concentration is not excessive.

- EMLA cream is a eutectic mixture of local anesthetics: 2.5% lidocaine and 2.5% prilocaine.

- Dose related side effects include drowsiness, dizziness, confusion, tremor, agitation, visual disturbances, nausea, vomiting, seizures.

- Students may refer to the Lund Browder diagram to determine the percentage of body surface area exposed.

Serum lidocaine levels for arrhythmias

Level	Traditional units	SI units
Therapeutic	2–6 µg/mL	6.4–25.6 µmol/L
Toxic: CNS and cardiovascular depression	6–8 µg/mL	25.6–34.2 µmol/L
Toxic: Seizures, obtundation, decreased cardiac output	>8 µg/mL	>32.2 µmol/L

Case study 1: A healthy, 23 year old man undergoing laser hair removal applied EMLA cream to his chest, abdomen, and back. Forty-five minutes later he developed nausea, vomiting confusion, brief loss of consciousness, and agitation. Physical exam revealed mild folliculitis on the torso; cerebrospinal fluid lidocaine level was 1 mg/L.

5. What percent of this patient's body surface area was covered with EMLA cream?
    (a) 18%
    (b) 27%
    (c) 36%
    (d) 45%
6. Aside from surface, what variable in the diffusion equation was altered in this patient?
    (a) Concentration of drug in the vehicle was excessive
    (b) Diffusion coefficient D was increased by occlusive wrap or base
    (c) Partitioning from the base into the skin was excessive
    (d) Thickness of the skin barrier h was thinner than usual

Case study 2: A 22 year old woman applied a topical gel containing 10% lidocaine, 10% tetracaine, and an unknown amount of phenylephrine to both legs under occlusion in preparation for laser hair removal. She developed convulsions, lapsed into a coma, and subsequently died.

7. What percent of this patient's body surface area was covered with gel?
    (a) 18%
    (b) 27%
    (c) 36%
    (d) 45%
8. Aside from surface, what variable in the diffusion equation was altered in this patient?
    (a) Concentration of drug in the vehicle was excessive
    (b) Diffusion coefficient D was increased by occlusive wrap or base
    (c) Partitioning from the base into the skin was excessive
    (d) Thickness of the skin barrier h was thinner than usual
9. A 5 year old child has poison oak dermatitis on his left arm and hand. The child's mother estimates that the surface is about equivalent to 4 of her palms. The physician has recommended that the family apply 10% zinc oxide ointment 6 times daily for 5 days. How many fingertip units (FTUs) should the parents be instructed to apply in each dose?

(a) 1 FTU

(b) 1.5 FTU

(c) 2 FTU

(d) 2.5 FTU

10. What size zinc oxide tube will the child require?

(a) 30 g

(b) 42.5 g

(c) 57 g

(d) 113 g

11. Urea is a keratolytic drug used to promote uptake of water by the stratum corneum for dry lesions. It has a water solubility of 545 g/L, is almost insoluble in ether and chloroform, melting point 132.7°C and a log P of −2.11. Calculate urea's penetrant polarity gap and the log P of the vehicle that would maximize urea's thermodynamic activity.

What kind of base is preferred for dry lesions. What kind of base do you recommend for this drug and why?

12. Terbinafine is an antifungal drug with log P 5.9, water solubility 0.00075 mg/L, and pK_a 7.1. If we put terbinafine in Dermabase (oil in water base), will it partition into the oil phase or the water phase?

Terbinafine

(a) Oil disperse phase

(b) Water disperse phase

(c) Oil continuous phase

(d) Water continuous phase

(e) Equally into oil and water

13. The preferred base for terbinafine for the treatment of a fungal infection would be:

(a) White ointment

(b) Eucerin cream

(c) Dermabase cream

(d) Methylcellulose gel

14. Which of the following approaches would enhance the solubility of terbinafine in a water-based gel?

(a) Use base to make salt

(b) Use acid to make salt

(c) Use cosolvents

(d) Derivatize

15. Assuming the pH of the stratum corneum is 5.1, what form of terbinafine predominates in the stratum corneum?

    (a) The uncharged form predominates, (B/BH$^+$ 100/1)

    (b) The charged form predominates, (B/BH$^+$ 1/100)

    (c) The uncharged form predominates, (A$^-$/HA 1/100)

    (d) The charged form predominates, (A$^-$/HA, 100/1)

    (e) The concentrations of the two forms are equal

16. Betamethasone has log P 1.94, water solubility of 66 mg/L, and FW 392.5. Which of the following approaches would enhance the solubility of betamethasone in a water-based gel?

Betamethasone

    (a) Use base to make salt

    (b) Use acid to make salt

    (c) Use cosolvents

    (d) Derivatize

17. You have a prescription for anthralin 0.3% for a patient with psoriasis. Anthralin has a MW of 226, log P 2.3, pK_a 9.06, and is soluble in ether, insoluble in water. Are there any stability concerns with anthralin?

Anthralin

    (a) Hydrolysis of the ketone

    (b) Photo-oxidation of the ketone

    (c) Hydrolysis of the phenol

    (d) Photo-oxidation of the phenol

18. Will you use anthralin or its sodium salt to prepare this prescription and why?

    (a) Sodium salt to increase partitioning into the stratum corneum

    (b) Sodium salt to enhance stability

    (c) Anthralin to increase partitioning into the stratum corneum

    (d) Anthralin to enhance stability

19. Where are the target receptors for anthralin and what layer(s) will the drug need to penetrate to reach them?

    (a) Receptors: viable epidermis; penetrate stratum corneum and dermis

    (b) Receptors: dermis; penetrate stratum corneum and viable epidermis

    (c) Receptors: viable epidermis; penetrate stratum corneum

    (d) Receptors: dermis; penetrate stratum corneum

20. Would an occlusive or nonocclusive base be preferred for psoriasis?

    (a) Occlusive base because psoriasis is a moist lesion

    (b) Non-occlusive base because psoriasis is a moist lesion

    (c) Occlusive base because psoriasis is a dry lesion

    (d) Non-occlusive base because psoriasis is a dry lesion

21. Calculate the PPG and log P for suitable vehicles for anthralin.

    (a) PPG = −1.5, log$P_{vehicle}$ lower than 0.8 or higher than 3.8

    (b) PPG = 1.5, log$P_{vehicle}$ lower than 0.8 or higher than 3.8

    (c) PPG = 1.5, log$P_{vehicle}$ lower than 3.8 or higher than 0.8

    (d) PPG = 1.5, log$P_{vehicle}$ lower than 0.8 or higher than 2.3

22. Which of the following would be most suitable given the disease state and the biopharmaceutical properties of anthralin?

    (a) White ointment

    (b) Eucerin cream

    (c) Dermabase cream

    (d) Methylcellulose gel

    (e) Polyethylene glycol ointment (non-occlusive)

23. Which of the following is suitable and likely to be preferred by the patient?

    (a) White ointment

    (b) Eucerin cream

    (c) Dermabase cream

    (d) Methylcellulose gel

    (e) Polyethylene glycol ointment (non-occlusive)

24. You have a prescription for tetracycline 2% topical for acne. Tetracycline has a MW of 444 (base), base solubility 0.4 mg/mL water, 20 mg/mL alcohol, log P (base) −0.304 HCl salt 100 mg/mL in water, 10 mg/mL alcohol, pK_a 9.7 (base) 7.7 (phenol).

Tetracycline

Are there any stability concerns with tetracycline?

(a) Hydrolysis of the ketone

(b) Photo-oxidation of the ketone

(c) Hydrolysis of the amide

(d) Photo-oxidation of the phenol

25. Will you use tetracycline or its hydrochloride salt to prepare this prescription and why?

(a) Hydrochloride salt to increase partitioning into the stratum corneum

(b) Hydrochloride salt to enhance stability

(c) Tetracycline to increase partitioning into the stratum corneum

(d) Tetracycline to enhance stability

26. Where are the targets for this drug and what layer(s) will the drug need to penetrate in order to reach them?

(a) Targets: viable epidermis; penetrate stratum corneum and dermis

(b) Targets: sebaceous glands; penetrate stratum corneum

(c) Targets: stratum corneum

(d) Targets: dermis; penetrate stratum corneum and viable epidermis

27. Would an occlusive or nonocclusive base be preferred for acne?

(a) Occlusive base because acne is a moist lesion

(b) Non-occlusive base because acne is a moist lesion

(c) Occlusive base because acne is a dry lesion

(d) Non-occlusive base because acne is a dry lesion

28. What specific base would you choose for this prescription?

(a) White ointment

(b) Eucerin cream

(c) Dermabase cream

(d) Methylcellulose gel

29. Does this base maximize?:

(a) Partition coefficient

(b) Concentration of drug in the vehicle

(c) Diffusion coefficient

Selected reading

1. KA Walters (ed.). Structure and Function of Skin in Dermatological and Transdermal Formulations, New York, Marcel Dekker: 2002.
2. RH Guy, J Hadgraft Physicochemical aspects of percutaneous penetration and its enhancement. Pharm Res 1988; 5: 753–758.
3. JW Wiechers, CL Kelly, TG Blease et al. Formulating for efficacy, Int J Cos Sci 2004; 26: 173–182.

4. JW Wiechers, AC Watkinson, SE Cross et al. Predicting skin penetration of actives from complex cosmetic formulations: an evaluation of inter formulation and inter active effects during formulation optimization for transdermal delivery. Int J Cos Sci 2012; 1–11.
5. S Heukchel, A Goebel, RHH Neubert. Microemulsions – Modern colloidal carrier for dermal and transdermal drug delivery. J Pharm Sci 2008; 97: 603–631.
6. SA Ibrahim, SK Li. Effects of chemical enhancers on human epidermal membrane: Structure-enhancement relationship based on maximum enhancement (E_{max}). J Pharm Sci 2009; 98: 926–944.
7. M Kalavala, CM Mills, CC Long et al. The fingertip unit: A practical guide to topical therapy in children. J Dermatol Treat 2007; 18: 319–320.
8. D Young. Student's death sparks concerns about compounded preparations. Am J Health-Sys Pharm 2005: 62: 450–454.
9. LV Allen. The Art, Science and Technology of Pharmaceutical Compounding, 4th edn, Washington, DC: American Pharmacists' Association, 2012.
10. JE Thompson, LW Davidow. A practical guide to contemporary pharmacy practice, 3rd edn, Philadephia, PA: Lippincott, Williams & Wilkins, 2009.

18

Transdermal dosage forms: Drug delivery to the blood stream

Learning objectives

Upon completion of this chapter, you should be able to:

- Differentiate between systemic and regional transdermal products and outline the movement of a drug from its site of administration to its target including:
  - release
  - absorption
  - distribution to target.
- Predict the effect of each of the following drug properties on the rate and extent of drug travel to its target from the skin: water solubility, log P, molecular weight, stability in solution, enzymatic degradation and location of drug receptors.
- Describe the design of matrix, drug in adhesive matrix and membrane (reservoir) patches and how the design impacts their performance.
- List the categories of ingredients that are added to compounded products for the skin and identify how the ingredient contributes to the manageable size of the dosage form, its palatability or comfort, its stability (chemical, microbial and physical), the convenience of its use and the release of drug from the dosage form.
- Given a drug proposed for transdermal use and adequate references, determine whether its physicochemical properties and dose are suitable, and design a formulation to maximize thermodynamic activity and a monitoring plan to determine patient response.
- Explain how the qualities of the dosage form are evaluated *in vitro* and/or *in vivo*.
- Explain to patients and/or other healthcare providers how to use, store and/or prepare dosage forms for administration via the skin for transdermal absorption.

Introduction

There is a great deal of excitement in the research and healthcare communities about the advent of transdermal drug delivery. For drugs with high first pass metabolism by mouth, the route offers an increase in

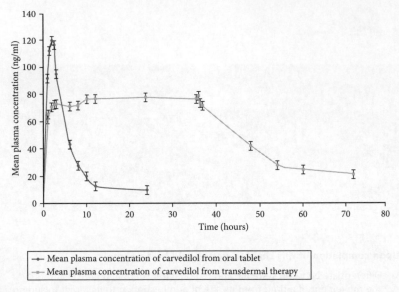

Figure 18.1 Transdermal carvedilol compared with oral carvedilol in healthy subjects. Transdermal drug therapy produces slow, consistent systemic levels of drug compared with oral tablets. (Reproduced with permission from Agrawal et al. J Contemp Clin Trial 2010; 31: 272–278.)

bioavailability. In addition the slow and consistent absorption from these dosage forms provides relatively constant plasma concentrations that result in good patient compliance, consistent therapeutic efficacy and lower incidence of adverse effects compared with the oral route (Figure 18.1). In addition drugs in transdermal dosage forms avoid the secretions, enzymes, drug interactions and variability in transit time of the gastrointestinal tract. The transdermal route is noninvasive and accessible, and available dosage forms are easy for the patient to administer. Drugs delivered by this route must be absorbed to a sufficient extent through the stratum corneum to reach the dermal capillaries and provide a therapeutic concentration at a systemic target. Presently the drugs used by this route are all small molecules of moderate lipophilicity that are active at very low doses. A number of newer products have been developed that are applied to the skin of an affected body part to provide 'regional' transdermal effects. Physical methods for modifying the permeability of the stratum corneum barrier such as microneedles or thermal ablation are currently under development and will allow the administration of larger drug molecules or small molecules with lower potency.

Characteristics of the transdermal route

Drug travel from dosage form to target

See Appendix Tables A.1 and A.4.

Release

Drugs are delivered through the skin from transdermal patches, gels and ointments.

- All are applied to a small area of healthy skin and release drug by diffusion while in contact with the stratum corneum.

There are three different patch designs: reservoir, matrix and drug in adhesive matrix devices.

- The reservoir patch designs are constructed with a semipermeable membrane that controls the rate at which the drug diffuses from the device.
  - Reservoir devices produce a constant (zero order) rate of release from the device for the length of time it is intended to be applied.
- The two matrix designs contain drug dissolved in a polymer or adhesive matrix along with a suspended reservoir of drug that keeps the matrix saturated.
  - Matrix devices produce a pattern of release that slowly declines as the concentration of drug in the device declines.
  - The skin controls the absorption rate of matrix devices producing a constant (zero order) rate for the length of time it is intended to be applied.

The ointment and gel products may be expected to produce a pattern of release that slowly declines as the concentration of drug in the vehicle declines.

- The skin controls the absorption rate of ointments and gels producing a constant (zero order) rate for the length of time it is intended to be applied.

Absorption

- *Systemic transdermal delivery* will require that drugs applied to the skin penetrate the intercellular lipids of the stratum corneum and partition into the hydrophilic media of the viable epidermis and dermis to be absorbed via dermal capillaries in quantities sufficient to produce a pharmacological response at systemic targets (Figure 18.2).
- As the drug moves from the dosage form into the skin, it accumulates in the skin producing a depot of drug that will continue to be absorbed once the patch is removed.
- Because these dosage forms need to produce therapeutic concentrations of drug in the systemic circulation, the performance of transdermal dosage forms can be evaluated by measuring the concentration in plasma over time and calculation of C_{max}, AUC and T_{max}.

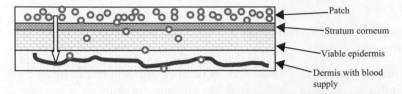

Figure 18.2 Movement of drug from transdermal dosage form to target

- The transdermal patch or gel should be applied to the site(s) specified in the product labeling to produce absorption consistent with its design.
  - The permeability of skin sites is not simply a function of stratum corneum thickness, but also number of skin appendages, surface area of corneocytes, lipid content and blood perfusion.
  - In general, the back and extremities are comparable in terms of permeability. The dose of a transdermal gel or ointment is determined by the weight of the vehicle applied to the skin.
- The main source of variability in drug absorption from transdermal products between patients may be the permeability of the patient's skin.
- The absorption route for *regional transdermal products* takes the drug through the layers of skin and past the dermal capillaries to neurons or inflamed joints beneath the area of application.
  - In this case absorption into the dermal capillaries would take the drug away from the site of action but to some extent is unavoidable.
- The activity of drug metabolizing enzymes in the skin is considerably lower than in the liver.
  - Nitroglycerin is approximately 20% metabolized in the skin while the first pass by the oral route is estimated to be greater than 90%.

Distribution

- The dermis contains the skin's vasculature with capillaries around the hair follicles and glands and just below the epidermis.
- The subcutaneous tissue located just below the dermis contains the arteries and veins that fill and drain the capillary system of the dermis and the base of the skin appendages, which extend upward into the other layers.
- The blood flow to the dermis is between 2 and 2.5 mL/minute per 100 g tissue, in the intermediate range.

Drug properties and the transdermal route

- Given the technology available, not all drugs have appropriate physicochemical properties for delivery through the skin in sufficient quantity to produce a systemic effect. In order to be effective transdermally, a drug must have:

Table 18.1 Drug properties for transdermal administration

Drug	Molecular weight	Log P	Melting point (°C)	Availability (TD %)
Fentanyl	336	3.9	83	92
Nicotine	162	1.1	−79	80–90
Selegiline	187	2.7	141–142	73
Nitroglycerin	227	1	13.5	72
Estradiol	272	4.2	176	
Levonorgestrel	312	3.8	235–237	
Norethindrone	298	3.2	203–204	
Ethinyl estradiol	296	4.3	141–146	
Norelgestromin	327	3.67 (pred)		
Scopolamine	303	0.8	59	
Buprenorphine	467.6	3.8	209	15
Capsaicin	305.4	4.0	65	
Clonidine	230	2.7	140	60
Diclofenac	296	3.9	156–158	10
Lidocaine	234	2.1	68–69	3±2
Granisetron	312	2.6		66
Methylphenidate	233	2.1	74–75	
Oxybutynin	357	4.3		
Rivastigmine	250	2.3	Liquid	43–49
Rotigotine	315	4.7		37
Testosterone	288	3.6	153	10
Average	285.1	3.267		

TD = transdermal.

   — low molecular weight (<500 Da)
   — adequate solubility in both water and oil (log P between 1 and 4)
   — low melting point (<200°C)
   — high potency (small amount required to cause an effect, <20 mg/dose).
- A quick survey of the properties of drugs commercially available for transdermal use reinforces the validity of these principles for drugs applied to the skin without the use of physical methods to alter their permeability (Table 18.1).

Table 18.2 Release rate, drug contents and surface area of fentanyl transdermal systems (Mylan)

Nominal release rate (µg/hour)	Fentanyl content (mg)	System size (cm²)
25	2.55	6.25
50	5.10	12.5
75	7.65	18.75
100	10.20	25

Reproduced with permission from Drugs@FDA.

Dosage forms for the transdermal route

Design

1. One dose in a manageable size unit
   - The dose of a transdermal patch is determined by the surface area of skin in contact with the patch. Thus as the size of patch area increases, the dose delivered will increase (Table 18.2).
2. Comfort
   - The occlusiveness of transdermal patches, in addition to the chemical irritant properties of drugs and adhesives, contribute to the high incidence of skin reactions to the devices.
   - Fortunately most reactions are mild and few patients discontinue patch therapy.
3. Stability
   - Individual transdermal patches are heat sealed into a multilaminate foil and polyethylene packaging since they may contain volatile solvents or hygroscopic substances. In general transdermal patches provide good stability.
4. Convenience/ease of use
   - Unit dose packaging is convenient for unit dose systems in healthcare institutions and enables the patient to easily pocket the devices to carry to work or leisure.
   - Transdermal products allow infrequent (daily, weekly) dosing that facilitates patient compliance.
   - Doses are relatively inflexible but may be cautiously reduced by modifying surface (see below).
5. Release of drug
   - The patches, gels and ointments release drug by diffusion. The dose of drug in a transdermal patch exceeds the amount delivered in order to provide consistent drug release over the period of application.

- The major challenge in their design has been to enhance the permeability of the skin sufficiently to permit adequate absorption for systemic action.
- All transdermal patches are constructed with occlusive backing that prevents the loss of water through the skin under the device resulting in hydration and swelling of the stratum corneum barrier.
  - Swelling of the keratinocytes of the membrane loosens the lipid mortar between the cells significantly increasing drug absorption.

Ingredients

Chemical penetration enhancement

- Many of the available transdermal patches and semisolids contain ingredients to modify the permeability of the skin in order to enhance the penetration of the active ingredient.
- The chemical penetration enhancers include small solvent molecules, fatty acids and their esters, and lipid mimics with bulky polar heads that create more fluidity in the intercellular lipids (Table 18.3).
- The chemical penetration enhancers are used to increase the diffusion of small drug molecules, and their use is limited by their tendency to cause skin irritation.
- Chemical penetration enhancers demonstrate only limited success in the permeation enhancement of large drugs like peptides and proteins.

Physical penetration enhancement

- Physical agents such as microneedles, heat, thermal ablation, iontophoresis and ultrasound can produce a greater magnitude of skin permeability enhancement than the chemical penetration enhancers (Table 18.4).
- Research indicates that physical methods can permit the permeation of large, polar and charged drugs such as peptides, proteins and oligonucleotides.
  - Macromolecular drugs are generally potent and have short biological half lives, and would benefit from the controlled release characteristics of transdermal devices.

Pharmaceutical example

Synera, a controlled, heat assisted, drug delivery (CHADD) patch contains a separate compartment of proprietary powder mixture above the drug reservoir that reacts with oxygen when the patch is opened to generate heat. The device, which was approved by the FDA in 2005, contains lidocaine and tetracaine and is used to anesthetize the patient's skin before the placement of catheters.

Table 18.3 Chemical penetration enhancers for the skin

Penetration enhancer	Description
Solvents	
Water	Produces swelling in the stratum corneum causing expansion of the intercellular lipid lamellae
DMSO, ethanol, Transcutol (diethyleneglycol monoethyl ether)	Solubilization and extraction of lipids from between cells promoting drug partitioning into the altered lipid
Propylene glycol, glycerin, polyethylene glycol	Provide adequate drug solubility at the skin surface after volatile solvents have evaporated
Lipid vesicles	
Liposomes	Lipid vesicles fuse with intercellular lipids delivering their cargo to the intercellular channel
Lipid mimics with bulky polar groups	
2-*n*-Nonyl-1,3-dioxolane (SEPAR 0009) and derivatives	Inserts hydrocarbon tail into the intercellular lipids creating both polar and nonpolar channels. Increases the thermodynamic activity of the drug in its vehicle
Cyclopentadecalactone (CPE-215)	Inserts hydrocarbon tail and large polar head group into the intercellular lipids, creating more fluid channel
Laurocapram,1-dodecylazacycloheptan-2-one (AzoneR) and derivatives	Inserts hydrocarbon tail and large polar head group into the intercellular lipids, creating more fluid channel
1-[2-(decylthio)ethyl]azacyclopentan-2-one (HPE-101)	Inserts hydrocarbon tail and large polar head group into the intercellular lipids, creating more fluid channel, more effective in polar vehicles
4-Decyloxazolid-2-one (Dermac SR-38)	Mimics natural skin lipids
Dodecyl-*N-N*-dimethylaminoisopropionate; DDAIP (NextAXT 88)	Interacts with keratin to increase stratum corneum hydration

Dosage forms

Patches

- The transdermal patches can be conveniently placed on the skin with administration frequency from daily to weekly that improves patient compliance.
- They contain a reservoir or matrix of drug designed to produce release of drug across a specific surface area and with sufficient occlusion to drive the drug through the skin into the dermal capillaries.

Table 18.4 Physical methods to enhance the permeability of the skin		
Method	**Description**	**Clinical/preclinical applications**
Heat	Separate compartment of proprietary powder mixture reacts with oxygen when the patch is opened to generate heat to warm the skin	Lidocaine/tetracaine controlled heat-assisted drug delivery patch (Synera)
Thermal ablation	Heating of the skin surface to hundreds of degrees for a very short period of time to create microchannels in the stratum corneum	Parathyroid hormone, interferon-alfa, hepatitis B antigen, erythropoietin, teriparatide
Microneedles	Hollow or solid device that is long enough to create a channel for drug solution through stratum corneum but short enough not to excite nerve endings in the dermis	Parathyroid hormone, insulin, immunoglobulin G, desmopressin, human growth hormone, influenza vaccine, hepatitis B vaccine
Iontophoresis	Continuous electric current to drive charged drug molecules across the skin	Fentanyl, lidocaine, zolmitriptan, parathyroid hormone, LHRH, insulin
Electroporation	Use of short, high voltage electrical pulses to create transient pore-like disruptions in the stratum corneum	Salmon calcitonin
Sonophoresis	Ultrasound applied to the skin to produce cavitational bubbles which oscillate and disrupt stratum corneum structure	Insulin, heparin, interferon gamma, erythropoietin

LHRH = luteinizing hormone releasing hormone.

- Controlled therapeutic drug levels produce consistent therapeutic efficacy and reduce side effects.
- When necessary the patient or caregiver can terminate drug action relatively quickly by removing the device from the skin.
- The basic design of various commercially available transdermal patches is presented in Tables 18.5 and 18.6.
- The reservoir devices enclose the active ingredient in a compartment separated from the skin by a polymeric membrane that controls the delivery of the drug (Figure 18.3a).
  - The reservoir contains an excess of drug in solution or suspension in a nonaqueous vehicle so that the membrane remains saturated with drug throughout the period of patch use.
  - The rate limiting membrane allows drug diffusion at a *slower* rate than the stratum corneum such that the control of drug absorption is provided by the patch.
  - The reservoir creates a bulkier device with only fair skin conformability, and if the rate limiting membrane is cut these patches will lose their ability to control release rate and surface exposed to drug.

Table 18.5 Design and therapeutic equivalence code of selected commercially available systemic transdermal systems

Product (generic) manufacturer	Design	Therapeutic equivalence code
Butrans (buprenorphine) Purdue Pharma	Matrix	None
Catapres-TTS (clonidine) Boehringer Ingleheim	Reservoir	AB
Clonidine (generic) Mylan	Matrix	AB
Alora (estradiol) Watson	Drug in adhesive matrix	BX
Climara (estradiol) Bayer	Drug in adhesive matrix	AB, AB2 depends on strength
Estraderm (estradiol) Novartis	Reservoir	BX
Menostar (estradiol) Bayer	Drug in adhesive matrix	none
Vivelle-Dot (estradiol) Novartis	Drug in adhesive matrix	BX, AB1
Estradiol (generic) Mylan	Matrix	AB, AB2 depends on strength
Duragesic (fentanyl)	Drug in adhesive matrix	AB
Fentanyl (generic) Mylan	Drug in adhesive matrix	AB
Fentanyl (generic) Watson	Reservoir	AB
Minitran (nitroglycerin) Graceway	Drug in adhesive matrix	AB1
Nitro-Dur (nitroglycerin) Key	Drug in adhesive matrix	AB1
Nitroglycerin (generic) Kremers Urban	Drug in adhesive matrix	AB1
Nitroglycerin (generic) Hercon	Drug in adhesive matrix	AB2
Nitroglycerin (generic) Mylan	Drug in adhesive matrix	AB2

Reproduced with permission from Drugs@FDA.

- — They create zero order release and absorption that is consistent from patient to patient.
- Matrix devices contain drug dispersed in a polymeric matrix (see Figure 18.3b).
  - — The labeling of some matrix patches may describe the matrix as a 'reservoir' because it represents an excess of drug, which maintains sustained drug release over the entire period of application.
  - — Diffusion through the matrix controls the release rate of the drug from the device while the absorption rate is controlled by the permeability of the skin over which it is applied.
  - — The matrix patch design is simpler to construct and the thinner construction provides good skin conformability.

Table 18.6 Selected transdermal products, their design and ingredients

Product	Stability	Release	Absorption (permeability enhancement)
Synera medicated lidocaine, tetracaine topical patch	Methylparaben, polyvinyl alcohol, propylparaben, sorbitan palmitate	Adhesive, distilled water	CHADD technology: utilizes controlled heat to enhance transdermal drug permeation
AndroGel testosterone gel	Carbomer 980	Ethanol, purified water	Isopropyl myristate
Vivelle-Dot estradiol drug in adhesive matrix TDS	Acrylic and silicone adhesive, povidone,	Oleyl alcohol, dipropylene glycol	Polyolefin film
Catapres TTS clonidine reservoir TDS	Mineral oil, polyisobutylene, colloidal silicon dioxide	Polypropylene microporous membrane	Polyester and aluminum backing
Exelon rivastigmine drug matrix patch	Silicone adhesive	Acrylic copolymer matrix	Polymer backing
Lidoderm lidocaine	Dihydroxyaluminum aminoacetate, disodium edetate, gelatin, glycerin, kaolin, methyl and propyl paraben, polyacrylic acid, polyvinyl alcohol, sodium carboxymethylcellulose, sodium polyacrylate, D-sorbitol, tartaric acid	Water, glycerin	Polymer backing and chemical penetration enhancers: urea and propylene glycol
Voltaren gel (diclofenac) Endo	Carbomer, strong ammonia solution (gels carbomer)	Isopropyl alcohol, mineral oil, purified water	Cocoyl caprylocaprate, polyoxyl 20 cetostearyl ether, propylene glycol

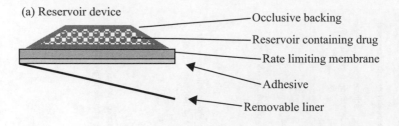

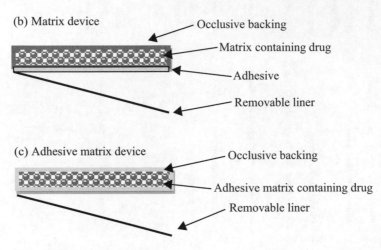

Figure 18.3 Transdermal patch designs.

- — Because the matrix is monolithic this patch design has low dose dumping potential. However, most manufacturers discourage cutting of patches to modify dose.
- The drug in adhesive matrix devices has been further simplified by incorporation of the active ingredient into the adhesive layer (see Figure 18.3c).
  - — Diffusion through the adhesive matrix controls the release rate of the drug from the device while the absorption rate is controlled by the permeability of the skin over which it is applied.
  - — The occlusive backing covers the drug adhesive matrix which may incorporate penetration enhancers to facilitate drug absorption.
  - — These systems are popular with patients because they are thin and conform well to the skin.
- The occlusive backing layer on all patch designs facilitates drug absorption by hydrating the skin beneath the patch after application.
- All patch designs have a liner that should be removed before use.

Gels and ointments

- Nitroglycerin, estradiol and testosterone have been formulated as topical semisolids for systemic delivery.

Pharmaceutical example

Nitroglycerin 2% ointment is a petrolatum based product containing lanolin and water. Testosterone 1% gel is a carbomer based product containing ethanol as a cosolvent and isopropyl myristate as a penetration enhancer, and estradiol is available as a transdermal gel, emulsion and aerosol spray.

- Because these dosage forms are less occlusive than trandermal patches, they usually contain a penetration enhancer in the formulation.
- While the dose is less exact than delivered by a patch, the semisolids may be a good choice for patients who experience allergic reactions to components of transdermal patches.

Quality evaluation and assurance

- There are a variety of methods to evaluate the drug release rate of transdermal drug delivery systems *in vitro*. These evaluations may also be used to guide the optimization of a formulation or design.
- They include the following.
  - The transdermal system is placed in the disc at the bottom of the USP Dissolution Apparatus 5 and suspended in a suitable medium at 32°C.
  - The transdermal system is placed between the receptor and donor compartment of a Franz diffusion cell with the releasing surface facing a suitable fluid in the receptor compartment.
- To receive approval for a generic transdermal drug product, the manufacturer must establish that its product has the same active ingredient, dosage form, strength, route of administration and conditions of use, and that the active ingredient of the proposed drug product is absorbed into the blood stream to the same extent and at the same rate as the innovator product.
- Note that reservoir patches may be pharmaceutically and bioequivalent to matrix or drug in adhesive matrix patches and therefore interchangeable.

Dosage forms: Regional transdermal systems (Table 18.7)

- Regional transdermal systems are available as patches, and as creams and gels that may be applied to the affected limb multiple times per day.

Table 18.7 Regional transdermal drug delivery systems

Product (generic) manufacturer	Design	Application site
Qutenza (capsaicin) NeurogesX	Drug in adhesive matrix	Painful area
Flector (diclofenac) King	Drug in adhesive matrix	Painful area
Lidoderm (lidocaine) Endo	Drug in adhesive matrix	Painful area
Synera (lidocaine, tetracaine) Zars	Reservoir	Site of procedure
Voltaren (diclofenac) gel Novartis	Gel	Painful area
Capsaicin cream	Cream	Painful area

- Regional transdermal systems do not differ in design from the systemic transdermal dosage forms but are intended to be applied to the affected body part.
- This method of use confines the absorption of drug to the affected area and reduces the volume through which the drug must distribute to produce the desired effect.
- The regional method of delivery may be used with drugs that have a daily dose larger than 20 mg when taken by mouth.
- Systemic exposure to the drug is lower resulting in fewer adverse effects.
- Regional transdermal therapy has been popular with nonsteroidal anti-inflammatory drugs as an approach to reduce gastrointestinal side effects.

Compounded formulations for transdermal applications

- The compounding pharmacist may be asked to prepare transdermal formulations for a patient who has difficulty taking medications by mouth or lacks the muscle mass to receive intramuscular injections.
- Because the skin is a significant barrier to drug absorption, the pharmacist's first concern should be for potential therapeutic failure of a compounded product whose biopharmaceutics cannot be rigorously evaluated.
- If in the pharmacist's judgment the preparation of a transdermal product is in the best interests of the patient, careful application of pharmaceutical principles coupled with close monitoring of the patient's response to the formulation are essential.
- The physicochemical properties of the drug should be examined to determine the likelihood of a favorable response.
- A suitable vehicle should be chosen that will maximize the thermo-dynamic activity of the drug in the formulation.
- The patient should be monitored for both therapeutic and adverse effects beginning on the first day of therapy.

Suitable vehicles

- To have the maximum concentration gradient from vehicle to the dermal capillaries, the vehicle must be saturated with dissolved drug.
- The compounding pharmacist is encouraged to use the permeant polarity gap method to develop a rational vehicle for transdermal use. For more details on this approach please see page 412 or the original paper (see selected reading 4).

Key Points

- Choose a base that will dissolve the drug and choose a drug form that will dissolve in the base. In the case of two phase systems (water-in-oil and oil-in-water emulsions) there must be an appreciable amount dissolved in the continuous phase.
- Use of a cosolvent ingredient with polarity similar to the stratum corneum enhances partitioning of the drug and cosolvent into the stratum corneum.
- Occlusive properties increase the hydration and permeability of the stratum corneum.
- Water from the formulation will be absorbed into the skin but will only have transient effects on hydration.
- Penetration enhancers may be required to enhance the mobility of large molecular weight drugs.

Choosing a transdermal dose

- Most compounding pharmacists will begin with the oral or rectal dose of a drug in a transdermal formulation and adjust the dose based on the patient's response.
- While an oral dose of a drug will include the amount of drug that is lost to hepatic metabolism on the first pass through the liver, we have little information about the metabolism of drugs by enzymes in the skin.
  - The bioavailability of transdermally applied drugs has a wide range from 10% to >90%. Thus despite the dissimilar comparison, this practice is reasonable given the information we have about the limited number of drugs available as transdermal gels.
- The pharmacist is advised to examine the proposed transdermal dose in light of Flynn's estimate that the maximum achievable limit for transdermal delivery is about 1 mg/cm^2 per day for a drug with 'model' physicochemical properties.

Monitoring the patient

- A comprehensive monitoring plan should be developed that will identify both therapeutic failure and adverse effects of the transdermal drug.

- The onset of effects for most drugs applied transdermally is substantially delayed with T_{max} between 8 and 24 hours, however the cautious practitioner will follow up with the patient within 4 hours of initiating treatment.
- The patient's response should be documented in writing and used to guide further formulation work with this patient and the drug in question.

Counseling healthcare providers and patients

Advising healthcare providers about patches

- While these dosage forms are generally easy to apply, there are a variety of issues relating to the use of transdermal dosage forms that may be addressed by the pharmacist with their background in dosage form design:
  — management of skin reactions
  — avoidance of burns
  — adhesive failures
  — writing on patches
  — reducing patch surface for dose adjustments.
- The incidence of skin reactions to transdermal patches is high, between 20% and 50%.
- The drug, the adhesive or the occlusiveness of the patch can all contribute to skin irritation.
- The majority of skin reactions are considered a form of irritant contact dermatitis, which can be managed by rotating the site of patch application.
- The following information on the management of skin reactions to transdermal patches is from selected reading 7.
  — Apply the patch to clean, relatively hairless area of the skin.
  — Remove the patch carefully to minimize damage to the skin and hairs.
  — Remove excess adhesive with a vegetable oil such as olive oil.
  — Clean the application area gently with water.
  — Rotate the area of application on a daily basis, preferably to an area drained into a different set of lymph nodes.
  — Do not use the same area for application for at least 7 days.
  — Do not use alcohol or detergents on the affected skin.
- The following is the suggested treatment of skin reactions.
  — Move the patch to another area of skin.
  — Apply moisturizers or calamine lotion.
  — Allergic contact dermatitis will respond to topical corticosteroids.
- The occlusive backing of transdermal patches may contain metal foil which creates a safety concern for patients undergoing magnetic resonance imaging and patients who may need external defibrillation during cardiac resuscitation.

   — These metals have the potential to conduct electrical currents and can lead to serious burns in patients wearing patches during these procedures.

   — One advantage of transdermal patch therapy is that the device can be easily removed and the reservoir of drug already transferred to the stratum corneum will continue to provide some absorption until the patch can be reapplied.

- A partially or completely detached patch may be reattached using medical tape or bandaging.

   — A few device manufacturers recommend the application of a new patch to a different site.

   — Patients should be advised to clean the new site with soap and water and carefully dry but they should not use alcohol or other organic solvents to clean the site before reapplication.

- Because the newer patches are transparent and may be difficult for the patient to locate for removal, or because caregivers need to communicate the time and date of patch placement to other people caring for the patient, it is common practice to write on patches in institutional and home settings.

   — Most manufacturers do not recommend writing on their patches primarily for theoretical reasons such as the pen may tear the patch or ink may interact with patch components.

   — Durand et al. (selected reading 9) suggest that caregivers write on medical tape and place the tape on or near the patch.

- Occasionally the pharmacist is asked how to reduce patch surface area for the purpose of reducing the dose delivered.

   — Cutting a reservoir patch to reduce its size would destroy its ability to control release rate and surface exposed to drug.

   — Theoretically a matrix patch could be cut without compromising its release characteristics but some of the manufacturers' labeling warns against this practice.

   — In any case with the advent of generic substitution for transdermal patches, it is preferred not to advise patients to cut their patches when the next week they may receive a reservoir device that would be compromised by this practice.

   — The currently recommended practice is to occlude a portion of the patch with wax paper or the patch liner allowing the barrier edges to protrude from the patch rather than trimming them.

   — The patch with barrier is covered with Tegaderm or other dressing and the date, time and fraction of exposed patch noted on the dressing.

Advising patients about the use of patches

The patient may be given the following advice.

- Wash area thoroughly with soap and warm water and pat dry.
- The location of placement is specified in the manufacturer's labeling. These locations (arm, chest, abdomen etc) have been studied by the manufacturer and are known to produce predictable absorption. Patients should use the manufacturer recommended sites unless their skin condition prohibits placement. General recommendations include: free of hair, no broken skin. Do not put where extremities bend or clothing will bind.
- *Rotate sites!*
- There is a large lag time before the onset of the effect. The scopolamine patch needs to be placed behind the ear 4 hours before needed. Patients need to understand that a fentanyl patch will not control pain for several hours. They need to have an immediate release pain reliever available until the fentanyl patch reaches therapeutic concentrations and for break through pain.
- Frequency varies with drug and device.
- Remember to take off the old patch before adding the new one. The old patch actually has substantial drug left in it when it is removed. (Nitroglycerin patch still contains around 80% of the original drug quantity after 24 hours, Duragesic patches around 50%.)
- Fold the patch to stick the adhesive surfaces together before disposal in a secure place so that children cannot access.
- Bathing, showering or swimming generally do not affect transdermal patches although it is prudent to avoid scrubbing the area.
- Remove before magnetic resonance imaging studies.

Questions and cases

1. Duragesic is a membrane-controlled patch applied every 72 hours for relief of severe pain. Which of the following features are true of membrane controlled (reservoir) patches?

    (a) Simple patch construction

    (b) Good skin conformability

    (c) Potential for dose dumping

    (d) First order release

    (e) Skin controlled absorption

2. Climara is a drug-in-adhesive matrix patch used for estrogen replacement therapy. Which of the following features is true of drug-in-adhesive matrix patches?

    (a) Good skin conformability

    (b) Zero order release

    (c) Thicker construction

(d) Complex formulation

(e) Potential for dose dumping

3. ZM is a burn patient who is experiencing significant pain but whose skin barrier is significantly compromised. Which transdermal patch design would be preferred for ZM?

(a) Reservoir (membrane) patch

(b) Matrix patch

(c) Drug in adhesive matrix patch

4. Drugs delivered transdermally are:

(a) More likely to cause side effects than oral products

(b) Less likely to cause side effects than oral products

(c) Unlikely to cause any side effects

(d) Equally likely to cause side effects as oral dosage forms

5. A patient receiving estrogen replacement therapy with the estradiol patch has developed severe nausea on 0.025 mg/day which is the lowest dose of the medication. The physician would like to restart at 0.015 mg/day and titrate her up to the 0.025 mg/day dose. If the 0.025 mg/day patch is 8 cm^2, how much surface would provide the patient with the lower dose?

(a) 3.9 cm^2

(b) 4.8 cm^2

(c) 5.2 cm^2

(d) 6 cm^2

6. When counseling patients about the use of transdermal patches, they should be advised to remove:

(a) The backing

(b) The matrix

(c) The adhesive

(d) The liner

7. Alora is a transdermal patch marketed by Watson Labs. It is described as follows in the package insert: Alora consists of three layers. Proceeding from the polyethylene backing film, the adhesive matrix drug reservoir that is in contact with the skin consists of estradiol, USP and sorbitan monooleate dissolved in an acrylic matrix. The polyester overlapped release liner protects the adhesive matrix during storage and is removed prior to application of the system to the skin. Which of the following designs applies to the Alora patch?

(a) Reservoir (membrane) patch

(b) Matrix patch

(c) Drug in adhesive matrix patch

8. Alora is labeled to provide 0.025 mg/24 hours. It is able to maintain a constant release rate over 4 days of use because:

(a) It has a rate controlling membrane

(b) It has a rate controlling matrix

(c) It has a rate controlling adhesive

(d) It has an excess of drug in the matrix, the delivery of which is controlled by the skin

9. You are counseling a patient who will be using Alora patches for postmenopausal symptoms. Which of the following counseling points applies?

   (a) Place the patch on the affected area(s)

   (b) Because the treatment is topical there are no side effects

   (c) Fold the old patch before discarding it

   (d) All of the above

10. Voltaren 1% gel is a non-steroidal anti-inflammatory indicated for relief of the pain of osteoarthritis. It is a 1% solution of diclofenac sodium in water-based carbomer gel with instructions to use 4 g to affected joint up to 4 times daily. (FW 296, log P 3.9, water solubility sodium salt >9 mg/mL.) Which of the following formulation strategies has been maximized by the manufacturer of Voltaren gel?

    (a) C_v, concentration of drug in the vehicle is maximized

    (b) S, surface of the skin exposed to the drug is maximized

    (c) D, diffusion coefficient is maximized

    (d) K, tissue/vehicle partition coefficient is maximized

Diclofenac

11. To reach its receptors, Voltaren gel must penetrate to:

    (a) Viable epidermis

    (b) Dermis

    (c) Dermal blood supply

    (d) Skin appendages

    (e) Tissues beneath the area applied

12. Diclofenac is also available as a 1.3% patch (Flector). You are counseling a patient who will be using Flector for osteoarthritis of the knees. Which of the following counseling points applies?

    (a) Remove while showering

    (b) Place the patch on the affected area(s)

    (c) Because the treatment is topical there are no side effects

    (d) All of the above

13. Which of the following is *not* required for two patches to be considered therapeutically equivalent?

    (a) Product must contain the same drug salt or derivative

    (b) Product must be the same design (matrix or reservoir)

    (c) Product must contain the same dose

    (d) Product must demonstrate a substantially similar rate and extent of absorption of the active ingredient as the reference listed drug

    (e) All of the above are required

14. An elderly patient receiving oxybutynin therapy for overactive bladder with a patch dosage form has developed severe agitation on 3.9 mg/day which is the only strength available. The physician would like to restart at 1.3 mg/day and titrate her up to the 3.9 mg/day dose. If the 3.9 mg/day patch is 39 cm², how much surface would provide the patient with the lower dose? Explain to the patient's caregiver how to limit the dose by altering the surface.

15. GB is a 57 year old woman on bioidentical hormone replacement therapy using a custom compounded product. Her prescription is for estradiol 0.1%, progesterone 5% cream, apply 1 g daily. She comes into the pharmacy for a refill only 1 week after receiving a 1 month supply of her transdermal hormone cream. She tells you that she liked the effect of the cream on her skin as well as on her menopausal symptoms so she has decided to apply it to her face, chest and arms. Counsel GB on the implications of applying the cream in this fashion.

16. An elderly patient has begun to refuse medications including an antidepressant medication and diltiazem for unstable angina. Her family is concerned about her untreated angina and has contacted you about preparing a transdermal product with this medication. Diltiazem is slightly soluble in water and soluble in alcohol, FW 451, log *P* 4.197; the hydrochloride salt is freely soluble in water and sparingly soluble in alcohol, melting point 187–188°C. The patient's daily dose is 120 mg. Comment on the suitability of diltiazem for transdermal administration.

Diltiazem

17. You have a prescription for transdermal lorazepam 0.5 mg/day for treatment of anxiety.

Lorazepam

Lorazepam has a molecular weight of 321, log P 2.959, water solubility 1 g/10 000 mL, melting point 166–168°C.

Discuss the suitability of lorazepam for transdermal drug delivery.

Calculate the penetrant polarity gap for lorazepam. What ingredients or base could be used to formulate the transdermal lorazepam?

How will you determine whether the patient's condition will be adequately treated?

Selected reading

1. KS Paudel, M Milewski, CL Swadley et al. Challenges and opportunities in dermal/transdermal delivery. Ther Deliv 2010: 1: 109–131.
2. H Kalluri, AK Banga. Transdermal delivery of proteins. AAPS PharmSci Tech 2011; 12: 431–441.
3. S Heukchel, A Goebel, RHH Neubert. Microemulsions – Modern colloidal carrier for dermal and transdermal drug delivery. J Pharm Sci 2008; 97: 603–631.
4. JW Wiechers, CL Kelly, TG Blease et al. Formulating for efficacy, Int J Cos Sci 2004; 26: 173–182.
5. J Jampilek, K Brychtova. Azone analogues: Classification, design, and transdermal penetration principles. Med Res Rev 2012: 5: 907–947.
6. SH Bariya, MC Gohel, TA Mehta et al. Microneedles: an emerging transdermal drug delivery system. J Pharm Pharmacol 2012; 64: 11–29.
7. I Ale, JM Lachapelle, HI Maibach. Skin tolerability associated with transdermal drug delivery systems: an overview. Adv Ther 2009; 26: 920–935.
8. A Mitchell, HS Smith. Applying partially occluded fentanyl transdermal patches to manage pain in pediatric patients. J Opioid Manag 2010; 6: 290–294.
9. C Durand, A Alhammad, KC Willett. Practical considerations for optimal transdermal drug delivery. Am J Health-Sys Pharm 2012; 69: 116–124.

Appendix

Table A.1 Membrane and drug permeability characteristics by route

Membrane	Paracellular resistance (Ω cm^2)	Optimal transcellular MW (Da)	Largest commercial product (MW in Da)	Optimal log P
Ophthalmic (cornea)	1012	≤500	Cyclosporin (1202.6)	1.0–2.0
Ophthalmic (RPE)	2000	≤400	Bevacizumab (149 000)	0–5.0
Nasal	261	≤1000	Salmon calcitonin (3432)	1.0–4.0
Pulmonary	266 (bronchial)	<500	Insulin (5808) <76 000	−1 to 4
Buccal	1803 (buccal)	<500	Buprenorphine (467.6)	2–4
Small intestine	211–266	≤500	Cyclosporin (1202.6)	1–5
Rectal	406	≤300	Ergotamine (581.6)	0–5
Intramuscular		<67 000	Immunoglobulins 145–160 kDa	NA
Subcutaneous		<67 000	Immunoglobulins 145–160 kDa	NA
Intravenous		<67 000	Immunoglobulins 145–160 kDa	NA
Skin	9703	<500	Tacrolimus (822)	1.0–4.0
Transdermal	9703	<400	Buprenorphine (467.6)	2.0–4.0
Capillaries (except testes, placenta and blood–brain barrier)	3–30	<67 000 (albumin)	Not applicable	Not applicable

(continues)

Table A.1 (continued)

Membrane	Paracellular resistance (Ω cm^2)	Optimal transcellular MW (Da)	Largest commercial product (MW in Da)	Optimal log P
Blood–brain barrier	2000	≤400	Amphotericin B (924)	0–5

MW = molecular weight; RPE = retinal pigmented epithelium.
Adapted from references 1–14.

Table A.2 Pharmaceutical excipients and their function

Excipient class	Function	Examples
Aerosol propellant	Agent that is gaseous under ambient conditions that provide the force to expel the product when the valve is opened. Delivers drug to the absorbing membrane in finely divided form	Apaflurane, butane, HFA-134a, HFA 227, isobutane, norflurane, propane
Air displacement agent	Displaces air in a hermetically sealed container to enhance product stability	Nitrogen
Anti-microbial preservative – oral liquids	Used in preparations containing water to prevent growth of bacteria and fungi. Effectiveness of parabens is usually enhanced by use in combination	Benzoic acid/sodium benzoate, butylparaben, ethylparaben, methylparaben, potassium sorbate/sorbic acid, propylparaben
Anti-microbial preservative – mucous membranes	Used in preparations containing water to prevent growth of bacteria and fungi. Effectiveness of parabens is usually enhanced by use in combination	Benzalkonium chloride, polyquaterium-1 (Polyquad), chlorobutanol, thimerosal, phenylmercuric nitrate, polixetonium, potassium sorbate, sodium perborate, benzododecinium bromide, methyl and propyl paraben, stabilized oxochloro complex (SOC)
Antioxidant	Used to prevent degradation of preparations by oxidation (chemical stability)	Ascorbic acid, ascorbyl palmitate, butylated hydroxyanisole, butylated hydroxytoluene, hypophosphorous acid, monothioglycerol, propyl gallate, sodium ascorbate, sodium bisulfite, sodium formaldehyde, sodium metabisulfite

(continues)

Table A.2 *(continued)*

Excipient class	Function	Examples
Buffers	Used to resist change in pH to stabilize, solubilize, or maintain at a pH close to the relevant body fluid	Acetic acid/sodium acetate, boric acid/sodium borate (eye only), citric acid/sodium citrate, lactic acid/sodium lactate, potassium phosphate, dibasic potassium phosphate, monobasic, sodium bicarbonate/carbonate
Bulking agent for freeze drying	Protects drug from freeze concentration during lyophilization. Protects chemical and physical stability	Alpha-lactalbumin, creatinine, glycine, mannitol, polydextrose, trehalose
Coloring agent	Used to impart color and distinctive appearance to liquid and solid preparations so that pharmacists do not mistake them for a different product	FD&C Red No. 3, FD&C Red No. 20, FD&C Red No. 40, FD&C Yellow No. 6, FD&C Blue No. 2, D&C Green No. 5, D&C Orange No. 5, D&C Red No. 8, Caramel, ferric oxide, red
Chelating agent	Substance that forms stable water soluble complexes with metals; used in some liquid pharmaceuticals to complex trace heavy metals that might promote instability	Edetic acid, edetate disodium
Cosolvent	Water miscible solvent used to increase the lipophilicity of a vehicle. Used to prepare solution dosage forms that contain the drug in an easily measured unit ready to move across membranes and can be easily swallowed	Alcohol, glycerin, isopropanol (topical only), polyethylene glycol 300, 400, propylene glycol
Cosurfactant cosolvents	Paired with surfactant emulsifier to enhance stability of emulsions	Alcohol, butanol, glycerin, isopropanol, polyethylene glycol 300, 400, propylene glycol, Transcutol (diethylene glycol monoethyl ether)
Emollient	Used in topical preparations ease spreading and soften skin. Prevent loss of water from skin increasing its permeability	Cetyl alcohol, cetyl esters wax, isopropyl myristate, lanolin, mineral oil, olive oil, petrolatum

(continues)

Table A.2 (continued)

Excipient class	Function	Examples
Emulsifying agent	Used to promote and maintain fine dispersions of one immiscible liquid in another (water-in-oil or oil-in-water). These ingredients maintain the physical stability of dispersed systems	Acacia (oral), carbomer, cetyl alcohol, cremophor, glyceryl monostearate, mono-oleate, labrasol (caprylocaproyl, polyoxylglycerides), lanolin alcohols, lecithin (oral, parenteral), polaxamer, polyoxyl ethers, polyoxyethylene fatty acid esters, polysorbates, stearic acid, trolamine
Flavorant	Provides a pleasant flavor and odor to a preparation	Anise oil, cinnamon oil, cocoa, menthol, orange oil, peppermint oil, vanillin
Flocculating agents	Reduce the tendency of particles to aggregate and loosen the density of the sediment	Acacia, methylcellulose, gelatin, tragacanth, lecithin, sodium acetate, sodium citrate, sodium phosphate, polysorbates, sorbitan esters
Humectant	Used to prevent drying of preparations, particularly ointments and creams	Glycerin, propylene glycol, sorbitol
Levigating agent	Liquid or semisolid used to facilitate the reduction of the particle size of a powder by grinding on an ointment slab	Mineral oil, glycerin, propylene glycol
Oil phase	Contributes to the emollient and occlusive effects of a topical product	Corn oil, cottonseed oil, decane, decanol, dodecane, dodecanol, ethyl oleate, isopropyl myristate, isopropyl palmitate, isostearyl isostearate, jojoba oil, lauryl alcohol, medium chain triglycerides, mineral oil, petrolatum, oleic acid, silicon oil, soybean oil
Oil vehicle	Dissolution of oil soluble drugs for oral or parenteral administration	Corn oil, cottonseed oil, olive oil, sesame oil, soybean oil
pH modifying agent	Used in liquid preparations to adjust pH. pH modifying agents can be used to improve the solubility of the drug in water or the ability of the drug to partition into the membrane	Citric acid, acetic acid, hydrochloric acid, nitric acid, sulfuric acid, ammonia solution, ammonium acetate, diethanolamine, potassium hydroxide, sodium bicarbonate, sodium borate, sodium carbonate, sodium hydroxide, trolamine

(continues)

Table A.2 (continued)

Excipient class	Function	Examples
Plasticizer	Provides flexibility to creams and ointments	Glycerin, lanolin alcohols, mineral oil, petrolatum, polyethylene glycol, propylene glycol
Respiratory surfactants	Dispersing aid to prevent growth of drug particles in respiratory suspensions	Lecithin, oleic acid, oligolactic acid, polyethylene glycol, sorbitan trioleate
Solubilizing agent	Used to increase the apparent solubility of a drug in water so that the dose is contained in a more manageable size unit and the drug is ready to move across membranes	Cyclodextrins, poloxamer, polyoxyethylene castor oil derivatives, polyoxyethylene stearates, polysorbates
Solvent	Used to dissolve another substance in preparation of a solution. Used to prepare solution dosage forms that contain the drug in an easily measured unit that is easily swallowed, ready to move across membranes	Alcohol, corn oil, cottonseed oil, glycerin, isopropyl alcohol (topical only), mineral oil (topical and oral only), oleic acid, peanut oil, purified water, sterile water for injection, sterile water for irrigation
Stiffening agent	Used to increase thickness or hardness of a semisolid preparation, such as ointments and creams	Cetyl alcohol, cetyl esters wax, microcrystalline wax, paraffin, stearyl alcohol, white wax
Suppository base	Carrying agent that melts or dissolves to release drug from suppositories. Provides drug in a manageable size unit	Cocoa butter, glycerinated gelatin, hydrogenated coconut, palm, vegetable oil, polyethylene glycols (PEGs)
Sweetening agent	Imparts sweetness and masks unpleasant flavors	Aspartame, acesulfame, dextrose, glycerin, mannitol, saccharin, saccharin sodium, sorbitol, sucralose, sucrose
Tonicity agent	Provides dissolved particles required to produce a solution isoosmotic with body fluids	Dextrose, sodium chloride, mannitol
Vehicle, oral flavored and sweetened	Carrying agent used in formulating liquids for oral administration. Palatability, used to prepare solution dosage forms that contain the drug in an easily measured unit ready to move across membranes and can be easily swallowed	Acacia syrup, aromatic syrup, aromatic elixir, cherry syrup, cocoa syrup, orange syrup, syrup

(continues)

Table A.2 (continued)

Excipient class	Function	Examples
Vehicle, sterile	Used in parenteral, ophthalmic, nasal and nebulized products. Include tonicity agent and solvent	Dextrose 5%, sterile water for injection, bacteriostatic sterile water for injection, sodium chloride 0.9%, bacteriostatic sodium chloride for injection
Viscosity increasing agent	Thickening or suspending agent used to reduce sedimentation rate of particles in a vehicle in which they are not soluble or to add body to a sugar free syrup reducing the contact of drug with the taste buds. In topicals to prepare water-based gels and prevent coalescence in oil-in-water emulsion bases	Acacia, carbomer, carboxymethylcellulose, colloidal silicon dioxide, guar gum, hydroxyethyl cellulose, hydroxypropyl cellulose, hypromellose (hydroxypropylmethylcellulose), methylcellulose, polaxamer, povidone, pregelatinized starch, xanthan gum
Wetting agent	Permits hydrophobic powders to displace air and be 'wetted' by water. In solid dosage forms a wetting agent will speed uptake of water and release. In dispersed liquids a wetting agent reduces aggregation and improves uniformity	See also cosolvents. Docusate sodium, polyoxyl 40 stearate, polyoxyl castor oil derivatives, polysorbates, sodium lauryl sulfate

Adapted from references 15–17.

Table A.3 Dosage form ingredients for oral solid dosage forms

Ingredient class	Purpose	Examples
Capsule shell	Form thin shells to enclose a solid or liquid drug to mask taste and contain the dose in an easily administered unit	Gelatin, gelatin, glycerin, gelatin, sorbitol
Tablet coating – immediate release	Outer surface layer added to tablet to protect against decomposition, facilitate ingestion, modify release patterns, mask taste or odor, or provide distinctive appearance. Thin, elastic surface that usually dissolves in stomach unless added to provide slow release (ethylcellulose)	Tablet coating, sugar, liquid glucose, sucrose, tablet coating, film, hydroxyethyl cellulose, hydroxypropyl cellulose, hydroxypropylmethylcellulose, methylcellulose, ethylcellulose, opaquing agent, titanium dioxide, tablet polishing agent, carnauba wax, white or yellow wax
Tablet coating – enteric	Outer tablet layer designed to dissolve after passing out of the stomach	Cellulose acetate phthalate, hypromellose phthalate, methacrylic acid copolymers (Eudragits), polyvinyl acetate phthalate, shellac
Tablet binders	Ensure that the tablet remains intact after compression	Acacia, alginic acid, carboxymethylcellulose sodium, ethylcellulose, gelatin, hydroxypropylmethylcellulose (HPMC), methylcellulose, microcrystalline cellulose, polyethylene glycol, povidone, polyvinylpyrrolidone, pregelatinized starch. Sugars: compressible sugar, lactose, glucose, liquid glucose
Mucoadhesive agents	Adhere to the mucus covering oral cavity	Carboxymethylcellulose, polycarbophil, carbomer, gelatin, pectin
Penetration enhancers	Increase fluidity of the membrane or extract lipids from intracellular space	Sodium caprate, sodium laurate, oleic acid, cyclopentadecalactone
Tablet and capsule diluents	Inert fillers to create desired bulk, flow properties and compression characteristics of tablets and capsules. Improved flow ensures that each tablet weighs the same amount and contains the same amount of drug	Calcium phosphate (direct compression), microcrystalline cellulose, powdered cellulose, kaolin, lactose, mannitol, sorbitol, starch

(continues)

Table A.3 (*continued*)

Ingredient class	Purpose	Examples
Tablet disintegrant	Promotes wicking of water into an oral solid so that it falls apart into particles. This causes the tablet to release drug faster because it exposes more surface to body fluids	Alginic acid/sodium alginate, carboxymethylcellulose (CMC, cellulose gum), croscarmellose, crospovidone, polacrilin potassium, sodium starch glycolate, starch
Tablet glidant	Improves flow properties and prevents caking of the powder mixture as it moves through tableting equipment so it can be accurately metered into the die and each tablet weighs the same amount and contains the same amount of drug	Colloidal silicon dioxide, corn starch, talc
Tablet lubricant	Prevents adhesion of the powder to tablet presses so powder can be accurately metered into the equipment and each tablet weighs the same amount and contains the same amount of drug	Calcium stearate, magnesium stearate, mineral oil, sodium stearyl fumarate, stearic acid, zinc stearate
Release-controlling membrane	Semipermeable membrane coating produces diffusion dependent release from drug core	Ethylcellulose, polyvinyl acetate, amino methacrylate copolymer, ammonio methacrylate copolymer, cellulose acetate, ethylacrylate methyl methacrylate copolymer, ethylene vinyl acetate, methacrylate copolymer
Plasticizer	Mixed with coating polymer to provide flexibility	Dibutyl phthalate, diethyl phthalate, dibutyl sebacate, acetylated monoglyceride, acetyl tributyl citrate, acetyl triethyl citrate, polyethylene glycol
Drug-containing core (used with release-controlling membrane)	Single unit systems: compressed tablet containing drug. Multiparticulate system: small spheres coated or embedded with drug and then coated with insoluble membrane	Sucrose, lactose, microcrystalline cellulose

(continues)

Table A.3 (continued)

Ingredient class	Purpose	Examples
Hydrophilic colloid matrix	Slowly bioerodible polymer forms a hydrated gel that slows the diffusion drug away from the eroding matrix. Produce combination diffusion and dissolution dependent release	Hydroxypropylmethylcellylose (HPMC or hypromellose), gelatin, hydroxypropylcellulose, pectin, hydroxyethylcellulose, carboxymethylcellulose, alginate, xanthan gum, chitosan, guar gum, locust bean gum, carrageenan, polyethylene oxide
Wax matrix	Non-eroding, matrix forming agents, produce diffusion dependent release	Carnauba wax, microcrystalline wax, hydrogenated vegetable oils, cetyl alcohol, glycerol palmitostearate, stearyl alcohol, beeswax, aluminum monostearate, glycerol monostearate
Inert matrix	Non-eroding (insoluble), polymer matrix, produce diffusion dependent release	Ethylcellulose, ammonio-methacrylate copolymers, polyvinyl chloride, polyethylene, polyvinyl acetate
Channeling agents	Dispersed in wax or insoluble matrices to dissolve in body fluids and produce channels through which drug can diffuse	NaCl, sucrose, sorbitol, hypromellose
Ion exchange resin	Insoluble polymer with acidic or basic functional groups that provide a counter ion to bind drug of interest	Styrene divinyl benzene copolymer, polymethacrylate
Osmotic pump membrane	Semipermeable membrane coating that allows water in. Produces diffusion dependent release from drug core through a laser drilled hole or pore produced by dissolution of a pore-forming agent	Cellulose acetate, cellulose diacetate, cellulose triacetate, cellulose propionate, cellulose butyrate, ethylcellulose, amylase triacetate
Osmotic agent	Draws water into the osmotic tablet to create pressure that pushes drug out	Potassium chloride, sodium chloride, mannitol
Pore-forming agents	Water soluble compound embedded in the membrane that dissolves in gastrointestinal fluids allowing drug diffusion out of the tablet	Sodium chloride, potassium chloride, polyvinyl pyrrolidone, polyvinyl alcohol, sucrose, glucose, sorbitol, mannitol, urea

Adapted from references 15–18.

Table A.4 Comparison of systemic routes of administration

Site	Surface area	Fluid volume available for drug dissolution, pH	Relative enzyme activity	Epithelial layers and permeability
Intravenous	Capillary bed in target tissue	95 mL/minute (median cubital vein), 7.4	Moderate	For extravascular drug targets single, endothelial layer, excellent permeability
Intramuscular	Capillary bed in muscle tissue	0.15–0.2 mL/g tissue, pH 7.4	Moderate	Single, endothelial layer, excellent permeability
Subcutaneous	Capillary bed in subcutaneous tissue	0.15–0.2 mL/g tissue, pH 7.4	Moderate	Single, endothelial layer, excellent permeability
Oral cavity	100–200 cm^2	0.9–1.1 mL, pH 5.8–7.4	Moderate	Multiple, moderate
Stomach	0.1–0.2 m^2	118 mL, pH 1–3.5	High	Single, moderate
Small intestine	100 m^2	212 mL, pH 5.0–7.0	High	Single, very good
Large intestine	0.5–1 m^2	187 mL, pH 6.4–7.0	Moderate	Single, moderate
Rectum	200–400 cm^2	2–3 mL, pH 7.0–7.4	Low	Single to multiple, moderate
Nose	160 cm^2	Airway surface liquid 0.7–7 µL/cm^2, pH 5.5–7.4	Moderate	Multiple to single, excellent
Lungs	>70 m^2	Airway surface liquid 0.7–7 µL/cm^2, alveolar surface liquid approx. 0.02 µL/cm^2, pH 6.6–6.9	Moderate	Single, excellent
Skin	1.73 m^2	Negligible (water is 10–20% of stratum corneum by weight), pH 4.2–5.6	Moderate	Multiple, Very low
Vagina	65–107 cm^2	1 mL/hour premenopausal	Moderate	Multiple, moderate

Table A.5 Pharmaceutically important pHs

Description	pH
Blood	7.4
Cerebral spinal fluid	7.35
Skeletal muscle	7.15 (lower with exercise)
Subcutaneous tissue	7.35
Interstitial fluid	7.35
Tears	7.0–7.4
Nasal mucosa	5.5–7.4
Respiratory tract	6.6–6.9
Breast milk	7.0
Skin (stratum corneum)	4.2–5.6
Saliva	5.8–7.4
Stomach	1.0–3.5
Small intestine	5.0–7.4 (5.0 fed, 6.5 fasted)
Large intestine	6.4–7.0
Rectum	7.0–7.4
Vagina	3.5–4.9
Bile	6.8–8.0
Urine	4.6–8.0

Adapted from references 2, 20, 22–24.

Table A.6 Buffers and their pK_a

Buffer pair	pK_a (pH)
Phosphoric acid/monosodium phosphate	2.12
Monosodium phosphate/disodium phosphate	7.21
Disodium phosphate/trisodium phosphate	12.67
Citric acid/monosodium citrate	3.15
Monosodium citrate/disodium citrate	4.78
Disodium citrate/trisodium citrate	6.4
Acetic acid/sodium acetate	4.74
Carbonic acid/sodium bicarbonate	6.4
Sodium bicarbonate/sodium carbonate	10.3

References

1. X Liu, B Testa, A Fahr. Lipophilicity and its relationship with passive drug permeation. Pharm Res 2011; 28: 962–977.
2. NR Mathias, MA Hussain. Non-invasive systemic drug delivery: developability considerations for alternate routes of administration. J PharmSci 2010; 99: 1–20.
3. N Washington, C Washington, CG Wilson. Physiological Pharmaceutics, 2nd edn, London: Taylor & Francis, 2001.
4. L Illum. Nasal drug delivery-possibilities, problems and solutions. J Contr Rel 2003; 87: 187–198.
5. AA Hussain. Intranasal drug delivery. Adv Drug Del Rev 1998; 29: 39–49.
6. NR Labiris, MB Dolovich. Pulmonary Drug Delivery. Part 1: Physiological factors affecting therapeutic effectiveness of aerosolized medications. Br J Clin Pharmacol 2003; 56: 588–599.
7. RH Guy, J Hadgraft. Physicochemical aspects of percutaneous penetration and its enhancement. Pharm Res 1988; 5: 753–758.
8. YB Choi, MR Praunitz. The Rule of Five for non-oral routes of drug delivery: Ophthalmic, inhalation and transdermal. Pharm Res; 2011; 28: 943–948.
9. JJ Lochhead, RG Thorne. Intranasal delivery of biologics to the central nervous system. Adv Drug Del Rev 2012; 64: 614–628.
10. A Tronde, B Norden, H Marchner, et al. Pulmonary absorption rate and bioavailability of drugs in vivo in rats: structure-absorption relationships and physicochemical profiling of inhaled drugs. J Pharm Sci 2003; 92: 1216–1233.
11. JW Wiechers, AC Watkinson, SE Cross, MS Roberts. Predicting skin penetration of actives from complex cosmetic formulation: an evaluation of inter formulation and inter active effects during formulation optimization for transdermal delivery. Int J Cos Sci 2012; 1–11.
12. S Geinoz, RH Guy, B Testa, et al. Quantitative structure-permeation relationships to predict skin permeation: A critical evaluation. Pharm Res 2004; 21: 83–92.
13. JF Holter, JE Weiland, ER Pacht, et al. Protein permeability in the adult respiratory distress syndrome. J Clin Invest 1986; 78: 1513–1522.
14. Y Rojanasakul, LY Wang, M Bhat, et al. The transport barrier of epithelia: a comparative study on membrane permeability and charge selectivity in the rabbit. Pharm Res 1992; 9: 1029–1034.
15. USP Convention. United States Pharmacopeia, 34th edn, and National Formulary, 29th edn. Baltimore, MD: United Book Press, 2011.
16. RC Rowe, PJ Sheskey, ME Quinn (eds). Handbook of Pharmaceutical Excipients, 6th edn, London: Pharmaceutical Press, 2009.
17. FDA Inactive ingredient database. http://www.accessdata.fda.gov/scripts/cder/iig/index.cfm
18. H Wen, K Park (eds). Oral Controlled Release Formulations Design and Drug Delivery: Theory to Practice, Hoboken, NJ: Wiley, 2010.
19. VF Patel, F Liu, MB Brown. Advances in oral transmucosal drug delivery. J Contr Rel 2011; 153: 106–116.
20. AW Ng, A Bidani, TA Heming. Innate host defense of the lung: effects of lung-lining fluid pH. Lung 2004; 182: 297–317.
21. PB Pendergrass, MW Belovicz, CA Reeves. Surface area of the human vagina as measured from vinyl polysiloxane casts. Gynecol Obstet Invest 2003; 55: 110–103.
22. P Arora, S Sharma, S Garg. Permeability issues in nasal drug delivery. Drug Discovery Today; 2002; 7: 967–975.
23. JE Hall. Guyton and Hall Textbook of Medical Physiology, 12th edn, Philadelphia: Saunders, 2011.
24. N Washington, C Washington, CG Wilson. Physiological Pharmaceutics, 2nd edn, London: Taylor & Francis, 2001.

Answers to questions and cases

Chapter 1

Q1 (c) An ingredient that enhances the size, stability, palatability or release of a drug

Q2 (b) Resistant to microbial growth

Q3 (c) Movement of the drug from interaction with elements of the dosage form to interaction with molecules of biological fluid

Q4 (c) Citalopram, log P 3.5, MW 324

Q5 (a) Need for rapid absorption

Q6 (d) Preparation as a buccal tablet

Q7 (d) Chemical instability

Q8 (c) Diffusion

Q9 (c) Sustained release

Q10 (c) Intravenous, sublingual, oral, transdermal

Q11

Ibuprofen		
Design issue	**Drug property**	**Analysis**
Drug must be dissolved before it can cross membranes	Water solubility of 0.049 mg/mL, maximum dose 800 mg	Drug has poor water solubility, would require over 16 000 mL to dissolve the 800 mg tablet. Need to work on solubility!
Drug must have sufficient lipophilicity to cross membranes	Log P of $3.6 = \frac{\log [\text{conc}]\text{oil}}{[\text{conc}]\text{water}}$ Antilog(3.6) = 3981	This drug will favor partitioning into lipid membranes over water by a 3981:1 ratio

(continues)

Ibuprofen (continued)

Design issue	Drug property	Analysis
Large molecular weight drugs have difficulty moving across membranes	Formula weight 206	Based on molecular weight, ibuprofen should move easily through membranes
Drugs may hydrolyze, oxidize, photolyze or otherwise degrade in solution	This drug is quite stable in solution	Stability in the dosage form should not be an issue for this drug
Some drugs may be substantially degraded by enzymes before they reach their target	Metabolized in the liver, 80% of the dose is delivered to the systemic circulation	Some drug is lost to enzyme deactivation but most makes it to the blood
Intracellular drug receptors or receptors in the brain or posterior eye are challenging to target	Receptors (cyclooxygenase) are in peripheral inflamed tissues such as joints and in the brain (hypothalamus)	ibuprofen's lipophilicity is excellent for reaching peripheral receptors as well as the brain. Its size (molecular weight) is within the optimal range
Some drugs cause serious toxicity as a result of poor selectivity of drug action	Not applicable	

Amoxicillin

Design issue	Drug property	Analysis
Drug must be dissolved before it can cross membranes	Water solubility of 3430 mg/L, maximum dose 250 mg	Drug is soluble in the contents of the gastrointestinal tract, even better if the patient takes 250 mL water with the tablet
Drug must have sufficient lipophilicity to cross membranes	$\text{Log } P \text{ of } 0 = \dfrac{\log [\text{conc}]\text{oil}}{[\text{conc}]\text{water}}$ $\text{Antilog}(0) = 1$	This drug is equally soluble in water and lipid membranes. References state that amoxicillin is well distributed
Large molecular weight drugs have difficulty moving across membranes	Formula weight 365	Based on molecular weight, amoxicillin should move easily through membranes
Drugs may hydrolyze, oxidize, photolyze or otherwise degrade in solution	Not stable in solution. Hydrolyzes	Must be made as a lyophilized powder for suspension and suspended by the pharmacist just prior to use. May stabilize in solid dosage form
Some drugs cannot be used by mouth because of substantial losses to enzyme degradation	10% metabolized in the liver, 74–92% of the dose is delivered to the systemic circulation. Stable to acid!	Some drug is lost to enzyme deactivation but most of the drug makes it to the blood

(continues)

Amoxicillin (*continued*)

Design issue	Drug property	Analysis
Intracellular drug receptors or receptors in the brain or posterior eye are challenging to target	Drug targets are in bacterial cell wall in the location of the infection	Amoxicillin should be able to distribute to infected tissues based on its log P and molecular weight. It is marginal for distribution to the brain
Some drugs cause serious toxicity as a result of poor selectivity of drug action	Not applicable	

Q12

Adrenaline chloride injection multidose vial (King, other manufacturers)

Ingredient	Category	Function
Sodium bisulfite	Antioxidant	Chemical stability
Chlorobutanol	Antimicrobial preservative	Microbial stability
Water for injection	Solvent	Size, release

Lumigan (bimatoprost) ophthalmic solution

Ingredient	Category	Purpose
Benzalkonium chloride	Antimicrobial preservative	Microbial stability
Sodium chloride	Tonicity agent	Comfort
Dibasic sodium phosphate	Buffer	Chemical stability, comfort
Citric acid	Buffer, antioxidant	Chemical stability
Purified water	Solvent	Size, release

Donepezil orally disintegrating tablet (Zydus)

Ingredient	Category	Purpose
Crospovidone	Disintegrant	Release
Magnasweet	Sweetener	Palatability
Magnesium stearate	Lubricant	Physical uniformity
Mannitol	Sweetener	Palatability
Silicon dioxide (colloidal)	Glidant	Physical uniformity
Sucralose	Sweetener	Palatability

Depakene (valproic acid) oral solution		
Ingredient	**Category**	**Purpose**
Cherry flavor	Flavor	Palatability
FD&C Red No. 40	Dye	To match flavor
Glycerin	Sweetener, cosolvent	Palatability, size, release
Methyl and propyl parabens	Antimicrobial preservative	Microbial stability
Sorbitol	Sweetener, cosolvent	Palatability, size, release
Purified water	Solvent	Size, release
Sucrose	Sweetener	Palatability

Q13

Skin (local treatment) route	
Answer	**Challenge or opportunity?**
Stratum corneum, very accessible	Opportunity
Multiple layers, very poor permeability	Challenge
Moderate, 2 m^2	Neither
Negligible (water is 10–25% of stratum corneum by weight), pH 4.2–5.6	Challenge
Usually but not always. Some dosage forms are lost onto clothing or pets before they are absorbed	Equivocal
There is moderate enzyme activity in the skin including: peptidases, esterases, cytochrome P450. Losses generally do not create a barrier to drug absorption	Opportunity
Intermediate, 60–90 minutes	Neither

Chapter 2

Q1 (a) High melting points due to strong attractive forces between ions

Q2 (c) A solid that forms distinct orderly arrangements between molecules

Q3 (b) A solid that takes up water from the environment

Q4 (b) Metaproterenol > terbutaline > isoproterenol

Q5 (b) Drug dissolution rate

Q6 (c) Slow dissolution due to orderly arrangement of molecules within the solid

Q7 (a) Milling

Q8 (a) A large spherical shaped solid

Q9 (d) Partial spreading on the solid surface

Q10 (b) A drug molecule that has been chemically altered to change its physical properties

Q11 (e) Smaller surface area, slower dissolution, better flow

Q12 (c) Crush the tablet first and take with lots of water

Q13 The amorphous form of warfarin will have a better water solubility than the crystalline form, which will mean that the amorphous form will dissolve more rapidly and to a greater extent. Use of the amorphous form could increase the amount of warfarin absorbed and the rate at which it is absorbed resulting in a greater anticoagulant effect and possible bleeding and bruising in patients who switch from the crystalline form.

Q14 A number of calcium and magnesium salts can be used as desiccants; generally they are used in granular form. Other desiccant materials include silica and clays. There is a rechargeable desiccant material available that goes from pink to blue when it needs to be dried. (Dry Brik 2.) Another possibility is to weigh the desiccant before you put it in the container and then reweigh periodically to determine how much water has been absorbed. When it exceeds a preset percent of the original weight, replace and dry the used desiccant in an oven. (2 hours at 107°C/225°F.)

Chapter 3

Q1 Both (a) and (b)

Q2 (b) The solid particles to pull together

Q3 (a) Use the emulsifying agent with the HLB of 5 because it will orient mostly in the oil phase

Q4 (c) The pressure on the propellant ingredient must be decreased

Q5 (a) The interfacial (surface) tension between water and ketoprofen is high

Q6 (c) Loosely packed sediment

Q7 (d) Will not sediment but will coalesce

Q8 (d) Viscosity

Q9 (b) Hydrophilic colloids are highly associated until shaken

Q10 (c) Pseudoplastic

Q11 Theoretically dilatant vehicles would be most suitable because their viscosity increases with increased shear rate when we blink in response to something in our eye.

Q12

Sugar-free suspension structured vehicle, NF		
Ingredient	**Category**	**Quality**
Xanthan gum	Viscosity agent	Physical stability (uniformity)
Saccharin sodium	sweetener	Palatability
Potassium sorbate	Antimicrobial preservative	Microbial stability
Citric acid	Buffer, antioxidant, flavoring agent	Chemical stability, palatability
Sorbitol	Sweetener	Palatability
Mannitol	Sweetener	Palatability
Glycerin	Vehicle	Size of dose, convenience
Purified water	Vehicle	Size of dose, convenience

Soft hand cream		
Ingredient	**Category**	**Quality**
Methylparaben	Antimicrobial preservative	Microbial stability
Propyl paraben	Antimicrobial preservative	Microbial stability
Stearic acid	Emulsifier	Physical stability (uniformity)
Triethanolamine	Emulsifier	Physical stability (uniformity)
Glycerin	Vehicle	Size of dose, convenience
Mineral oil	Vehicle	Size of dose, convenience
Purified water	Vehicle	Size of dose, convenience

Carbomer aqueous jelly		
Ingredient	**Category**	**Quality**
Carbomer 934	Viscosity agent	Physical stability (uniformity)
Triethanolamine	pH adjuster, gelling agent	Physical stability (uniformity)

(continues)

Carbomer aqueous jelly (*continued*)		
Ingredient	**Category**	**Quality**
Methylparaben	Antimicrobial preservative	Microbial stability
Propyl paraben	Antimicrobial preservative	Microbial stability
Purified water	Vehicle	Size of dose, convenience

Q13 By convention we present the independent variable, that is, the variable that we manipulate in a study, on the x axis and the dependent variable, the variable that changes as a function of the independent variable, on the y axis. This allows us to visualize how flow is affected by changes in force. The advantage of using a presentation counter to this convention is that it allows us to visualize how viscosity changes with increasing force because it is reflected in the slope.

Chapter 4

Q1

Molarity of lidocaine HCl 20 mg/mL:

$$\frac{(20\ \text{g})(1\ \text{mole})}{(\text{L})} = \frac{0.069\ \text{mol/L lidocaine}}{(288.82\ \text{g})}$$

Calculating the osmolarity contributed by lidocaine HCl:

Osmolarity $= (1.79)(0.069\ \text{mol/L}) = 0.124$ osmol/L or 124 mosmol/L

Molarity of NaCl 6 mg/mL:

$$\frac{(6\ \text{g})(1\ \text{mole})}{(\text{L})} = \frac{0.1025\ \text{mol/L NaCl}}{(58.5\ \text{g})}$$

Calculating the osmolarity contributed by NaCl:

Osmolarity $= (1.823)(0.1025\ \text{mol/L}) = 0.187$ osmol/L or 187 mosmol/L

The osmolarities are additive:

124 mosmol/L + 187 mosmol/L $= 311$ mosmol/L

Q2 Vancomycin HCl

From the table:

1 g Vancomycin HCl equals 0.05 g sodium chloride

$$\frac{0.05 \text{ g NaCl}}{1 \text{ g VHCl}} (0.1 \text{ g VHCl}) = 0.005 \text{ g NaCl}$$

NaCl required to make 10 mL isotonic:

$$\frac{0.9 \text{ g NaCl}}{100 \text{ mL soln}} (10 \text{ g mL soln}) = 0.09 \text{ g NaCl}$$

Need to add 0.09 g – 0.005 g = 0.085 g = 85 mg sodium chloride

Q3

Methods of manipulating solubility					
Drug	pH	Cosolvents	Solubilizing agents	Amorphous solid	Derivatization
Carisoprodol	No	Yes	Yes	Yes	No
Digoxin	No	Yes	Yes	Yes	Yes
Ethambutol	Yes	Yes	Yes	Yes	Yes
Hydrocodone	Yes	Yes	Yes	Yes	No
Ibuprofen	Yes	Yes	Yes	Yes	Yes
Loratidine	Yes	Yes	Yes	Yes	No
Penicillin G	Yes	Yes	Yes	No*	yes
Prednisone	No	Yes	Yes	Yes	Yes
Spironolactone	No	Yes	Yes	Yes	No

*The Merck Index lists penicillin G as an amorphous solid so presumably its solubility could not be improved in this fashion because it is not crystalline.

Q4 *Bases*: amines. *Acids*: carboxylic acids, imides, phenols, sulfonamides, thiols

Q5

$$\frac{\text{Moles B:}}{\text{L}} = \frac{(0.7 \text{ mg})(1 \text{ g})}{(1 \text{ mL})(1000 \text{ mg})} \frac{(1000 \text{ mL})}{(1\text{L})} \frac{(1 \text{ mol})}{(645)} = \frac{(0.001085 \text{ mol/L B:})}{\text{L}}$$

$$\frac{\text{Moles B:H}^+}{\text{L}} = \frac{(50 \text{ mg})(1 \text{ g})}{(\text{mL})(1000 \text{ mg})} \frac{(1000 \text{ mL})}{(1 \text{ L})} \frac{(1 \text{ mol})}{(681.8 \text{ g})}$$

$$= \frac{(0.073 \text{ mol/L B:H}^+)}{\text{L}}$$

$$\text{pHP} = 6.56 + \frac{\log[0.001085]}{[0.073 - 0.001085]} = 6.56 - 1.828 = 4.73$$

Cannot be mixed with phenytoin injection.

Q6

$$\frac{\text{Moles HA}}{\text{L}} = \frac{(0.0159 \text{ mg})(1 \text{ g})}{(\text{mL})(1000 \text{ mg})} \frac{(1000 \text{ mL})}{(1 \text{ L})} \frac{(1 \text{ mol})}{(230 \text{ g})}$$

$$= \frac{(0.000069 \text{ mol/L HA})}{\text{L}}$$

$$\frac{\text{Moles A}^-}{\text{L}} = \frac{(220 \text{ mg})(1 \text{ g})}{(5 \text{ mL})(1000 \text{ mg})} \frac{(1000 \text{ mL})}{(1 \text{ L})} \frac{(1 \text{ mole})}{(252 \text{ g})} = \frac{(0.175 \text{ mol/L A}^-)}{\text{L}}$$

$$\text{pHP} = 4.2 + \log \frac{[0.175 - 0.000069]}{[0.000069]} = 4.2 + 3.4 = 7.6$$

Aromatic eriodictyon syrup with a pH of 8.0 is the only vehicle that would keep this drug in solution.

Q7 (b) B:H$^+$

Q8 (d) Chloride, sulfate or citrate

Q9 (d) The pH of fentanyl injection would be acidic because the B:H$^+$ donates protons to water creating hydronium ion

Q10 (c) Acetic acid/sodium acetate, pK_a 4.74

Q11 (c) Decrease the solubility owing to the common ion effect

Q12 (a) B:

Q13 (d) B:H$^+$ which favors water solubility

Q14 We want to acidify the urine so that it is in its B:H$^+$ form in the urine and cannot be reabsorbed from urine back into blood.

Q15 (d) A$^-$

Q16 (b) Sodium, potassium or calcium

Q17 (b) The pH·of dinoprostone injection would be basic because the A^- picks up protons from water creating hydroxyl ion

Q18 (b) Monosodium phosphate/disodium phosphate, pK_a 7.21

Q19 (c) HA

Q20 (c) The pH of dinoprostone gel would be acidic because the HA donates protons to water creating hydronium ion

Q21 (e) HA and A^- are approximately equal at this pH

Q22 In blood:

$$74 = 4.6 + \frac{\log [A^-]}{[HA]} = \frac{7.4-4.6}{[HA]} = \frac{2.8}{[HA]} = \frac{\log[A^-][A^-] \sim 630}{1}$$

In breast milk:

$$70 = 4.6 + \frac{\log [A^-]}{[HA]} = \frac{7.0-4.6}{[HA]} = \frac{2.4}{[HA]} = \frac{\log[A^-][A^-] \sim 251}{1}$$

There would be more dinoprostone in the blood than in the breast milk because the amount of A, which cannot diffuse across the membrane, is higher in blood than in the breast milk.

Q23 Calculate slope:

$$Slope = \frac{\Delta y}{\Delta x} = \frac{y_1 - y_2}{x_1 - x_2} = \frac{-1.93 - -4.097}{1-0} = 2.167$$

Our equation is:

$$-3.7 = -4.097 + 2.167fc$$

$$4097 - 3.7 = -4.097 + 4.097 + (2.167)fc$$

$$0397 = (2.167)fc$$

Dividing both sides by 2.167:

$$\frac{0397}{2167} = \frac{(2.167)fc}{(2.167)} = 0.183 = 18.3\%$$

Q24 (b) The three hydroxyls

Q25 (d) All of the above

Q26 (a) The hydroxyl

Q27 (b) Benzoate

Q28 (a) The hydroxyl

Q29 (d) All of the above

Chapter 5

Q1

Cloxacillin: Hydrolysis of lactam

Betamethasone: Oxidation of keto alcohol

Tetracycline: Oxidation of phenol or oxidation of keto alcohol

Pilocarpine: Hydrolysis of lactone

Sufentanil: Photolysis of benzylic N

Simvastatin: Hydrolysis of lactone, photo-oxidation of conjugated alkene

Q2 Hydrolysis of the ester

Q3

$$t_{90} = \frac{0.105}{k_1} = \frac{875 \text{ s } (1 \text{ minute})}{1.2 \times 10^{-4}} = \frac{14.6 \text{ minutes at pH 8}}{(60 \text{ s})}$$

$$t_{90} = \frac{0.105}{k_1} = \frac{0.105}{3.4 \times 10^{-10}}$$

$$= \frac{308\,823\,529 \text{ seconds}(1 \text{ minute})(1 \text{ hour})(1 \text{ day})(1 \text{ year})}{(60 \text{ seconds})(60 \text{ minutes})(24 \text{ hours})(365 \text{ day})}$$

$$= 9.8 \text{ years at pH 4.5}$$

Q4 Acetic acid/sodium acetate (pK_a 4.74), monosodium citrate/disodium citrate (pK_a 4.78) or benzoic acid/sodium benzoate (pK_a 4.2) could all produce suitable buffers around pH 4.5 which is the pH at which cocaine is more stable.

Q5 The solution with a concentration of 40 mg/mL would degrade faster than the solution with the concentration of 20 mg/mL because the rate of degradation in solutions is concentration dependent. This means that the higher the concentration of drug, the higher the rate of degradation.

Q6

$$\Delta T = T_2 - T_1 = -20°C - 20°C = -40°$$

$$\frac{k_2}{k_1} = Q_{\Delta T} - 2^{\Delta T/10} = 2^{-40/10} = 2^{-4} = 0.0625$$

$$\frac{T_{90T2} = T_{90T1}}{Q_{\Delta T}} = \frac{24 \text{ hours}}{0.0625} = \frac{384 \text{ hours}(1 \text{ day})}{24 \text{ hours}} = 16 \text{ days}$$

Q7

$$\Delta T = T_2 - T_1 = 25°C - 5°C = 20°C$$

$$\frac{k_2}{k_1} = Q_{\Delta T} - 3^{\Delta T/10} = 3^{20/10} = 3^2 = 9$$

6 hours (9) = 54 hours or about 2.25 days lost from the shelf life

Q8

$$T_1 = 25°C \quad T_2 = 5°C \quad \Delta T = -20°C$$

For E_a of 19.4 kcal/mol:

$$\frac{k_2}{k_1} = Q_{\Delta T} = Q_{10}^{\Delta T/10} = 3^{-20/10} = 3^{-2} = -0.111$$

The new shelf life would be:

$$T_{90T2} = \frac{T_{90T1}}{Q_{\Delta T}} = \frac{7 \text{ days}}{0.111} = 63 \text{ days}$$

Q9

$$t_{90} = \frac{0.105}{k_1} = \frac{0.105}{1.65 \times 10^{-9}}$$

$$= \frac{63\,636\,364 \text{ seconds(1 minute)(1 hour)(1 day)(1 year)}}{(60 \text{ seconds})(60 \text{ minutes})(24 \text{ hours})(365 \text{ day})} = 2.02 \text{ years}$$

Q10

$$\Delta T = T_2 - T_1 = 35°C - 5°C = 30°C$$

$$\frac{k_2}{k_1} = Q_{\Delta T} = 4^{\Delta T/10} = 4^{30/10} = 4^3 = 64$$

1 hour (64) = 64 hours or about 2.67 days lost from the shelf life

Q11

Venlafaxine

Oxidation (photo) of the benzylic ether

Formulation:

- Lower pH with buffers or HCl
- Use antioxidants such as sodium bisulfite or ascorbic acid
- Make suspension

Storage:

- Store in refrigerator before opening
- Store under nitrogen

Packaging:

- Pack in amber or other light-excluding material
- Pack in glass or polyvinyl chloride

Q12

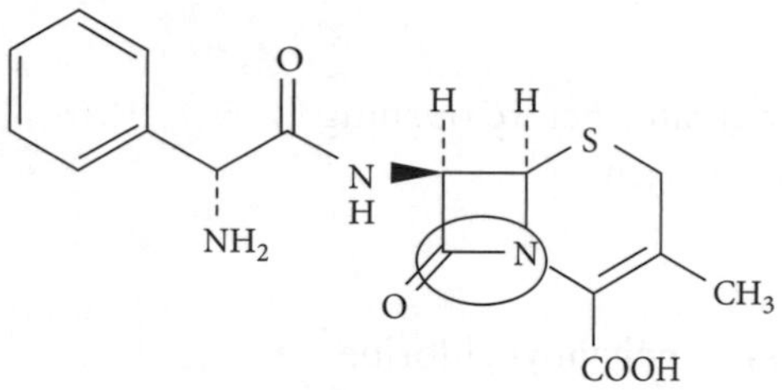

Cephalexin

Hydrolysis of lactam

Formulation:

- Use buffers
- Make a suspension
- Make powder for reconstitution

Storage:

- Refrigerate liquid forms
- Do not refrigerate solid forms

Packaging:

- Package in glass or HDPE

Q13 Hydrolysis of the imide

Formulation:

- Use buffers
- Make suspension
- Make powder for reconstitution

Storage:

- Refrigerate liquid forms
- Do not refrigerate solid forms

Packaging:

- Package in glass or HDPE

Q14

Formulation:

- Lower pH with buffers or HCl
- Use antioxidants such as ascorbic acid
- Make suspension
- Make powder for reconstitution

Storage:

- Store in refrigerator before opening
- Store under nitrogen

Packaging:

- Pack in glass or polyvinyl chloride

Q15 Hydrolysis of the lactam

Formulation:

- Use buffers
- Make suspension
- Make powder for reconstitution

Storage:

- Refrigerate liquid forms
- Do not refrigerate solid forms

Packaging:

- Package in glass or HDPE

Q16 Oxidation of the thiol

Formulation:

- Lower pH with buffers or HCl
- Use antioxidants such as sodium bisulfite or ascorbic acid
- Remove oxygen from vehicle and head space

Storage:

● Store under nitrogen

Packaging:

● Pack in glass

Q17 Hydrolysis of the ester, photolysis of the benzylic carbon

Buffer with citric acid, sodium citrate, pK_a 3.15, store under refrigeration in amber HDPE bottle.

Chapter 6

Q1 (b) The small intestine pH 6.5 – 7.6

Q2 (a) Absorption will be good because log P < 5, MW < 500, there are fewer than 5 hydrogen bond donors and 10 hydrogen bond acceptors

Q3 (b) Unbound drug will move into the brain via passive diffusion

Q4 (b) Cephalexin reaches its site of elimination before it reaches the site of the infection

Q5 (a) Easily crosses the blood–brain barrier

Q6 (a) The amount of felodipine in the blood increases

Q7 (c) Atropine is more likely to distribute to the CNS because it is more lipophilic

Q8 (b) The amount of nifedipine decreases because the number of metabolic enzymes increases

Q9 (b) The amount of methotrexate increases because amoxicillin decreases its secretion

Q10 (c) Water would reduce the viscosity of the dissolution medium (D)

Q11 (d) The salt form would increase the saturation solubility of the drug (C_s)

Q12 (b) The ingredient would decrease the area exposed to solvent (A)

Q13 (a) The condition would increase the width of the boundary layer (h)

Q14 (a) Cannot cross membranes

Q15 (a) Percentage of drug absorbed will remain constant because diffusion is not saturable

Q16 (c) Percentage of drug absorbed will go down because carrier mediated influx is saturable

Q17 (c) Percentage of drug absorbed will go down because carrier mediated efflux is saturable

Q18

Drug	Log CaCO$_2$ permeability coefficient	Log partition coefficient
Amoxicillin	−6.1	0
Acyclovir	−6.15	−1.56
Clonidine	−4.59	2.7
Diazepam	−4.32	2.9
Enalapril	−5.64	2.1
Felodipine	−4.64	3.8
Furosemide	−6.5	1.4
Hydrochlorothiazide	−6.06	−0.5
Ibuprofen	−4.28	3.6
Labetalol	−5.03	2.7
Methotrexate	−5.92	−2.2
Metoprolol	−4.59	1.6
Naproxen	−4.83	2.8
Phenytoin	−4.57	2.2
Ranitidine	−6.31	1.3
Verapamil	−4.58	4.7
Warfarin	−4.68	3

Q19 Decreased OCT1 activity decreases the movement of metformin from blood to active site and reduces the therapeutic effect (blood sugar and hemoglobin A1C will remain elevated) compared with a diabetic patient with a normal transporter. You will need to increase the patient's dose to get the same therapeutic effect compared with a diabetic patient with a normal transporter.

Q20 Rifampin is an inducer of CYP1A2, which deactivates warfarin. The patient's warfarin dose was carefully adjusted while she was taking the inducer drug and now that she has discontinued the rifampin she will need to have her warfarin dose adjusted downward.

Substrate drug → lower blood levels → reduced effect → higher dose required → higher blood level → toxicity

Drug inducer → faster metabolism → discontinue inducer → normal metabolism

Q21 Decreased albumin increases the unbound fractions of furosemide, valproic acid, and lovastatin. This will increase access to each drug's receptor and to elimination. Increased access to the receptor will increase the therapeutic effect but increased access to eliminating organs will likely offset (decrease) the therapeutic effect. Because you cannot predict which effect will predominate, monitor each of the bound drugs for both efficacy and toxicity.

Drug	Bound (%)	Target location	Elimination	Monitoring
Furosemide	97%	Kidney	Kidney	Blood pressure, dizziness
Atenolol	10%	Heart, vascular smooth muscle	Kidney	Not applicable
Valproic acid	90%	Brain	Liver	Agitation, sedation
Lovastatin (active metabolite)	>95%	Liver	Liver	Fasting lipids, muscle pain

Q22 Dilacor XR is an extended release capsule containing an entire day's medication that should be swallowed whole and not opened, crushed or chewed. When Mrs Lawson opens the capsule into apple sauce this releases a full day's medication at once resulting in high diltiazem levels and dizziness. Diltiazem is not available as an oral liquid which would be easier to swallow. It is available as an immediate release tablet that could be crushed and administered in apple sauce; however, the daily dose would need to be divided into three doses and (80 mg) given every 8 hours.

Chapter 7

Q1 (b) AB_1

Q2 (b) The product is judged bioequivalent based on *in vivo* data submitted

Q3 (a) Wellbutrin SR

Q4 (c) Zyban 150 mg ER tablets by GSK

Q5 (d) Bupropion HCl 150 mg ER tabs by Watson (A079094)

Q6 (c) Provera

Q7 (d) The product has potential bioequivalence problems

Q8 (c) Coumadin contains different inactive ingredients than the Barr warfarin product but is otherwise equivalent

Q9 (c) No, one is a capsule and one is a tablet and (d) one is immediate release and one is extended release

Q10 (c) No, one is a succinate and the other is a tartrate and (d) one is extended release and the other is immediate release

Q11 (b) Dilt-CD and diltiazem HCl by Valeant (A075116)

Q12 Mylan levothyroxine sodium has been rated equivalent to all of the reference listed drugs available: AB1, AB2, AB3, AB4. While hospitals are not bound by the same substitution laws as community pharmacies, it is preferable to substitute a product that will produce equivalent blood levels and therapeutic effect.

Q13

Drug	Highest strength oral dosage form	Water solubility	Water soluble by USP criteria	Water soluble by BCS criteria
Prednisone	50 mg	312 mg/L	No	Yes
Levonorgestrel	1.5 mg	2.05 mg/L	No	No
Griseofulvin	500 mg	8.64 mg/L	No	No
Simvastatin	80 mg	0.76 mg/L	No	No
Digoxin	0.25 mg	0.0648 mg/mL	No	Yes

Q14

No: $\dfrac{(22.2 \text{ mg MPA})}{(1000 \text{ mL})}(250 \text{ mL}) = 5.55$ mg MPA should dissolve in 250 mL

Solubility will not be different at different pH because the drug is not an acid or base.

The log P compared with the class defining metoprolol (1.88) indicates that it is highly permeable.

This drug is highly lipophilic so likely to be permeable and highly absorbed. It has low bioavailability because it is highly metabolized. Permeability and fraction dose absorbed take into account the parent drug and metabolites.

The drug is BCS Class 2, low solubility, high permeability. It is likely dissolution rate limited.

The FDA will require *in vivo* bioequivalence data for this drug.

Q15

No: $\dfrac{(2.5 \text{ mg A})(250 \text{ mL})}{(\text{mL})} = 625 \text{ mg A should dissolve in } 250 \text{ mL}$

The solubility is likely to be better at 1.2 when the amine is charged and at 6.5 when there is more enolate ion because this drug is both a weak base and a weak acid.

The fraction absorbed (0.2) compared with the class-defining metoprolol (0.95) indicates that acyclovir is a low permeability drug. The log P_{app} (−6.15) compared with the class-defining metoprolol (−4.59) indicates that acyclovir has low permeability.

Rank highest accuracy to lowest.

P_{eff} from human intestinal perfusion experiments, fraction absorbed, P_{eff} from human excised intestinal membrane, bioavailability with metabolite data, P_{app} from monolayers or P_{eff} from animal perfusion studies, log D and log P.

This drug is BCS Class IV. It is limited by both dissolution rate and permeability.

The drug is BDDCS Class IV. It is likely eliminated as unchanged drug in the urine or bile.

Q16

Yes: $\dfrac{(500 \text{ mg M})(250 \text{ mL})}{(\text{mL})} = 125\,000 \text{ mg M should dissolve in } 250 \text{ mL}$

The solubility is likely to be better at 1.2, 4.5 and 6.5 when the amine is charged because this drug is a weak base.

The fraction absorbed (0.53) compared with the class-defining metoprolol (0.95) indicates that metformin is a low permeability drug.

This drug is BCS Class III. It is limited by permeability.

The drug is BDDCS Class III. It is likely eliminated as unchanged drug in the urine or bile.

Q17 (c) Side effects may occur due to higher drug levels

Q18 (b) T_{max}

Chapter 8

Q1 (c) Predictable drug effects

Q2 (b) Phlebitis

Q3 (d) Solubility decreases, onset is slower and duration is longer

Q4 (c) Zinc and protamine are added to increase its tendency to aggregate

Q5 (c) Sterile saline

Q6 (b) The drug is poorly soluble in water

Q7 (a) The drug is not stable in water

Q8 (b) Reduces particle count per cubic foot to no more than 100

Q9 (c) Jet injector

Q10 (d) Insert the needle at a 45° angle

Diprovan:

Q11 Class of dosage form: parenteral emulsion

Q12 Prompt or sustained release: prompt, semiprompt

Q13 Route(s) of administration: IV, IM, SC

Q14

Diprivan injection	
	Purpose of ingredient
Soybean oil 100 mg	Oil phase vehicle
Glycerin 22.5 mg	Tonicity agent
Lecithin 12 mg	Emulsifier
Disodium EDTA 0.05 mg	Chelating agent
Water for injection 1 mL	Vehicle/solvent

Procaine penicillin:

Q11 Class of dosage form: parenteral suspension

Q12 Prompt or sustained release: sustained

Q13 Route(s) of administration: IM, SC

Q14

Procaine penicillin injection	
	Purpose of ingredient
Procaine penicillin 300 000 U	Active ingredient
Disodium citrate	Buffer
Trisodium citrate	Buffer
Lecithin 5 mg	Flocculating agent
Carboxymethylcellulose 5 mg	Suspending agent
Povidone 5 mg	Suspending agent
Propyl paraben 0.01 mg	Antimicrobial preservative
Methyl paraben 1 mg	Antimicrobial preservative
Water for injection qs 1 mL	Solvent/vehicle

Haldol:

Q11 Class of dosage form: oily parenteral solution

Q12 Prompt or sustained release: sustained

Q13 Route(s) of administration: IM

Q14

Haldol deconate injection	
	Purpose of ingredient
Haldoperidol deconate 70.52 mg (50 mg haldol)	Active ingredient
Benzyl alcohol 12 mg	Local anesthetic or preservative
Sesame oil 1 mL	Solvent/vehicle

Herceptin:

Q11 Class of dosage form: dry powder for reconstitution

Q12 Prompt or sustained release: prompt

Q13 Route(s) of administration: IV, IM, SC

Q14

Herceptin vials	
	Purpose of ingredient
Trastuzumab 440 mg	Active ingredient
L histidine HCl 9.9 mg	Buffer
L histidine 6.4 mg	Buffer
Trehalose 400 mg	Protectant
Polysorbate 20 1.8 mg	Wetting agent

Q15 Consider each of the drugs with a separate calculation. Acyclovir is a weak acid, HA:

$$\frac{\text{Moles HA}}{\text{L}} = \frac{(2.5 \text{ mg})(1 \text{ g})}{(\text{mL})(1000 \text{ mg})} \frac{(1000 \text{ mL})}{(1 \text{ L})} \frac{(1 \text{ mole})}{(225.2 \text{ g})} = \frac{(0.011 \text{ M HA})}{\text{L}}$$

$$\frac{\text{Moles A}^-}{\text{L}} = \frac{(50 \text{ mg})(1 \text{ g})}{(\text{mL})(1000 \text{ mg})} \frac{(1000 \text{ mL})}{(1 \text{ L})} \frac{(1 \text{ mole})}{(248.2 \text{ g})} = \frac{(0.201 \text{ M A}^-)}{\text{L}}$$

$$\text{pH}_\text{P} = 9.25 + \log\frac{[0.201 - 0.011]}{[0.011]} = 9.25 + 1.24 = 10.49$$

Acyclovir will precipitate at pH 10.49 and below.

Granisetron is a weak base, B:

$$\frac{\text{Moles B}}{\text{L}} = \frac{(0.434 \text{ mg})(1 \text{ g})}{(\text{mL})(1000 \text{ mg})} \frac{(1000 \text{ mL})}{(1 \text{ L})} \frac{(1 \text{ mole})}{(312.4)} = \frac{(0.00139 \text{ M B})}{\text{L}}$$

$$\frac{\text{Moles B:H+}}{\text{L}} = \frac{(1.12 \text{ mg})(1 \text{ g})}{(\text{mL})(1000 \text{ mg})} \frac{(1000 \text{ mL})}{(1 \text{ L})} \frac{(1 \text{ mole})}{(348.87 \text{ g})}$$

$$= \frac{(0.00321 \text{ M B:H}^+)}{\text{L}}$$

$$\text{pH}_\text{P} = 9.79 + \frac{\log[0.00139]}{[0.0032 - 0.00139]} = 9.79 - 0.115 = 9.56$$

Granisetron will precipitate at pH 9.56 and above. Both drugs are likely to precipitate if mixed. If data are not available to permit calculations, the pharmacist may mix small amounts of the injections in a syringe to determine the result.

Q16 *Dopamine:*

Formulation: sodium metabisulfite, citric acid, sodium citrate, water for injection; pH 3.3

Sodium bicarbonate:

Formulation: sodium bicarbonate in water for injection; pH 8.0

There are no cosolvents or surfactants.

Yes, sodium bicarbonate is a carbonate buffer.

The amine group on dopamine is a base in uncharged form and the phenols are acids. Sodium bicarbonate in uncharged form is carbonic acid and, therefore, an acid.

Because dopamine has acidic and basic function groups it will be most soluble at a range of pHs. As an acid sodium bicarbonate should be more soluble at basic pH.

Neither product is a powder for reconstitution.

The potential degradative pathway for dopamine is photooxidation – slowest rate in acidic pH. Sodium bicarbonate will release carbon dioxide gas at acidic pH.

Dopamine is most stable at pH 5 or below. Sodium bicarbonate is most stable at the pH it is formulated (pH of greatest stability not specified).

The potential problems are oxidation of dopamine owing to increased pH and degradation of sodium bicarbonate by conversion to carbon dioxide.

Q17 Ganciclovir formulation: Powder for reconstitution, reconstitute with sterile water for injection, pH 11.

Gemcitabine formulation: Powder for reconstitution containing mannitol, sodium acetate. Reconstitute with 0.9% sodium chloride, pH 3.0.

There are no cosolvents or surfactants.

There are no carbonate buffers.

Ganciclovir has an enol (acid) and an amine (basic), gemcitabine has an amine (base).

Ganciclovir can be soluble at either acidic (protonated amine) or basic (sodium enolate) pH, gemcitabine is more soluble at acidic pH.

Yes, both products are powders for reconstitution.

The potential degradative pathway for ganciclovir is hydrolysis of cyclic amide and photolysis of conjugated system/benzylic NH. For gemcitabine it is hydrolysis of cyclic carbamide and photolysis of conjugated system.

The pH at which the drug is most stable is not specified in references. Photolysis of both drugs is slower at acidic pH, hydrolysis not predicable in this case.

The potential problem are: chemical instability; both drugs are powders for reconstitution; loss of solubility for gemcitabine owing to increase in the pH of the mixture.

Q18 Amiodarone formulation: Polysorbate 80, benzyl alcohol in water for injection, pH 4.08.

Furosemide formulation: Sodium chloride, water for injection, pH 8.6.

No cosolvents; amiodarone contains a surfactant (polysorbate 80).

There are no carbonate buffers.

Amiodarone is a base (amine), furosemide is an acid (carboxylic acid, sulfonamide) and base (amine).

Amiodarone will be most soluble in acidic pH, furosemide will be soluble at acidic or basic pH.

Neither product is a powder for reconstitution.

The potential degradative pathway for amiodarone is photooxidation; for furosemide, photooxidation.

The pH at which the drug is most stable is not specified for amiodarone, but it is likely most stable at the pH of formulation; furosemide is most stable above pH 5.5.

The potential problems are: precipitation due to dilution of surfactant solubilizing amiodarone; loss of chemical stability of both owing to change in pH (photocatalyzed oxidation of benzylic positions in amiodarone and furosemide).

Q19

- Wash hands. Remove from refrigerator, swab diluent and powder vials, remove 1.1 mL. Inject the diluent into the powder vial and roll gently until dissolved. Draw contents of vial (assuming 33 μg dose) into syringe.
- Point out the different injection sites and the need to rotate the site used.
- Gently tap the site to be injected to stimulate nerve endings and minimize initial pain.
- Cleanse area with alcohol swab.
- Stretch the skin taut with one hand for easier needle insertion.
- Using the other hand, insert the needle at a 90° angle using a quick, dart-like thrust.
- Inject medication at a slow, even rate.

- Withdraw needle rapidly and press area with cotton ball.
- Dispose of syringe and needle in sharps container.

It is a single dose vial and must be discarded.

Q20 The onset could be as soon as 5–10 minutes but may be as long as 20 minutes depending primarily on where the patient injects the drug. Most patients give Imitrex into the thigh which is slower than the deltoid. He should wait at least an hour because it may be anywhere from 20 minutes to 1 hour before he can know the maximum effects of the drug.

Q21 Regular insulin formulation: glycerin, phenol, water for injection, pH 7.4.

Lantus formulation: cresol, water for injection, pH 4.0

Lantus should not be mixed with regular (or any other) insulin because its solubility is pH dependent. At pH 7.4 it will form microprecipitates in body tissues that slowly dissolve and produce its characteristic long duration. If mixed with regular insulin it will precipitate in the syringe. A study by Kaplan et al. (Effects of mixing glargine and short acting insulin analogs on glucose control. Diabetes Care 2004; 27: 2739–2740) explored the effect of mixing glargine and insulin aspart and insulin lispro, and found that while these combinations indeed resulted in precipitation, their use did not affect the patients' glucose control. The study could be used to discuss why precipitation of glargine in the syringe would not necessarily interfere with its therapeutic effect.

Q22 The best product would be the 100 µg/mL single dose vials. The other sizes would last the patient 10 or 30 months and could be destabilized chemically and microbially by multiple additions of air.

Chapter 9

Q1 (b) This reduces their elimination

Q2 (c) Attachment of monoclonal antibodies

Q3 (a) Attachment of polyethylene glycol

Q4 (c) Polyethylene glycol

Q5 (c) Reduce the frequency of dosing

Q6 (a) Is faster from small particles than large particles

Q7 (d) All of the above

Q8 (b) Lentivirus because its duration of gene transfer is stable

Q9 The liver has sinusoidal capillaries that allow it to take up very large particles into the blood supply. Order liver function panel (alkaline phosphatase, aspartate aminotransferase (AST), alanine aminotransferase (ALT), albumin and bilirubin) to monitor therapy.

Q10 The use of colloidal formulations produces prolonged levels of drug in the body as evidenced by the longer half lives of the drugs in the colloidal products. The colloidal (liposomal) doxorubicin product provides a larger extent of exposure to drug as evidenced by the larger AUC. We might predict that the colloidal product would produce greater anticancer activity than the conventional product. The colloidal (Depo) morphine has a lower peak concentration although the overall extent of exposure (AUC) to morphine from the colloidal product is roughly comparable with the conventional product. We could predict that the pain relieving effects would last longer and the side effects of the drug would be less prominent.

Q11 Reduction in the peak concentration without affecting the area under the curve should decrease the toxicity of the drug compared with delivery in its conventional dosage form.

Chapter 10

Q1 (c) A large proportion of drug applied to the eye is lost before it reaches the receptors

Q2 (a) Adequate lipid and water solubility

Q3 (e) Answers (b) and (c) only

Q4 (d) Wait 5 minutes before applying a second drop to the same eye

Q5

Timolol maleate for administration to the anterior portion of the eye		
Design issue	**Drug property**	**Analysis**
Drug must be dissolved before it can cross membranes	Water solubility: 2.74 mg/mL Maximum concentration: 0.5%	This water solubility must be for timolol and not timolol maleate
Drug must have sufficient lipophilicity to cross membranes	Log P 1.2	Good range for the cornea
Large molecular weight drugs have difficulty moving across membranes	Formula weight: 316	Good for corneal permeability

(continues)

Timolol maleate for administration to the anterior portion of the eye (continued)

Design issue	Drug property	Analysis
Drugs may hydrolyze, oxidize, photolyze, or otherwise degrade in solution	Stability in solution: sensitive to light in solution	Protect solution from light
Is the enzymatic degradation of this drug substantial enough to prevent its use by this route?	Enzymatic degradation: Primarily hepatic (80%) via the cytochrome P450 2D6 isoenzyme. Oral bioavailability is approximately 60%	Because enzyme activity in the eye is only about 4% of that in the liver, this drug will have good availability in the anterior eye
Intracellular drug receptors or receptors in the brain or posterior eye are challenging to target	Receptor location: beta adrenergic receptor on the surface of ciliary body cells	Cell surface receptor, local treatment
Some drugs cause serious toxicity as a result of poor selectivity of drug action	Systemic toxicities: low heart rate, dizziness, hypoglycemia, bronchospasm, constipation	Not applicable because the toxicity of timolol is not serious and the availability to the systemic circulation will be low

Q6

Timolol maleate ophthalmic gel (Timpoptic XE)

Ingredient	Category	Purpose of ingredient
Benzododecinium bromide	Antimicrobial preservative	Microbial stability
Mannitol	Tonicity agent	Comfort
Tromethamine	Buffer, pH modifier	Chemical stability, comfort
Gellan gum	Cation activated gelling agent	Retention
Water for injection	Solvent	Size, release

Q7

Ophthalmic route – anterior segment

Questions	Answer	Challenge or opportunity?
What is the membrane that will absorb the drug from this route and how accessible is it?	Drug is absorbed through the cornea and sclera. It is accessible.	Opportunity
Is the absorbing membrane single or multiple layers and how permeable is it?	Multiple layers of stratified squamous epithelium with moderate permeability	Challenge
How large is its surface?	Very small, 1.3–1.5 cm^2	Challenge

(continues)

Ophthalmic route – anterior segment (*continued*)		
Questions	**Answer**	**Challenge or opportunity?**
What is the nature of the body fluids that bathe the absorbing membrane?	Tears: pH 7.0–7.4, small volume: 7–10 µL	Challenge
Will the dosage form remain at the absorbing membrane long enough to release its drug?	This is a big problem for ophthalmic medications administered to the front of the eye	Challenge
Will the drug encounter enzymes or extremes of pH that can alter its chemical structure before it is distributed to its target?	There are a few cytochrome P450s in the eye but their activity is much lower than in the liver	Opportunity
What is the blood flow primacy to the absorbing membrane and the distribution time from it?	Drugs are not distributed in the eye via the blood so this does not apply to this route	Neither

Q8　The liquid will be shear thinning – this is actually preferred for ophthalmic products in that Newtonian vehicles do not shear thin and become uncomfortable with blinking.

Q9　This product is made with hydroxypropylmethylcellulose, a gelling agent that is liquid at room temperature but gels at body temperature. The temperature inside a car in the summer can easily exceed body temperature thus the product has gelled in the container. It should return to normal viscosity if stored at room temperature. Perhaps the drops should be kept in the cooler or in the tent while camping. Be sure to wash your hands!

Q10　Products made for the eye are sterile and ointments have strict standards on the number of particles greater than 50 µm in diameter per tube. In addition a tube of ophthalmic ointment will have a narrow opening that facilitates the extrusion of a thin ribbon of ointment. Free clinics will differ from location to location but the pharmacist should make herself aware of what options are available in the area where she practices. Many of the larger drug manufacturers have patient-assistance programs which can be accessed through the website NeedyMeds.org (http://www.needymeds.org/). Alcon Laboratories has a program that provides several antibiotic ophthalmic ointments without charge to qualifying patients. However, because it may take as long as 4 weeks for the patient to prepare, submit and be approved for free medication, this approach may not be practical in this situation. More information on medication assistance programs is available from the following references:

TM Felder, NR Palmer, LS Lal, et al. What is the evidence for pharmaceutical patient assistance programs? A systematic review. J Health Care Poor Underserved. 2011; 22: 24–49. Available from NIH Public Access at http://www.ncbi.nlm.nih.gov/pmc/articles/PMC3065996/pdf/nihms276308.pdf.

PE Johnson. Patient assistance programs and patient advocacy foundations: Alternatives for obtaining prescription medications when insurance fails. Am J Health-Sys Pharm 2006; 63: Suppl 7: S13–S17.

Chapter 11

Q1

Design issue	Drug property	Analysis
Drug must be dissolved before it can cross membranes	Hydrochloride salt is freely soluble in water	This drug has good water solubility for rapid dissolution in the mouth
Drug must have sufficient lipophilicity to cross membranes	Log P 2.7	This drug has good lipophilicity for absorption through the oral cavity
Large molecular weight drugs have difficulty moving across membranes	Molecular weight 296.41	This drug is within the molecular weight range for use in the oral cavity
Drugs may hydrolyze, oxidize, photolyze or otherwise degrade in solution	This drug has a benzylic carbonyl that is potentially photo-oxidizable	Will need to protect from light. Perhaps use antioxidants
Some drugs cannot be used by mouth because of substantial losses to enzyme degradation	This drug is about 50% metabolized by enzymes in the liver. Oral bioavailability is unknown because the drug is only available IV	Administration in the oral cavity should delay the exposure of the drug to hepatic enzymes
Intracellular drug receptors or receptors in the brain or posterior eye are challenging to target	Cell surface 5HT3 receptors in the gastrointestinal tract.	Not an issue
Some drugs cause serious toxicity as a result of poor selectivity of drug action	Not applicable	

From ASHP, Lexicomp.

Q2

Ingredient	Category	Quality
Gelatin	Binder	Contributes to physical stability
Mannitol	Diluent, sweetener	Contributes to size of dose. Enhances palatability
Sucralose	Sweetener	Enhances palatability
Black cherry flavor	Flavor	Enhances palatability

Q3

Oral cavity route		
Questions	**Answer**	**Challenge or opportunity?**
What is the membrane that will absorb the drug from this route and how accessible is it?	Drug is absorbed through the sublingual or the buccal epithelia. Both of these areas are easily accessible to the placement of dosage forms	Opportunity
Is the absorbing membrane single or multiple layers and how permeable is it?	Multiple, modest permeability	Challenge
How large is its surface?	Relatively small, 100–200 cm^2	Neither
What is the nature of the body fluids that bathe the absorbing membrane?	Oral cavity fluids: pH 5.8–7.4, relatively small volume (0.9–1.1 mL)	Challenge
Will the dosage form remain at the absorbing membrane long enough to release its drug?	Fluids in the mouth are rapidly swallowed creating the need to have the dosage form dissolve rapidly or stick to the oral mucosa	Challenge
Will the drug encounter enzymes or extremes of pH that can alter its chemical structure before it is distributed to its target?	There are mainly esterases and peptidases in the mouth so metabolism is not a significant problem for drugs administered to the oral cavity	Opportunity
What is the blood flow primacy to the absorbing membrane and the distribution time from it?	Highly perfused, < 2 minutes	Opportunity

Q4

Drug, strength	Oral bioavailability (%)	Plan for using sublingually
Lorazepam 2 mg/mL	90	Start with 0.5 mL (1 mg). Evaluate response after 5–10 minutes. Monitor for respiratory depression
Haloperidol 2 mg/mL	60–70	Use 2 mg × 0.7 = 1.4 mg or 0.7 mL sublingually. Evaluate after 5–10 minutes, monitor vital signs
Diazepam 5 mg/mL	85–100	Start with 0.5 mL (2.5 mg). Evaluate response after 5–10 minutes. Monitor for respiratory depression

Q5

Reduced salivary volume will reduce the dissolution rate of the tablet resulting in delayed absorption. The patient could rinse her mouth with water just prior to placing the tablet under her tongue. However, sublingual tablets should not be taken with water since that increases the likelihood that they will be swallowed.

Parameter	Intravenously	Sublingual pH 4	Sublingual pH 9
C_{max} (ng/mL)	185	64.88	95.24
T_{max} (hours)	0	0.33	0.33
$T_{1/2}$ (hours)	1.13	1.00	1.11
AUC (ng · minute/mL)	12 791	5807	8965.3
AUC_{SL}/AUC_{IV}	1.00	0.454	0.701

Data from AAPS PharmSciTech 2006; 7: 1.

Q6 (c) Absorption was greater from the pH 9 formulation because there was more membrane soluble form at this pH

Q7 (d) Orajel Severe Pain Formula because it contains a mucoadhesive agent and oral pain relief topical gel does not

Q8 (b) Rapid, sustained release

Q9 (d) Answers (b) and (c)

Q10 (d) Mucoadhesion

Q11 (c) The gums are the least mobile surface

Q12

Lorazepam 0.5 mg sublingual troches		
Ingredient	Category	Quality
Gelatin base	Base, mucoadhesive	Provides physical stability, size of dose, retention of dosage form
Silica gel	Dispersing aid	Enhances uniformity
Aspartame	Sweetener	Enhances palatability
Acacia powder	Thickener	Provides physical stability
Citric acid monohydrate	Tartness, antioxidant	Provides palatability, chemical stability
Orange oil	Flavor	Enhances palatability

Q13

Palonosetron 0.25 sublingual troches		
Ingredient	**Category**	**Quality**
Gelatin base	Base, mucoadhesive	Provides physical stability, size of dose, retention of dosage form
Silica gel	Dispersing aid	Enhances uniformity
Aspartame	Sweetener	Enhances palatability
Acacia powder	Thickener	Provides physical stability
Citric acid monohydrate	Tartness, antioxidant	Provides palatability, chemical stability
Orange oil	Flavor	Enhances palatability
Edetate disodium	Antioxident	Provides chemical stability

Chapter 12

Q1 (c) The solubility will increase because there will be more charged form in a patient with gastric bypass

Q2 (d) The rate may not be affected but the extent is likely to decrease

Q3

Celexa (citalopram) tablets		
Ingredient	**Category**	**Quality**
Copolyvidone	Binder	Physical stability of tablet
Corn starch	Binder, disintegrant	Physical stability, release
Croscarmellose sodium	Disintegrant	Release
Glycerin	Binder	Physical stability
Lactose	Binder	Physical stability
Magnesium stearate	Lubricant	Physical stability (uniformity) of tablet
Hypromellose	Film coat	Palatability
Microcrystalline cellulose	Binder	Physical stability
Polyethylene glycol	Binder	Physical stability
Titanium dioxide	Opaquing agent	Chemical stability

Zyprexa Zydis orally disintegrating tablets		
Ingredient	**Category**	**Quality**
Gelatin	Binder	Physical stability
Mannitol	Protectant, sweetener	Chemical stability, palatability
Aspartame	Sweetener	Palatability
Methylparaben	Antimicrobial preservative	Microbial stability
Propyl paraben	Antimicrobial preservative	Microbial stability

Q4 Granisetron's benzylic nitrogen may photolyze. No, this product's stability problems must be addressed with light occluding packaging rather than formulation.

Granisetron HCl 20 mg	**Active ingredient**
Purified water 50 mL	Solvent
Cherry syrup qs 100 mL	Sweetened, flavored vehicle

Package in tight light resistant container, store at refrigerator temperature.

Q5 Tiagabine: stable, possible oxidation of the alkene

Tiagabine hydrochloride 100 mg	**Active ingredient**
Sodium carboxymethylcellulose 0.25 g	Suspending agent
Methylparaben 200 mg	Antimicrobial preservative
Glycerin 10 mL	Wetting agent
Cherry flavor qs	Flavor
Syrup 40 mL	Sweetened vehicle
Purified water qs 100 mL	Solvent

Tight packaging with head space for shaking. Under refrigeration.

Q6 Oxidation of the keto hydroxyl functional group. The citric acid may keep the pH low enough to suppress oxidation. This product should be given a very short beyond-use date (14 days). Store in refrigerator. Package in tight container.

Betamethasone 12 mg	Active ingredient
Alcohol 0.75 mL	Cosolvent
Methylcellulose 1500 300 mg	Viscosity enhancing agent
Citric acid, anhydrous 500 mg	Adds tartness, lowers pH
Cherry flavor qs	Flavor
Orange flavor qs	Flavor
Propylene glycol 10 mL	Cosolvent
Sodium benzoate 250 mg	Antimicrobial preservative
Sodium chloride 100 mg	Flavor enhancer
Sorbitol solution 70% 20 mL	Sweetened vehicle
Sucrose 50 g	Sweetener
Purified water qs 100 mL	Solvent

Q7

(a) Possible hydrolysis of the ester group, photolysis of the benzylic oxygen, oxidation of conjugated alkene.

(b) This would need to be prepared as a suspension.

(c) Ingredients: wetting agent (alcohol or glycerin), suspending agent, (xanthan gum or methylcellulose), sweetener (sucrose or sorbitol), flavor (fruit with citric acid to add tartness and lower pH), antimicrobial preservative (sodium benzoate), purified water (solvent). Storage: refrigeration. BUD: 14 days. Packaging: tight, light resistant container.

(d) Show the parent where the dose will measure on the dosi-spoon or medicine cup. Shake well! If the suspension is particularly difficult to resuspend, the pharmacist should let the patient know how many vigorous shakes the suspension will require to redisperse properly.

Store in refrigerator and discard after 14 days.

Q8

Metformin HCl 250 mg/5mL for an extemporaneous oral liquid		
Design issue	Drug property	Analysis
Drug must be dissolved before it can cross membranes	Water solubility freely soluble, maximum dose 1000 mg	Drug is soluble in the contents of the gastrointestinal tract, even better if the patient takes 250 mL water with the tablet

(continues)

Metformin HCl 250 mg/5mL for an extemporaneous oral liquid (*continued*)

Design issue	Drug property	Analysis
Drug must have sufficient lipophilicity to cross membranes	Log *P* of −0.5	This drug will not partition well into lipid membranes from body water
Large molecular weight drugs have difficulty moving across membranes	Formula weight 129	Based on molecular weight, this drug probably diffuses between cells in membranes rather than passing through lipid bilayer
Drugs may hydrolyze, oxidize, photolyze or otherwise degrade in solution	Stability in solution, stable to hydrolysis and oxidation. Could be susceptible to light	Protect from light
Some drugs cannot be used by mouth because of substantial losses to enzyme degradation	Not metabolized, 50–60% of the dose is delivered to the systemic circulation	The drug is incompletely but sufficiently absorbed orally to allow use by mouth
Intracellular drug receptors or receptors in the brain or posterior eye are challenging to target	Receptors are in the liver	The drug should move easily into the liver because its MW is small and the liver's capillaries are leaky
Some drugs cause serious toxicity as a result of poor selectivity of drug action	Not applicable	

Q9

Ingredient	Category	Quality
Cherry flavor, other flavor	Flavor	Palatability
Saccharin sodium or sucralose	Sweetener	Palatability
Benzoic acid	Antimicrobial preservative	Microbial stability
Methyl cellulose (or other viscosity agent)	Viscosity agent	Palatability (mouth feel of syrup)
Purified water	Solvent, vehicle	Size of dose, dissolution of drug for absorption

Q10 Gastric mixing promotes drug disintegration and ultimately dissolution from oral solid dosage forms. Restriction of gastric motility will reduce the rate of disintegration, dissolution and absorption from oral solid dosage forms. The extent of absorption would also be affected if the dosage form was unable to completely dissolve before it left the absorptive surface of the small intestine. One solution to this would be to give these patients medication in the form of solutions and suspensions.

Q11 For hydrophobic drugs loss of solubilization by bile acids can substantially reduce their apparent solubility and dissolution rate. This would reduce both the rate and the extent of absorption of these drugs. In addition the reduction in the absorbing surface that the drug is exposed to will reduce the extent of absorption. This problem may require increase in the patient's dose.

Chapter 13

Q1

Design issue	Drug property	Analysis
Drug must be dissolved before it can cross membranes	5 mg/mL	This drug has good water solubility for rapid dissolution in the gastrointestinal tract
Drug must have sufficient lipophilicity to cross membranes	Log P −1.8 or −2.3	This drug is too hydrophilic to move through the lipid bilayer well. It is transported via the oligopeptide and monocarboxylate transporters in the small intestine
Large molecular weight drugs have difficulty moving across membranes	Molecular weight 197	This drug may be able to diffuse through paracellular channels in the small intestine
Drugs may hydrolyze, oxidize, photolyze or otherwise degrade in solution	This drug has two phenolic groups that are photo-oxidizable	This drug is available in solid dosage forms only: ODT, IR and ER tablets
Some drugs cannot be used by mouth because of substantial losses to enzyme degradation	This drug is substantially metabolized by enzymes in the periphery	Levodopa is formulated with carbidopa to prevent its metabolism in the periphery
Intracellular drug receptors or receptors in the brain or posterior eye are challenging to target	Levodopa's receptors are in the brain	Brain uptake of levodopa is provided by large amino acid transporter 1
Some drugs cause serious toxicity as a result of poor selectivity of drug action	Not applicable	

From DrugBank.

Q2 Levodopa is a drug that is absorbed via oligopeptide and monocarboxylate transporters in the small intestine. There are two possible consequences of food on drugs that are transported via this mechanism. 1. Amino acids in the meal can compete with levodopa for the carrier and reduce its absorption. 2. Food can slow the transit of the Sinemet dosage form so that it remains in the absorption window for levodopa longer resulting in better absorption. The data indicate that AUC and peak C_{max} of levodopa after a single dose of SINEMET CR 50–200 increased by about 50% and 25%, respectively, when administered with food. It would appear that the latter mechanism is operating and this dosage form would be better administered with food. The information for patients in the package insert says: 'The patient should be informed that a change in diet to foods that are high in protein may delay the absorption of levodopa and may reduce the amount taken up in the circulation. Excessive acidity also delays stomach emptying, thus delaying the absorption of levodopa. Iron salts (such as in multivitamin tablets) may also reduce the amount of levodopa available to the body. The above factors may reduce the clinical effectiveness of the levodopa or carbidopa-levodopa therapy.'

Q3

Ingredient	Category	Quality
FD&C Blue No. 2	Coloring agent	Makes the dosage form distinctive
FD&C Red No. 40	Coloring agent	Makes the dosage form distinctive
Hydroxypropyl cellulose	Hydrocolloid matrix	Release
Hypromellose	Hydrocolloid matrix	Release
Magnesium stearate	Lubricant	Uniformity (physical stability)

This tablet is a hydrophilic matrix design.

Q4

Questions	Answer	Challenge or opportunity?
What is the membrane that will absorb the drug from this route and how accessible is it?	Drug is intended for local absorption on the large intestinal epithelium. The dosage form must be swallowed and be emptied from the stomach and move through the small intestine before absorption starts	Challenge
Is the absorbing membrane single or multiple layers and how permeable is it?	Single, modest permeability	Neither
How large is its surface?	Moderate, 0.5–1 m^2	Neither

(continues)

(continued)		
Questions	**Answer**	**Challenge or opportunity?**
What is the nature of the body fluids that bathe the absorbing membrane?	Large intestinal fluids: pH 6.4–7.0, relatively large volume	Opportunity
Will the dosage form remain at the absorbing membrane long enough to release its drug?	It is difficult to design dosage forms to release precisely at the beginning of the large intestine and continue to do so until eliminated in the feces. Some drugs for colonic delivery rely partially on absorption from the small intestine.	Challenge
Will the drug encounter enzymes or extremes of pH that can alter its chemical structure before it is distributed to its target?	Yes, extremely low pH in the stomach, lots of enzymes: peptidases, esterases, cytochrome P450	Challenge
What is the blood flow primacy to the absorbing membrane and the distribution time from it?	Highly perfused <2 minutes	Neither, drug intended for colon

Q5 The extended release tablets are all rated AB and can be used interchangeably. The capsule products are not rated and therefore should not be used interchangeably.

Q6 Extended release tablets and capsule granules are intended to control the release rate and therefore absorption rate of the drug rather than permitting variables within the gastrointestinal tract such as gastric emptying or permeability to control rate. Thus for the most part the rate and extent of drug release and absorption from extended release dosage forms are not affected in a clinically meaningful way by administration with food.

Q7 (a) They must leave the stomach intact.

Q8

Dosage form	Design	Cut or opened?
Dilacor XR	Microparticular reservoir	No
Seroquel XR	Hydrophilic matrix	No
Opana ER	Hydrophilic matrix	No
Oxycontin	Inert matrix	No
Klor-Con	Microparticulate reservoir	No

(continues)

(continued)		
Dosage form	**Design**	**Cut or opened?**
Toprol XL	Hybrid	Yes
Ultram ER	Single unit reservoir	No
Glucotrol XL	Osmotic	No
Procardia XL	Osmotic	No

There does not appear to be any trend in terms of a design that can be opened or cut.

Q9 Mixing an enteric coated tablet or capsule pellets with apple sauce should allow the coat of the dosage form to remain intact and deliver the drug to the small intestine without release in the stomach. If an enteric coated tablet or capsule pellets are mixed with pudding the coat may begin to dissolve in the pudding resulting in release in the stomach, and loss of drug due to instability or irritation of the stomach may occur. Nurses should be advised to mix enteric coated dosage forms with apple sauce but not with pudding. The student may be interested in the original study: KA Wells, WG Losin. In vitro stability, potency, and dissolution of duloxetine enteric-coated pellets after exposure to applesauce, apple juice and chocolate pudding. Clin Ther 2008; 30: 1300–1308.

Q10 To titrate a patient to the correct dose an immediate release tablet will provide more rapid onset and more rapid achievement of steady state and therefore more rapid feedback to the prescriber about whether the dose is appropriate. Extended release products take much longer to reach steady state.

Q11 (c) Sulfate

Q12 (d) pH 9.95, which does not occur in the gastrointestinal tract

Q13 (b) Sodium

Q14 (c) Mucoadhesive agents

Chapter 14

Q1

Thiopental prepared extemporaneously for rectal administration		
Design issue	**Drug property**	**Analysis**
Drug must be dissolved before it can cross membranes	Soluble in water (1 g/10 to 1 g/30) at basic pH 10–11 (pK_a 7.6)	The requested concentration appears to be at or above the solubility of thiopental sodium. The dosage form may need to be a suspension
Drug must have sufficient lipophilicity to cross membranes	Log P of 2.3	This drug will favor partitioning into lipid membranes over water by a 199:1 ratio
Large molecular weight drugs have difficulty moving across membranes	Formula weight 242	Based on molecular weight, oxycodone should move easily through membranes
Drugs may hydrolyze, oxidize, photolyze, or otherwise degrade in solution	Thiopental sodium is available as a powder for injection indicating that it has poor stability in solution, probably because of hydrolysis of the imide	This drug will need to be made just prior to use
Some drugs cannot be used by mouth because of substantial losses to enzyme degradation	Not applicable	
Intracellular drug receptors or receptors in the brain or posterior eye are challenging to target	Receptors are in the brain	Thiopental's lipophilicity is in a good range for reaching the brain. Its size (molecular weight) is within the optimal range
Some drugs cause serious toxicity as a result of poor selectivity of drug action	Not applicable	

Q2

Ingredient	Category	Quality
Methylcellulose (or other viscosity enhancing agent for oral use)	Viscosity agent	Physical stability, retention
Glycerin	Wetting agent	Physical stability
Purified or sterile water	Solvent, vehicle	Size, release

The dosage form should not require a preservative because it will need to be used immediately owing to chemical stability issues. If students research published formulas they will find 'bare bones' formulations prepared from the powder for injection and sterile water. The powder for injection contains sodium carbonate as a buffer. (See Nguyen et al Radiology 2001; 221: 760–762.)

Q3

Questions	Answer	Challenge or opportunity?
What is the membrane that will absorb the drug from this route and how accessible is it?	Drug is absorbed through the rectal epithelium. It is easily though not conveniently accessible	Opportunity
Is the absorbing membrane single or multiple layers and how permeable is it?	Single, moderate to good permeability	Neither
How large is its surface?	Small, 200–400 cm^2	Challenge
What is the nature of the body fluids that bathe the absorbing membrane?	Rectal fluids: pH 7.0–7.4, small volume	Neither
Will the dosage form remain at the absorbing membrane long enough to release its drug?	Usually but not always. Sometimes the dosage form triggers the need to defecate and the patient cannot retain it	Challenge
Will the drug encounter enzymes or extremes of pH that can alter its chemical structure before it is distributed to its target?	No	Opportunity
What is the blood flow primacy to the absorbing membrane and the distribution time from it?	Highly perfused <2 minutes	Opportunity

Q4 The solubility characteristics and molecular weight of the drug will determine its ability to be absorbed systemically. Hydrophilic drugs with larger molecular weight would remain in the vaginal cavity better than lipophilic drugs with small molecular weight. If the drug is basic, it would be more likely to be in its charged form in the acidic vaginal fluids and therefore more likely to remain local.

Q5 (a) Hydrophilic, basic, larger molecular weight

Q6 (a) Amount of diazepam that dissolves in rectal fluids

Q7 (b) Hydromorphone hydrochloride to provide a higher concentration of water soluble form for dissolution

Q8 (d) A fat soluble base to provide better partitioning out of the base

Chapter 15

Q1 (c) Be poorly absorbed due to the keratinized squamous cell lining

Q2 (d) Reach the nasal vasculature

Q3 (a) Candesartan (FW 440)

Q4 (b) Phenylephrine (log P −0.31)

Q5 (a) Head upright

Q6 (b) The anterior turbinate region

Q7

Momentasone furoate nasal suspension		
Ingredient	Category	Quality
Benzalkonium chloride	Antimicrobial preservative	Microbial stability
Carboxymethylcellulose	Viscosity agent	Retention, physical stability
Citric acid	Buffer	Chemical stability
Sodium citrate	Buffer	Chemical stability
Glycerin	Wetting agent, solvent	Size of dose, physical stability
Microcrystalline cellulose	Viscosity agent	Retention, physical stability
Polysorbate 80	Wetting agent, dispersing aid	Physical stability, release
Purified water	Vehicle, solvent	Size of dose, release

Q8

Terbinafine HCl nasal liquid		
Design issue	Drug property	Analysis
Drug must be dissolved before it can cross membranes	Water solubility slightly soluble, 1 g in 100 mL requested	Drug is not soluble in dosage form unless the HCl salt has much better solubility than the uncharged form
Drug must have sufficient lipophilicity to cross membranes	Log P of 5.9	This drug will partition well into lipid membranes from body water and to get into fungal cells
Large molecular weight drugs have difficulty moving across membranes	Formula weight 291	Based on molecular weight, this drug should diffuse easily through the nasal lipid bilayer
Drugs may hydrolyze, oxidize, photolyze, or otherwise degrade in solution	Stability in solution, unsaturated section could be susceptible to oxidation	Protect from light

(continues)

Terbinafine HCl nasal liquid (continued)

Design issue	Drug property	Analysis
Some drugs cannot be used by mouth because of substantial losses to enzyme degradation	Hepatic	Metabolism in the nose has not been characterized
Intracellular drug receptors or receptors in the brain or posterior eye are challenging to target	Inhibits squalene epoxidase in the fungal cell	Drug should move into the site of infection and into the fungal cells relatively easily
Some drugs cause serious toxicity as a result of poor selectivity of drug action	Not applicable	

Q9

Ingredient	Category	Quality
Propylene glycol or glycerin	Wetting agent, vehicle	Physical stability, size of dose, dissolution of drug for absorption
Benzalkonium chloride or other antimicrobial preservative	Antimicrobial preservative	Microbial stability
Sodium chloride	Tonicity agent	Comfort
Methyl cellulose (or other viscosity agent)	Viscosity agent	Retention in the nose, if the product is a suspension it will maintain the uniformity of the suspension while the patient administers the dose
Purified water	Solvent, vehicle	Size of dose, dissolution of drug for absorption

Q10

Nasal route

Questions	Answer	Challenge or opportunity?
What is the membrane that will absorb the drug from this route and how accessible is it?	Drug is absorbed through the respiratory epithelium. This area is behind the nasal valve which means that some drug is wasted by application too far anterior on the nasal vestibule or too far posterior in the nasopharynx	Challenge
Is the absorbing membrane single or multiple layers and how permeable is it?	Single, excellent permeability	Opportunity

(continues)

Nasal route (continued)		
Questions	Answer	Challenge or opportunity?
How large is its surface?	Modest, 160 cm^2	Neither
What is the nature of the body fluids that bathe the absorbing membrane?	Relatively small volume, mucus is viscous pH 5.5–6.5, epithelial cells pH 7.4	Challenge
Will the dosage form remain at the absorbing membrane long enough to release its drug?	Mucociliary clearance fairly quickly moves drug back to the nasopharynx such that drug may not be completely released by the time it is swallowed	Challenge
Will the drug encounter enzymes or extremes of pH that can alter its chemical structure before it is distributed to its target?	There are many different enzymes in the nose: peptidases, esterases, cytochrome P450. We do not know to what extent they remove drug before distribution	Opportunity
What is the blood flow primacy to the absorbing membrane and the distribution time from it?	Highly perfused <2 minutes	Opportunity

Q11 A bad taste after use of nasal sprays indicates that the product has flowed into the nasopharynx and made contact with the taste buds. This is undesirable not just from the aesthetic standpoint of taste but from the pharmaceutical standpoint that the drug should be absorbed in the nose and not in the gastrointestinal tract. The patient should be questioned about whether she uses one puff or two in each nostril and if two should space these more to prevent overloading the mucosa with more than 100 µL drug solution at a time. In addition ensure that the dose is delivered more towards the anterior turbinate region than the back of the nose, which will rapidly carry the drug to the nasopharynx.

Chapter 16

Q1 (a) The alveoli have no mucociliary clearance

Q2 (c) The aerosol deposits in the alveoli where there is no smooth muscle

Q3 (d) Bronchioles and alveoli

Q4 (b) Drug metabolizing enzymes in lung tissue

Q5 (b) Beclomethasone dipropionate log P 4.07

Q6 (d) Becoming a gas as they are released into atmospheric pressure

Q7 (c) The drug powder is carried on a larger, inert solid

Q8   (b) Oxidation

Q9   (e) Answers (a) and (b)

Q10   (d) Answer (b)

Q11   (d) Rinse your mouth after using your inhaler

Q12   A possible explanation for this is that most of the 7.7 μm particle aerosol is deposited in the oropharyngeal region and larger airways and therefore does not reach the small airways, where the majority of beta2 receptors are located, in sufficient quantity to induce an equivalent bronchodilator response to the finer particle aerosol.

Q13   The particle size emitted from the MDI with built in spacer would be smaller, there would be less drug deposited in the mouth and more drug deposited in the lung.

Q14   The HFA formulation would be deposited primarily in the terminal bronchi, respiratory bronchioles and alveoli while the CFC formulation would be deposited in the mouth, upper and lower respiratory tract. The HFA product should have fewer side effects because less drug is deposited in the mouth and swallowed.

Q15   This child is young enough that he may need to use a nebulizer for his drug therapy. Both albuterol and budesonide are available as nebulized medications. His parents need to be instructed in how to have the child rinse and spit after using the nebulizer. Assemble the nebulizer according to its instructions. Fill the medicine cup or reservoir with the prescription, according to the instructions. Connect to the power source or compressor.

Place the mouthpiece in the child's mouth. Tell the child to breathe through his mouth until all the medicine is used, about 10–15 minutes. Wash the medicine cup and mouthpiece with water, and air-dry until next treatment. Once or twice a week after follow the manufacturer's instructions on disinfection.

Another possibility is an MDI with valved chamber. The pharmacist would need to train the family and have the parents and child demonstrate that he is able to use these devices. There are several corticosteroid MDIs and albuterol is also available as a MDI. Shake the inhaler and place the mouthpiece into the open end of the chamber ensuring a tight fit. Place the facemask over the nose and mouth. Actuate one dose into the chamber.

Allow the child to inhale and exhale normally into the chamber at least 10 times.

Clean the face after the mask is removed in case medication was deposited on the skin.

Q16 Either a MDI with valved chamber or a nebulizer would be appropriate with this patient assuming there are no barriers to explaining the use of the MDI and valved chamber. If the patient is fearful that would favor the nebulizer. Albuterol is available as a nebulizer solution and MDI.

Q17 This patient would benefit from the use of a nebulizer that requires little explanation since his hearing represents a barrier to educating him on the use of the MDI and valved chamber.

Q18 This will depend on the patient's life style. If he is very active, it would be preferred to provide him with training on the use of an MDI or the new Combivent Respimat soft mist inhaler. If he is relatively sedentary he may be well served with the use of a nebulizer and the combination albuterol and ipratropium nebulizer solution. MDI use: Shake the inhaler and remove the mouthpiece cap. Exhale fully and place the mouthpiece 2.5–5 cm from the open mouth. As you take a deep breath, squeeze the canister and mouthpiece together at the same time. Draw the spray from the inhaler into your lungs with a slow, deep breath. Hold your breath as long as comfortable. Exhale slowly. If you did not use a space or a spacer, rinsing your mouth before you swallow drug solution that may be on the sides of your mouth can reduce the chances of side effects. The rinse water should be spat into the sink. Wipe or rinse the mouthpiece clean. Rinse the plastic actuator once a week to prevent clogging. Take out the drug filled canister and run the plastic actuator under warm water and allow to air dry before reassembling.

Chapter 17

Q1

Nifedipine topical		
Design issue	**Drug property**	**Analysis**
Drug must be dissolved before it can cross membranes	Water solubility of 1 g/10 000 mL, 2 g in 100 mL requested	Because this dosage form needs to be nonocclusive, the HCl salt with better solubility than the uncharged form, will need to be used
Drug must have sufficient lipophilicity to cross membranes	Log P 2.343	This drug will partition well into intercellular lipids from a dosage form and yet is not too lipophilic to partition into viable cells

(continues)

Nifedipine topical (*continued*)

Design issue	Drug property	Analysis
Large molecular weight drugs have difficulty moving across membranes	Formula weight 346	Based on molecular weight, this drug should diffuse fairly well; its molecular weight is <500
Drugs may hydrolyze, oxidize, photolyze or otherwise degrade in solution	Stability in solution, potentially hydrolyzable esters and unsaturated section could be susceptible to oxidation	Protect from water and light; will need to formulate in a water free base
Some drugs cannot be used by mouth because of substantial losses to enzyme degradation	Extensively metabolized in the liver	Metabolism in the skin has not been characterized
Intracellular drug receptors or receptors in the brain or posterior eye are challenging to target	Inhibits the influx of extracellular calcium through vascular membrane pores to increase blood flow to the wound	The drug has good biopharmaceutical characteristics even for the skin and further the stratum corneum is not intact which provides the drug good access to the capillaries of the dermis
Some drugs cause serious toxicity as a result of poor selectivity of drug action	Not applicable	

Q2

Ingredient	Category	Quality
Propylene glycol or glycerin	Levigating agent	Physical stability, dissolution of drug for absorption
Polyethylene glycol ointment	Base, vehicle	Size of dose, chemical stability, dissolution of drug for absorption

Q3 (b) Betamethasone 0.05% cream (water in oil) by Schering

Q4

(a) Protect from light.

(b) Histamine receptors on sensory nerve endings in the dermis.

(c) This may be a debatable point but children do scratch chicken pox, creating little wounds that would benefit from a non-occlusive base.

(d) $\text{PPG} = |\log P_{\text{penetrant}} - \log P_{\text{stratum corneum}}| = |4.682 - 0.8| = 3.882$.

The $\log P$ of the vehicle chosen for cyproheptadine should be either $4.682 + 3.882 = 8.564$ or higher, *or* $4.682 - 3.882 = 0.8$ or lower.

(e) If we need to use a non-occlusive base for this indication, we would need to use the hydrochloride salt to maximize its solubility in a polar base.

(f) The pharmacist could use either a water-based gel or an oil-in-water cream. The latter may be preferred for providing some emolliency which can be helpful with itching. Note that although we do not have a log P value for the available oil-in-water cream bases, the active ingredient would be dissolved in the water continuous phase which has a log P of −4.0.

Q5 (c) 36%

Q6 (d) Thickness of the skin barrier h was thinner than usual.

Q7 (c) 36%

Q8 (a) Concentration of drug in the vehicle was excessive and (b) diffusion coefficient D was increased by occlusive wrap or base.

Q9 (c) 2 FTU

Q10 (a) 30 g

Q11

$$PPG = |\log P_{penetrant} - \log P_{stratum\ corneum}| = |-2.11 - 0.8| = 2.91$$

The log P of the vehicle chosen for urea should be either −2.11 + 2.91 = 0.8 or higher *or* −2.11 − 2.91 = −5.02 or lower.

The preferred base for dry lesions would be an occlusive one but it is unlikely that urea would be soluble in such a base making an oleaginous base a poor choice. A better choice for urea would be a water based gel which would provide a high C_v. The addition of alcohol or propylene glycol to the gel may create some increased partitioning by decreasing the polarity of the gel.

Q12 (a) Oil disperse phase

Q13 (d) Methylcellulose gel

Q14 (b) Use acid to make salt and (c) use cosolvents

Q15 (b) The charged form predominates, (B/BH^+ 1/100).

Q16 (c) Use cosolvents and (d) derivatize

Q17 (d) Photo-oxidation of the phenol

Q18 (c) Anthralin to increase partitioning into the stratum corneum and (d) to enhance stability

Q19 (c) Receptors: viable epidermis; penetrate stratum corneum

Q20 (c) Occlusive base because psoriasis is a dry lesion

Q21 (b) PPG $= 1.5$, $\log P_{\text{vehicle}}$ lower than 0.8 or higher than 3.8

Q22 (a) White ointment

Q23 (b) Eucerin cream

Q24 (d) Photo-oxidation of the phenol

Q25 (b) Hydrochloride salt to enhance stability and (c) tetracycline to increase partitioning into the stratum corneum

Q26 (b) Targets: sebaceous glands; penetrate stratum corneum

Q27 (b) Non-occlusive base because acne is a moist lesion

Q28 (d) Methylcellulose gel

Q29 (b) Concentration of drug in the vehicle

Chapter 18

Q1 (c) Potential for dose dumping

Q2 (a) Good skin conformability

Q3 (a) Reservoir (membrane) patch

Q4 (b) Less likely to cause side effects than oral products

Q5 (b) 4.8 cm^2

Q6 (d) The liner

Q7 (c) Drug in adhesive matrix patch

Q8 (d) It has an excess of drug in the matrix, the delivery of which is controlled by the skin.

Q9 (c) Fold the old patch before discarding it.

Q10 (a) C_v, concentration of drug in the vehicle is maximized.

Q11 (e) Tissues beneath the area applied.

Q12 (b) Place the patch on the affected area(s).

Q13 (b) Product must be the same design (matrix or reservoir).

Q14 The patient's skin should be exposed to 13 cm^2 or 33% of the patch surface to produce the lower dose. This can be done by occluding a portion of the patch with wax paper or the patch liner allowing the barrier edges to protrude from the patch rather than trimming them. The patch with barrier is covered with Tegaderm or other dressing.

Q15 GB is likely to receive excessive estradiol and progesterone doses if she applies the cream to a larger surface than recommended. This could cause side effects including serious problems such as clot formation in vessels.

Q16 Diltiazem has poor water solubility and its log P is slightly high for good transdermal absorption. Its melting point and formula weight are within the suitable range; however, the daily dose is high and will not be possible to provide 120 mg per day transdermally.

Q17 Lorazepam's weight is below 500, log P between 1 and 4 and melting point below 200°C, all of which should confer adequate mobility in the stratum corneum. It has poor water solubility so there is some concern about its ability to partition into and through the viable epidermis and dermis. It is active at a relatively low dose (0.5 to 2 mg/day).

$$PPG = |2.959 - 0.8| = 2.159$$

The log P of the vehicle chosen for lorazepam should be either $2.959 + 2.159 = 5.118$ or higher *or* $2.959 - 2.159 = 0.8$ or lower.

The lorazepam could be formulated in a water-based gel with ethanol cosolvent (below log P 0.8) or in a lecithin isopropyl palmitate organogel without water. (log P higher than 5.118)

Efficacy: monitor anxiety level, agitated behaviors

Adverse effects: monitor sedation, dizziness, gait

Index